Shigeji Fujita and Akira Suzuki

Electrical Conduction in Graphene and Nanotubes

Related Titles

Jiang, D.-E., Chen, Z.

Graphene Chemistry
Theoretical Perspectives

2013
ISBN: 978-1-119-94212-2

Malic, E., Knorr, A.

Graphene and Carbon Nanotubes
Ultrafast Relaxation Dynamics and Optics

2013
ISBN: 978-3-527-41161-0

Monthioux, M.

Carbon Meta-Nanotubes
Synthesis, Properties and Applications

2011
ISBN: 978-0-470-51282-1

Jorio, A., Dresselhaus, M. S., Saito, R., Dresselhaus, G.

Raman Spectroscopy in Graphene Related Systems

2011
ISBN: 978-3-527-40811-5

Delhaes, P.

Solids and Carbonated Materials

2010
ISBN: 978-1-84821-200-8

Saito, Y. (ed.)

Carbon Nanotube and Related Field Emitters
Fundamentals and Applications

2010
ISBN: 978-3-527-32734-8

Akasaka, T., Wudl, F., Nagase, S. (eds.)

Chemistry of Nanocarbons

2010
ISBN: 978-0-470-72195-7

Krüger, A.

Carbon Materials and Nanotechnology

2010
ISBN: 978-3-527-31803-2

Guldi, D. M., Martín, N. (eds.)

Carbon Nanotubes and Related Structures
Synthesis, Characterization, Functionalization, and Applications

2010
ISBN: 978-3-527-32406-4

Reich, S., Thomsen, C., Maultzsch, J.

Carbon Nanotubes
Basic Concepts and Physical Properties

2004
ISBN: 978-3-527-40386-8

Shigeji Fujita and Akira Suzuki

Electrical Conduction in Graphene and Nanotubes

Verlag GmbH & Co. KGaA

The Authors

Prof. Dr. Shigeji Fujita
University of Buffalo
SUNY, Dept. of Physics
329 Fronczak Hall
Buffalo, NY 14260
USA

Prof. Dr. Akira Suzuki
Tokyo University of Science
Dept. of Physics
Shinjuku-ku
162-8601 Tokyo
Japan

Library of Congress Card No.:
applied for

British Library Cataloguing-in-Publication Data:
A catalogue record for this book is available from the British Library.

Bibliographic information published by the Deutsche Nationalbibliothek
The Deutsche Nationalbibliothek lists this publication in the Deutsche Nationalbibliografie; detailed bibliographic data are available on the Internet at http://dnb.d-nb.de.

Print ISBN 978-3-527-41151-1

Composition le-tex publishing services GmbH, Leipzig
Printing and Binding Markono Print Media Pte Ltd, Singapore
Cover Design Adam-Design, Weinheim

Printed in Singapore
Printed on acid-free paper

Contents

Preface *XI*

Physical Constants, Units, Mathematical Signs and Symbols *XV*

1 Introduction *1*
1.1 Carbon Nanotubes *1*
1.2 Theoretical Background *4*
1.2.1 Metals and Conduction Electrons *4*
1.2.2 Quantum Mechanics *4*
1.2.3 Heisenberg Uncertainty Principle *4*
1.2.4 Bosons and Fermions *5*
1.2.5 Fermi and Bose Distribution Functions *5*
1.2.6 Composite Particles *6*
1.2.7 Quasifree Electron Model *6*
1.2.8 "Electrons" and "Holes" *7*
1.2.9 The Gate Field Effect *7*
1.3 Book Layout *8*
1.4 Suggestions for Readers *9*
1.4.1 Second Quantization *9*
1.4.2 Semiclassical Theory of Electron Dynamics *9*
1.4.3 Fermi Surface *9*
References *10*

2 Kinetic Theory and the Boltzmann Equation *11*
2.1 Diffusion and Thermal Conduction *11*
2.2 Collision Rate: Mean Free Path *12*
2.3 Electrical Conductivity and Matthiessen's Rule *15*
2.4 The Hall Effect: "Electrons" and "Holes" *17*
2.5 The Boltzmann Equation *19*
2.6 The Current Relaxation Rate *21*
References *25*

3 Bloch Electron Dynamics *27*
3.1 Bloch Theorem in One Dimension *27*
3.2 The Kronig–Penney Model *30*

3.3 Bloch Theorem in Three Dimensions *33*
3.4 Fermi Liquid Model *36*
3.5 The Fermi Surface *37*
3.6 Heat Capacity and Density of States *40*
3.7 The Density of State in the Momentum Space *42*
3.8 Equations of Motion for a Bloch Electron *46*
References *51*

4 Phonons and Electron–Phonon Interaction *53*
4.1 Phonons and Lattice Dynamics *53*
4.2 Van Hove Singularities *57*
4.2.1 Particles on a Stretched String (Coupled Harmonic Oscillators) *57*
4.2.2 Low-Frequency Phonons *59*
4.2.3 Discussion *61*
4.3 Electron–Phonon Interaction *65*
4.4 Phonon-Exchange Attraction *71*
References *75*

5 Electrical Conductivity of Multiwalled Nanotubes *77*
5.1 Introduction *77*
5.2 Graphene *78*
5.3 Lattice Stability and Reflection Symmetry *81*
5.4 Single-Wall Nanotubes *84*
5.5 Multiwalled Nanotubes *85*
5.6 Summary and Discussion *87*
References *89*

6 Semiconducting SWNTs *91*
6.1 Introduction *91*
6.2 Single-Wall Nanotubes *93*
6.3 Summary and Discussion *98*
References *98*

7 Superconductivity *99*
7.1 Basic Properties of a Superconductor *99*
7.1.1 Zero Resistance *99*
7.1.2 Meissner Effect *100*
7.1.3 Ring Supercurrent and Flux Quantization *101*
7.1.4 Josephson Effects *102*
7.1.5 Energy Gap *104*
7.1.6 Sharp Phase Change *104*
7.2 Occurrence of a Superconductor *105*
7.2.1 Elemental Superconductors *105*
7.2.2 Compound Superconductors *106*
7.2.3 High-T_c Superconductors *107*
7.3 Theoretical Survey *107*
7.3.1 The Cause of Superconductivity *107*

7.3.2 The Bardeen–Cooper–Schrieffer Theory *108*
7.3.3 Quantum Statistical Theory *110*
7.4 Quantum Statistical Theory of Superconductivity *111*
7.4.1 The Generalized BCS Hamiltonian *111*
7.5 The Cooper Pair Problem *114*
7.6 Moving Pairons *116*
7.7 The BCS Ground State *119*
7.7.1 The Reduced Generalized BCS Hamiltonian *119*
7.7.2 The Ground State *121*
7.8 Remarks *126*
7.8.1 The Nature of the Reduced Hamiltonian *126*
7.8.2 Binding Energy per Pairon *127*
7.8.3 The Energy Gap *127*
7.8.4 The Energy Gap Equation *128*
7.8.5 Neutral Supercondensate *130*
7.8.6 Cooper Pairs (Pairons) *130*
7.8.7 Formation of a Supercondensate and Occurrence of Superconductors *130*
7.8.8 Blurred Fermi Surface *131*
7.9 Bose–Einstein Condensation in 2D *133*
7.10 Discussion *137*
References *139*

8 Metallic (or Superconducting) SWNTs *141*
8.1 Introduction *141*
8.2 Graphene *147*
8.3 The Full Hamiltonian *149*
8.4 Moving Pairons *151*
8.5 The Bose–Einstein Condensation of Pairons *153*
8.6 Superconductivity in Metallic SWNTs *157*
8.7 High-Field Transport in Metallic SWNTs *159*
8.8 Zero-Bias Anomaly *161*
8.9 Temperature Behavior and Current Saturation *162*
8.10 Summary *162*
References *164*

9 Magnetic Susceptibility *165*
9.1 Magnetogyric Ratio *165*
9.2 Pauli Paramagnetism *167*
9.3 The Landau States and Levels *170*
9.4 Landau Diamagnetism *171*
References *176*

10 Magnetic Oscillations *177*
10.1 Onsager's Formula *177*
10.2 Statistical Mechanical Calculations: 3D *181*

10.3 Statistical Mechanical Calculations: 2D *184*
10.4 Anisotropic Magnetoresistance in Copper *189*
10.4.1 Introduction *189*
10.4.2 Theory *192*
10.4.3 Discussion *194*
10.5 Shubnikov–de Haas Oscillations *196*
References *201*

11 Quantum Hall Effect *203*
11.1 Experimental Facts *203*
11.2 Theoretical Developments *206*
11.3 Theory of the Quantum Hall Effect *208*
11.3.1 Introduction *208*
11.3.2 The Model *210*
11.3.3 The Integer QHE *212*
11.3.4 The Fractional QHE *217*
11.4 Discussion *218*
References *219*

12 Quantum Hall Effect in Graphene *221*
12.1 Introduction *221*
References *227*

13 Seebeck Coefficient in Multiwalled Carbon Nanotubes *229*
13.1 Introduction *229*
13.2 Classical Theory of the Seebeck Coefficient in a Metal *232*
13.3 Quantum Theory of the Seebeck Coefficient in a Metal *235*
13.4 Simple Applications *239*
13.5 Graphene and Carbon Nanotubes *240*
13.6 Conduction in Multiwalled Carbon Nanotubes *242*
13.7 Seebeck Coefficient in Multiwalled Carbon Nanotubes *243*
References *246*

14 Miscellaneous *247*
14.1 Metal–Insulator Transition in Vanadium Dioxide *247*
14.1.1 Introduction *247*
14.2 Conduction Electrons in Graphite *249*
14.3 Coronet Fermi Surface in Beryllium *250*
14.4 Magnetic Oscillations in Bismuth *251*
References *251*

Appendix *253*
A.1 Second Quantization *253*
A.1.1 Boson Creation and Annihilation Operators *253*
A.1.2 Observables *256*
A.1.3 Fermion Creation and Annihilation Operators *257*
A.1.4 Heisenberg Equation of Motion *259*

A.2 Eigenvalue Problem and Equation-of-Motion Method *261*
A.2.1 Energy-Eigenvalue Problem in Second Quantization *261*
A.2.2 Energies of Quasielectrons (or "Electrons") at 0 K *264*
A.3 Derivation of the Cooper Equation (7.34) *267*
A.4 Proof of (7.94) *270*
A.5 Statistical Weight for the Landau States *271*
A.5.1 The Three-Dimensional Case *271*
A.5.2 The Two-Dimensional Case *272*
A.6 Derivation of Formulas (11.16)–(11.18) *273*
References *274*

Index *275*

Preface

Brilliant diamond and carbon black (graphite) are both made of carbon (C). Diamond is an insulator while graphite is a good conductor. This difference arises from the lattice structure. Graphite is a layered material made up of sheets, each forming a two-dimensional (2D) honeycomb lattice, called *graphene*. The electrical conduction mainly occurs through graphene sheets. Carbon nanotubes were discovered by Iijima[1] in 1991. The nanotubes ranged from 4 to 30 nm in diameter and were microns (μm) in length, had scroll-type structures, and were called *Multi-walled Nanotubes* (MWNTs) in the literature. *Single-Wall Nanotubes* (SWNTs) have a size of about 1 nm in diameter and microns in length. This is a simple two-dimensional material. It is theorists' favorite system. The electrical transport properties along the tube present, however, many puzzles, as is explained below. Carbon nanotubes are very strong and light. In fact, carbon fibers are used to make tennis rackets. Today's semiconductor technology is based on silicon (Si) devices. It is said that carbon chips, which are stronger and lighter, may take the place of silicon chips in the future. It is, then, very important to understand the electrical transport properties of carbon nanotubes. The present book has as its principal topics electrical transport in graphene and carbon nanotubes.

The conductivity σ in individual carbon nanotubes varies, depending on the tube radius and the pitch of the sample. In many cases the resistance decreases with increasing temperature while the resistance increases in the normal metal. Electrical conduction in SWNTs is either semiconducting or metallic, depending on whether each pitch of the helical line connecting the nearest-neighbor C-hexagon centers contains an integral number of hexagons or not. The second alternative occurs more often since the pitch is not controlled in the fabrication process. The room-temperature conductivity in metallic SWNTs is higher by two or more orders of magnitude than in semiconducting SWNTs. Currents in metallic SWNTs do not obey Ohm's law linearity between current and voltage. Scanned probe microscopy shows that the voltage does not drop along the tube length, implying a superconducting state. The prevailing theory states that electrons run through the one-dimensional (1D) tube ballistically. But this interpretation is not the complete story. The reason why the ballistic electrons are not scattered by impurities and

1) Iijima, S. (1991) *Nature (London)*, **354**, 56.

phonons is unexplained. We present a new interpretation in terms of the model in which superconducting Bose-condensed Cooper pairs (bosons) run as a supercurrent. In our text we start with the honeycomb lattice, construct the Fermi surface, and develop Bloch electron dynamics based on the rectangular unit cell model. We then use kinetic theory to treat the normal electrical transport with the assumption of "electrons," "holes," and Cooper pairs as carriers.

To treat the superconducting state, we assume that the phonon-exchange attraction generates Cooper pairs (pairons). We start with a Bardeen–Cooper–Schrieffer (BCS)-like Hamiltonian, derive a linear dispersion relation for the moving pairons, and obtain a formula for the Bose–Einstein Condensation (BEC) temperature

$$k_B T_c = 1.24 \hbar v_F n^{1/2} , \quad (2D)$$

where n is the pairon density and v_F the Fermi speed. The superconducting temperature T_c given here, is distinct from the famous BCS formula for the critical temperature: $3.53 k_B T_c = 2\Delta_0$, where Δ_0 is the zero-temperature electron energy gap in the weak coupling limit. The critical temperature T_c for metallic SWNTs is higher than 150 K while the T_c is much lower for semiconducting SWNTs.

MWNTs have open-ended circumferences and the outermost walls with greatest radii, contribute most to the conduction. The conduction is metallic (with no activation energy factor) and shows no pitch dependence.

In 2007 Novoselov *et al.* [2] discovered the room-temperature *Quantum Hall Effect* (QHE) in graphene. This was a historic event. The QHE in the GaAs/AlGaAs heterojunction is observed around 1 K and below. The original authors interpreted the phenomenon in terms of a Dirac fermion moving with a linear dispersion relation. But the reason why Dirac fermions are not scattered by phonons, which must exist at 300 K, is unexplained. We present an alternative explanation in terms of the *composite bosons* traditionally used in QHE theory. The most important advantage of our bosonic theory over the Dirac fermion theory is that our theory can explain why the plateau in the Hall conductivity (σ_{xy}) is generated where the zero resistivity ($\rho_{xx} = 0$) is observed.

This book has been written for first-year graduate students in physics, chemistry, electrical engineering, and material sciences. Dynamics, quantum mechanics, electromagnetism, and solid state physics at the senior undergraduate level are prerequisites. Second quantization may or may not be covered in the first-year quantum course. But second quantization is indispensable in dealing with phonon-exchange, superconductivity, and QHE. It is fully reviewed in Appendix A.1. The book is written in a self-contained manner. Thus, nonphysics majors who want to learn the microscopic theory step-by-step with no particular hurry may find it useful as a self-study reference.

Many fresh, and some provocative, views are presented. Experimental and theoretical researchers in the field are also invited to examine the text. The book is based on the materials taught by Fujita for several courses in quantum theory of solids and quantum statistical mechanics at the University at Buffalo. Some of the

2) Novoselov, K.S. *et al.* (2007) *Science*, **315**, 1379.

book's topics have also been taught by Suzuki in the advanced course in condensed matter physics at the Tokyo University of Science. The book covers only electrical transport properties. For other physical properties the reader is referred to the excellent book *Physical Properties of Carbon Nanotubes*, by R. Saito, G. Dresselhaus and M.S. Dresselhaus (Imperial College Press, London 1998).

The authors thank the following individuals for valuable criticisms, discussions and readings: Professor M. de Llano, Universidad Nacional Autonoma de México; Professor Sambandamurthy Ganapathy, University at Buffalo, Mr. Masashi Tanabe, Tokyo University of Science and Mr. Yoichi Takato, University at Buffalo. We thank Sachiko, Keiko, Michio, Isao, Yoshiko, Eriko, George Redden and Kurt Borchardt for their encouragement, reading and editing of the text.

Buffalo, New York, USA
Tokyo, Japan
December, 2012

Shigeji Fujita
Akira Suzuki

Physical Constants, Units, Mathematical Signs and Symbols

Useful Physical Constants

Quantity	Symbol	Value
Absolute zero temperature		$0\,\mathrm{K} = -273.16\,°\mathrm{C}$
Avogadro's number	N_A	$6.02 \times 10^{23}\,\mathrm{mol^{-1}}$
Bohr magneton	$\mu_B = e\hbar/(2m_e)$	$9.27 \times 10^{-24}\,\mathrm{J\,T^{-1}}$
Bohr radius	$a_B = 4\pi\varepsilon_0\hbar^2/(m_e e^2)$	$5.29 \times 10^{-11}\,\mathrm{m}$
Boltzmann's constant	$k_B = R/N_A$	$1.38 \times 10^{-23}\,\mathrm{J\,K^{-1}}$
Coulomb's constant	$k_0 = 1/(4\pi\varepsilon_0)$	$8.988 \times 10^{9}\,\mathrm{N\,m\,C^{-2}}$
Dirac's constant (Planck's constant/(2π))	$\hbar = h/(2\pi)$	$1.05 \times 10^{-34}\,\mathrm{J\,s}$
Electron charge (magnitude)	e	$1.60 \times 10^{-19}\,\mathrm{C}$
Electron rest mass	m_e	$9.11 \times 10^{-31}\,\mathrm{kg}$
Gas constant	$R = N_A k_B Z$	$8.314\,\mathrm{J\,K^{-1}\,mol^{-1}}$
Gravitational constant	G	$6.674 \times 10^{-11}\,\mathrm{N\,m^2\,kg^{-2}}$
Gravitational acceleration	g	$9.807\,\mathrm{m\,s^{-2}}$
Magnetic flux quantum	$\Phi_0 = h/(2e)$	$2.068 \times 10^{-15}\,\mathrm{Wb}$
Mechanical equivalent of heat		$4.184\,\mathrm{J\,cal^{-1}}$
Molar volume (gas at STP)		$2.24 \times 10^{4}\,\mathrm{cm^3} = 22.4\,\mathrm{L}$
Permeability of vacuum	μ_0	$4\pi \times 10^{-7}\,\mathrm{H\,m^{-1}}$
Permittivity of vacuum	ε_0	$8.85 \times 10^{-12}\,\mathrm{F\,m^{-1}}$
Planck's constant	h	$6.63 \times 10^{-34}\,\mathrm{J\,s}$
Proton mass	m_p	$1.67 \times 10^{-27}\,\mathrm{kg}$
Quantum Hall conductance	e^2/h	$3.874 \times 10^{-6}\,\mathrm{S}$
Quantum Hall resistance	$R_H = h/e^2$	$25\,812.81\,\Omega$
Speed of light	c	$3.00 \times 10^{8}\,\mathrm{m\,s^{-1}}$

Subsidiary Units

newton $1\,\mathrm{N} = 1\,\mathrm{kg\,m\,s^{-2}}$
joule $1\,\mathrm{J} = 1\,\mathrm{N\,m}$

coulomb	$1\,\mathrm{C} = 1\,\mathrm{A\,s}$
hertz	$1\,\mathrm{Hz} = 1\,\mathrm{s}^{-1}$
pascal	$1\,\mathrm{Pa} = 1\,\mathrm{N\,m}^{-2}$
bar	$1\,\mathrm{bar} = 10^5\,\mathrm{Pa}$

Prefixes Denoting Multiples and Submultiples

10^3	kilo (k)
10^6	mega (M)
10^9	giga (G)
10^{12}	tera (T)
10^{15}	peta (P)
10^{-3}	milli (m)
10^{-6}	micro (μ)
10^{-9}	nano (n)
10^{-12}	pico (p)
10^{-15}	femto (f)

Mathematical Signs

$\mathbb{N}$	set of natural numbers
$\mathbb{Z}$	set of integers
$\mathbb{Q}$	set of rational numbers
$\mathbb{R}$	set of real numbers
$\mathbb{C}$	set of complex numbers
$\forall x$	for all x
$\exists x$	existence of x
$\mapsto$	maps to
$\therefore$	therefore
$\because$	because
$=$	equals
$\simeq$	approximately equals
$\neq$	not equal to
$\equiv$	identical to, defined as
$>$	greater than
$\gg$	much greater than
$<$	smaller (or less) than
$\ll$	much smaller than
$\geq$	greater than or equal to
$\leq$	smaller (or less) than or equal to
$\propto$	proportional to
$\sim$	represented by, of the order
$\mathcal{O}(x)$	order of x

$\langle x \rangle$, $\overline{x}$	the average value of x
ln	logarithm of base e (natural logarithm)
Δx	increment in x
dx	infinitesimal increment in x
$z^* = x - iy$	complex conjugate of complex number; $z = x + iy$ ($x, y \in \mathbb{R}$, i = imaginary unit = $\sqrt{-1}$)
$\alpha^\dagger$	Hermitian conjugate of operator α
$\boldsymbol{\alpha}^\dagger$	Hermitian conjugate of matrix $\boldsymbol{\alpha}$
P^{-1}	inverse of P
$\delta(x)$	Dirac's delta function
$\delta_{ij} = \begin{cases} 1 & \text{if } i = j \\ 0 & \text{if } i \neq j \end{cases}$	Kronecker's delta
$\Theta(x) = \begin{cases} 1 & \text{if } x \geq 0 \\ 0 & \text{if } x < 0 \end{cases}$	Heaviside's step function
$\text{sgn}x = \begin{cases} 1 & \text{if } x > 0 \\ -1 & \text{if } x < 0 \end{cases}$	sign of x
$\dot{x} = dx/dt$	time derivative
$\text{grad}\phi \equiv \nabla\phi$	gradient of ϕ
$\text{div}\boldsymbol{A} \equiv \nabla \cdot \boldsymbol{A}$	divergence of $\boldsymbol{A}$
$\text{curl}\boldsymbol{A} \equiv \text{rot}\boldsymbol{A} \equiv \nabla \times \boldsymbol{A}$	curl (or rotation) of $\boldsymbol{A}$
∇	Nabla (or del) operator
$\Delta \equiv \nabla^2$	Laplacian operator

List of Symbols

The following list is not intended to be exhaustive. It includes symbols of frequent occurrence or special importance in this book.

Å	ångstrom ($= 10^{-8}$ cm $= 10^{-10}$ m)
$\boldsymbol{A}$	vector potential
a_0	lattice constant
$\boldsymbol{a}_1$, $\boldsymbol{a}_2$	nonorthogonal base vectors
$\boldsymbol{B}$	magnetic field (magnetic flux density)
C_V	specific heat at constant volume
c	heat capacity per particle
$c_{\mathbb{V}}$	heat capacity per unit volume
c	speed of light
$\mathcal{D}(\varepsilon)$	density of states in energy space
$\mathcal{D}(\omega)$	density of states in angular frequency
$\mathcal{D}(p)$, $\mathcal{D}(k)$	density of states in momentum space
E	total energy

E	internal energy
E_{F}	Fermi energy
$\boldsymbol{E}$	electric field vector
$\boldsymbol{E}_{\mathrm{H}}$	electric field vector due to the Hall voltage
e	electronic charge (absolute value)
$\hat{\boldsymbol{e}}_x, \hat{\boldsymbol{e}}_y, \hat{\boldsymbol{e}}_z$	orthogonal unit vectors
F	Helmholtz free energy
f	one-body distribution function
f_{B}	Bose distribution function
f_{F}	Fermi distribution function
f_0	Planck distribution function
G	Gibbs free energy
g	g-factor
$\mathcal{H}$	Hamiltonian
$\boldsymbol{H}_{\mathrm{a}}$	applied magnetic field vector
H_{c}	critical magnetic field (magnitude)
h	Planck's constant
h	single-particle Hamiltonian
$\hbar$	Planck's constant divided by 2π
I	magnetization
$\mathrm{i} \equiv \sqrt{-1}$	imaginary unit
$\boldsymbol{i}, \boldsymbol{j}, \boldsymbol{k}$	Cartesian unit vectors
$\boldsymbol{J}$	total current
$\boldsymbol{j}$	single-particle current
$\boldsymbol{j}$	current density
K	thermal conductivity
$\boldsymbol{k}$	wave vector (k-vector)
k_0	Coulomb's constant
k_{B}	Boltzmann constant
$\mathcal{L}$	Lagrangian function
L	normalization length
ℓ	mean free path
$\boldsymbol{l}$	angular momentum
M	molecular mass
M^*	magnetotransport mass
M	(symmetric) mass tensor
m	electron mass
m^*	cyclotron mass
m^*	effective mass
N	number of particles
$\mathcal{N}$	number operator
N_{L}	Landau level
n	particle number density
n_{c}	number density of the dressed electrons
n_{p}	number density of pairons

P	pressure
$\boldsymbol{P}$	total momentum
$\boldsymbol{p}$	momentum vector
p	momentum (magnitude)
Q	quantity of heat
$\boldsymbol{q}$	heat (energy) current
q	charge
R	resistance
$\boldsymbol{R}$	Bravais lattice vector
$\boldsymbol{R}$	position vector of the center of mass
R_{H}	Hall coefficient
r	radial coordinate
$\boldsymbol{r}$	position vector
S	entropy
S	Seebeck coefficient
T	absolute temperature
T_0	transition temperature
T_{c}	critical temperature
T_{F}	Fermi temperature
$\mathcal{T}$	kinetic energy
TR	grand ensemble trace
Tr	many-particle trace
tr	one-particle trace
V, $\mathbb{V}$	volume
V_{H}	Hall voltage
$\mathcal{V}$	potential energy
v	speed (magnitude of $\boldsymbol{v}$)
$\boldsymbol{v}$	velocity
$\boldsymbol{v}_{\mathrm{thermal}}$	thermal velocity
$\boldsymbol{v}_{\mathrm{d}}$	drift velocity
$v_{\mathrm{d}}(= \lvert\boldsymbol{v}_{\mathrm{d}}\rvert)$	drift speed
$\boldsymbol{v}_{\mathrm{F}}$	Fermi velocity
$v_{\mathrm{F}}(= \lvert\boldsymbol{v}_{\mathrm{F}}\rvert)$	Fermi speed
W	work
$\boldsymbol{w}$	wrapping vector
Z	partition function
$\alpha = -e/(2m)$	magnetogyric (magnetomechanical) ratio
$\mathrm{e}^{\alpha} \equiv z$	fugacity
$\beta \equiv 1/(k_{\mathrm{B}} T)$	reciprocal temperature
χ	magnetic susceptibility
ε	single-particle energy
ε_{F}	Fermi energy
ε_{g}	energy gap
ε_{p}	pairon energy
$\Theta(x)$	step function

θ	polar angle
λ	wavelength
λ	penetration depth
$\lambda (\equiv e^{\beta\mu})$	fugacity
κ	curvature
κ	quantum state
μ	chemical potential
$\boldsymbol{\mu}$	magnetic moment
μ_B	Bohr magneton
ν	frequency
ν	Landau level occupation ratio (filling factor)
Ξ	grand partition function
ξ	dynamical variable
ξ	coherence length
ρ	mass density
ρ	density operator
ρ	many-particle distribution function
ρ	resistivity
$\rho(B)$	magnetoresistivity
ρ_H	Hall resistivity
σ	total cross section
σ	electrical conductivity
σ_H	Hall conductivity
$\sigma_x, \sigma_y, \sigma_z$	Pauli spin matrices
τ	relaxation time
τ_c	collision time, average time between collision
τ_d	duration of collision
φ	distribution function
ϕ	azimuthal angle
ϕ	scalar potential
$\boldsymbol{\Phi}$	magnetic flux
Φ_0	flux quantum
$\boldsymbol{\Psi}$	quasiwavefunction for many condensed bosons
ψ	wavefunction for a quantum particle
$d\Omega = \sin\theta \, d\theta \, d\phi$	element of solid angle
$\omega \equiv 2\pi\nu$	angular frequency
ω_c	cyclotron frequency
ω_c	rate of collision (collision frequency)
ω_D	Debye frequency
$\langle \vert$	bra vector
$\vert \rangle$	ket vector
$(hkl), [hkl], \langle hkl \rangle$	crystallographic notation
$[\, , \,]$	commutator brackets
$\{\, , \,\}$	anticommutator brackets
$\{\, , \,\}$	Poisson brackets

Units

In much of the literature quoted, the unit of magnetic field $\boldsymbol{B}$ is the gauss. Electric fields are frequently expressed in $\mathrm{V\,cm^{-1}}$ and resistivity in Ω cm.

$$1\ \text{tesla (T)} = 10^4\ \text{gauss (G, (Gs))} \quad 1\ \Omega\,\text{m} = 10^2\ \Omega\,\text{cm}$$

The Planck constant h over 2π, $\hbar \equiv h/(2\pi)$, is used in dealing with an electron. The original Planck constant h is used in dealing with a photon.

Crystallographic Notation

This is mainly used to denote a direction, or the orientation of a plane, in a cubic metal. A plane (hkl) intersects the orthogonal Cartesian axes, coinciding with the cube edges, at a/h, a/k, and a/l from the origin, a being a constant, usually the length of a side of the unit cell. The direction of a line is denoted by $[hkl]$, the direction cosines with respect to the Cartesian axes being h/N, k/N, and l/N, where $N^2 = h^2 + k^2 + l^2$. The indices may be separated by commas to avoid ambiguity. Only occasionally will the notation be used precisely; thus, [100] or [001] usually means any cube axis and [111], any diagonal.

B and *H*

When an electron is described in quantum mechanics, its interaction with a magnetic field is determined by $\boldsymbol{B}$ rather than $\boldsymbol{H}$; that is, if the permeability μ is not unity, the electron motion is determined by $\mu \boldsymbol{H}$. It is preferable to forget $\boldsymbol{H}$ altogether and use $\boldsymbol{B}$ to define all field strengths. The $\boldsymbol{B}$ is connected with a vector potential $\boldsymbol{A}$ such that $\boldsymbol{B} = \nabla \times \boldsymbol{A}$. The magnetic field $\boldsymbol{B}$ is effectively the same inside and outside the metal sample.

List of Abbreviations

1D	one dimensional
2D	two dimensional
3D	three dimensional
ARPES	angle-resolved photoemission spectroscopy
bcc	body-centered cubic
BCS	Bardeen–Cooper–Schrieffer
BEC	Bose–Einstein condensation
c-	composite-
c.c.	complex conjugate
C-	carbon-
CM	center of mass

CNT	carbon nanotube
cub, cub	cubic
dHvA	de Haas–van Alphen
dia	diamond
DOS	density of states
DP	Dirac picture
"electron"	see p. 7
EOB	Ehrenfest–Oppenheimer–Bethe
f (c-)	fundamental (composite-)
fcc	face-centered cubic
h.c.	Hermitian conjugate
hcp	hexagonal closed packed
"hole"	see p. 7
hex	hexagonal
HP	Heisenberg picture
HRC	high-resistance contacts
HTSC	high-temperature superconductivity
KP	Kronig–Penney
lhs	left-hand side
LL	Landau level
LRC	low-resistance contacts
mcl, mcl	monoclinic
MIT	metal-insulator transition
MR	magnetoresistance
MWNT	multiwalled (carbon) nanotube
NFEM	nearly free electron model
NT	nanotube
orc	orthorhombic
QH	quantum Hall
QHE	quantum Hall effect
rhs	right-hand side
rhl	rhombohedral
sc	simple cubic
SdH	Shubnikov–de Haas
SP	Schrödinger picture
sq	square
SQUID	superconducting quantum interference device
SWNT	single-wall (carbon) nanotube
tcl	triclinic
tet	tetragonal
vrh	variable range hopping
WS	Wigner–Seitz
ZBA	zero-bias anomaly

1
Introduction

1.1
Carbon Nanotubes

Graphite and diamond are both made of carbons. They have different lattice structures and different properties. Diamond is brilliant and it is an insulator while graphite is black and it is a good conductor.

In 1991 Iijima [1] discovered carbon nanotubes (CNTs) in the soot created in an electric discharge between two carbon electrodes. These nanotubes ranging from 4 to 30 nm in diameter were found to have helical multiwalled structures as shown in Figures 1.1 and 1.2 after electron diffraction analysis. The tube length is about 1 μm.

The scroll-type tube shown in Figure 1.2 is called a *multiwalled carbon nanotube* (MWNT). A *single-wall nanotube* (SWNT) was fabricated by Iijima and Ichihashi [2] and by Bethune *et al.* [3] in 1993. Their structures are shown in Figure 1.3.

The tube is about 1 nm in diameter and a few micrometers in length. The tube ends are closed as shown. Because of their small radius and length-to-diameter ratio $> 10^4$, they provide an important system for studying two-dimensional (2D)

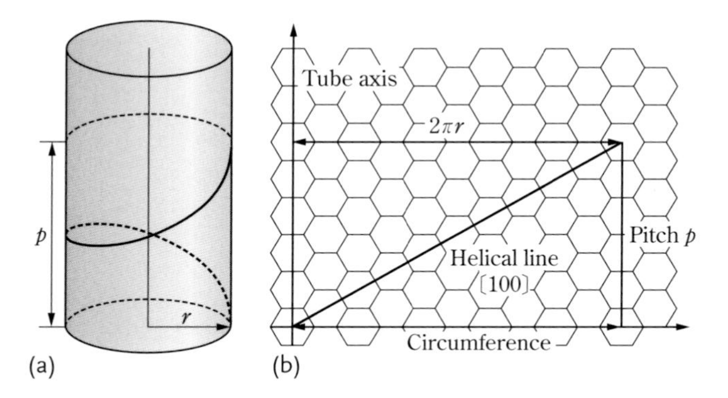

Figure 1.1 Schematic diagram showing (a) a helical arrangement of graphitic carbons and (b) its unrolled plane. The helical line is indicated by the heavy line passing through the centers of the hexagons.

Electrical Conduction in Graphene and Nanotubes, First Edition. S. Fujita and A. Suzuki.

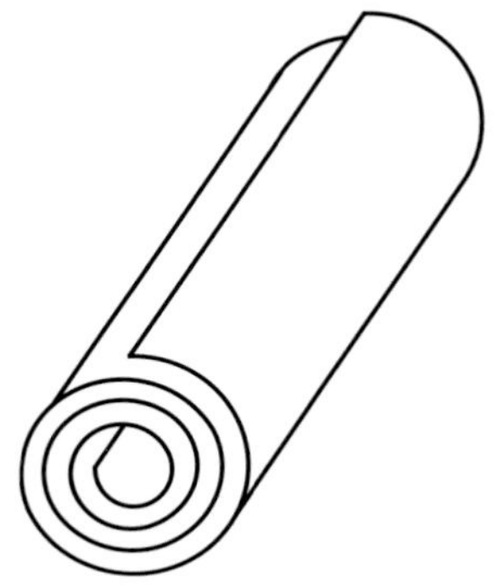

Figure 1.2 A multiwalled nanotube. The tube diameter ranges from 4 to 30 nm and its length is about 1 μm. (Original figure, Iijima [1])

physics, both theoretically and experimentally. Unrolled carbon sheets are called *graphene*.[1] They have a honeycomb lattice structure as shown in Figure 1.1b.

A SWNT can be constructed from a slice of graphene (that is a single planar layer of the honeycomb lattice of graphite) rolled into a circular cylinder.

Carbon nanotubes are light since they are entirely made of the light element carbon (C). They are strong and have excellent elasticity and flexibility. In fact, carbon fibers are used to make tennis rackets, for example. Their main advantages in this regard are their high chemical stability as well as their strong mechanical properties.

Today's semiconductor technology is based mainly on silicon (Si). It is said that carbon-based devices are expected to be as important or even more important in the future. To achieve this purpose we must know the electrical transport properties of CNTs, which are very puzzling, as is explained below. The principal topics in this book are the remarkable electrical transport properties in CNTs and graphene on which we will mainly focus in the text.

The conductivity σ in individual CNTs varies, depending on the tube radius and the pitch of the sample. In many cases the resistance decreases with increasing temperature. In contrast the resistance increases in the normal metal such as copper (Cu). The electrical conduction properties in SWNTs separates samples into two classes: *semiconducting* or *metallic*. The room-temperature conductivities are higher for the latter class by two or more orders of magnitude. Saito *et al.* [6] proposed a model based on the different arrangements of C-hexagons around the circumference, called the *chiralities*. Figure 1.3a–c show an *armchair*, *zigzag*, and a general *chiral* CNT, respectively. After statistical analysis, they concluded that semiconducting SWNTs should be generated three times more often than metallic SWNTs. Moriyama *et al.* [7] fabricated 12 SWNT devices from one chip, and observed that

1) Graphene is the basic structural element of some carbon allotropes including graphite, CNTs, and fullerenes. The name comes from graphite + -ene; graphite itself is composed of many graphene sheets stacked together. Graphene as a name indicates a single, two-dimensional layer of three-dimensional graphite which contains many layers of carbon hexagons. Two-dimensional graphene can exist in nature as the spacing between layers (3.35 Å) is longer than the distance to neighboring atoms $a_{C\text{–}C}$ (1.42 Å) within the same plane. It has been challenging to isolate one layer from bulk graphite.

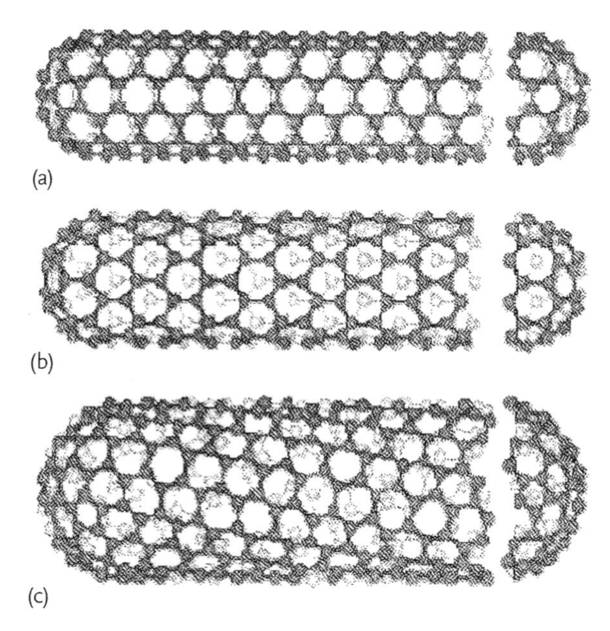

Figure 1.3 SWNTs with different chiralities and possible caps at each end: (a) shows a so-called *armchair* carbon nanotube (CNT), (b) a *zigzag* CNT, and (c) a general *chiral* CNT. One can see from the figure that the orientation of the C-hexagon in the honeycomb lattice relative to the tube axis can be taken arbitrarily. The terms "armchair" and "zigzag" refer to the arrangement of C-hexagons around the circumference. (From [4, 5]).

two of the SWNT samples were semiconducting and the other ten were metallic, a clear discrepancy between theory and experiment. We propose a new classification. The electrical conduction in SWNTs is either *semiconducting* or *metallic* depending on whether each pitch of the helical line connecting the nearest-neighbor C-hexagon contains an integral number of hexagons or not. The second alternative (metallic SWNT) occurs more often since the helical angle between the helical line and the tube axis is not controlled in the fabrication process. In the former case the system (semiconducting SWNT) is periodic along the tube length and the "holes" (and not "electrons") can travel along the wall. Here and in the text "electrons" ("holes"), by definition, are *quasi*electrons which are excited above (below) the Fermi energy *and* which circulate clockwise (counterclockwise) when viewed from the tip of the external magnetic field vector. "Electrons" ("holes") are generated in the negative (positive) side of the Fermi surface which contains the negative (positive) normal vector, with the convention that the positive normal points in the energy-increasing direction. In the Wigner–Seitz (WS) cell model [7] the primitive cell for the honeycomb lattice is a rhombus. This model is suited to the study of the ground state of graphene. For the development of the electron dynamics it is necessary to choose a *rectangular unit cell* which allows one to define the effective masses associated with the motion of "electrons" and "holes" in the lattice.

Silicon (Si) (germanium (Ge)) forms a diamond lattice which is obtained from the zinc sulfide (ZnS) lattice by disregarding the species. The electron dynamics of Si are usually discussed in terms of cubic lattice languages. Graphene and graphite have hexagonal lattice structures. Silicon and carbon are both quadrivalent materials but because of their lattice structures, they have quite different physical properties.

1.2
Theoretical Background

1.2.1
Metals and Conduction Electrons

A metal is a conducting crystal in which electrical current can flow with little resistance. This electrical current is generated by moving electrons. The electron has mass m and charge $-e$, which is negative by convention. Their numerical values are $m = 9.1 \times 10^{-28}$ g and $e = 4.8 \times 10^{-10}$ esu $= 1.6 \times 10^{-19}$ C. The electron mass is about 1837 times smaller than the least-massive (hydrogen) atom. This makes the electron extremely mobile. It also makes the electron's quantum nature more pronounced. The electrons participating in the transport of charge are called *conduction electrons*. The conduction electrons would have orbited in the outermost shells surrounding the atomic nuclei if the nuclei were separated from each other. Core electrons which are more tightly bound with the nuclei form part of the metallic ions. In a pure crystalline metal, these metallic ions form a relatively immobile array of regular spacing, called a *lattice*. Thus, a metal can be pictured as a system of two components: mobil electrons and relatively immobile lattice ions.

1.2.2
Quantum Mechanics

Electrons move following the quantum laws of motion. A thorough understanding of quantum theory is essential. Dirac's formulation of quantum theory in his book, *Principles of Quantum Mechanics* [9], is unsurpassed. Dirac's rules that the quantum states are represented by *bra* or *ket* vectors and physical observables by Hermitian operators are used in the text. There are two distinct quantum effects, the first of which concerns a single particle and the second a system of identical particles.

1.2.3
Heisenberg Uncertainty Principle

Let us consider a simple harmonic oscillator characterized by the Hamiltonian

$$\mathcal{H} = \frac{p^2}{2m} + \frac{kx^2}{2} , \qquad (1.1)$$

where m is the mass, k the force constant, p the momentum, and x the position. The corresponding energy eigenvalues are

$$\varepsilon_n = \hbar\omega_0\left(n + \frac{1}{2}\right), \quad \omega_0 \equiv \left(\frac{k}{m}\right)^{1/2}, \quad n = 0, 1, 2, \ldots \tag{1.2}$$

The energies are quantized in (1.2). In contrast the classical energy can be any positive value. The lowest quantum energy $\varepsilon_0 = \hbar\omega_0/2$, called the *energy of zero-point motion*, is not zero. The most stable state of any quantum system is not a state of *static equilibrium* in the configuration of lowest potential energy, it is rather a *dynamic equilibrium* for the zero-point motion [10, 11]. Dynamic equilibrium may be characterized by the minimum total (potential + kinetic) energy under the condition that each coordinate q has a range Δq and the corresponding momentum p has a range Δp, so that the product $\Delta q \Delta p$ satisfies the *Heisenberg uncertainty relation*:

$$\Delta q \Delta p > h \,. \tag{1.3}$$

The most remarkable example of a macroscopic body in dynamic equilibrium is liquid helium (He). This liquid with a boiling point at 4.2 K is known to remain liquid down to 0 K. The zero-point motion of He atoms precludes solidification.

1.2.4
Bosons and Fermions

Electrons are fermions. That is, they are indistinguishable quantum particles subject to the *Pauli exclusion principle*. Indistinguishability of the particles is defined by using the permutation symmetry. According to Pauli's principle no two electrons can occupy the same state. Indistinguishable quantum particles not subject to the Pauli exclusion principle are called bosons. Bosons can occupy the same state with no restriction. Every elementary particle is either a *boson* or a *fermion*. This is known as the *quantum statistical postulate*. Whether an elementary particle is a boson or a fermion is related to the magnitude of its spin angular momentum in units of $\hbar$. Particles with integer spins are bosons, while those with half-integer spins are fermions [12]. This is known as Pauli's *spin-statistics theorem*. According to this theorem and in agreement with all experimental evidence, electrons, protons, neutrons, and μ-mesons, all of which have spin of magnitude $\hbar/2$, are fermions, while photons (quanta of electromagnetic radiation) with spin of magnitude $\hbar$, are bosons.

1.2.5
Fermi and Bose Distribution Functions

The average occupation number at state $\boldsymbol{k}$, denoted by $\langle n_k \rangle$, for a system of free fermions in equilibrium at temperature T and chemical potential μ is given by the

Fermi distribution function:

$$\langle n_k \rangle = f_F(\varepsilon_k) \equiv \frac{1}{\exp((\varepsilon_k - \mu)/(k_B T)) + 1} \quad \text{for fermions,} \tag{1.4}$$

where ε_k is the single-particle energy associated with the state $\boldsymbol{k}$. The average occupation number at state $\boldsymbol{k}$ for a system of free bosons in equilibrium is given by the *Bose distribution function*:

$$\langle n_k \rangle = f_B(\varepsilon_k) \equiv \frac{1}{\exp((\varepsilon_k - \mu)/(k_B T)) - 1} \quad \text{for bosons.} \tag{1.5}$$

1.2.6
Composite Particles

Atomic nuclei are composed of *nucleons* (protons, neutrons), while atoms are composed of nuclei and electrons. It has been experimentally demonstrated that these composite particles are indistinguishable quantum particles. According to *Ehrenfest–Oppenheimer–Bethe's rule* [12, 13], the center of mass (CM) of a composite moves as a fermion (boson) if it contains an odd (even) number of elementary fermions. Thus, He^4 atoms (four nucleons, two electrons) move as bosons while He^3 atoms (three nucleons, two electrons) move as fermions. Cooper pairs (two electrons) move as bosons.

1.2.7
Quasifree Electron Model

In a metal at the lowest temperatures conduction electrons move in a nearly stationary periodic lattice. Because of the Coulomb interaction among the electrons, the motion of the electrons is correlated. However, each electron in a crystal moves in an extremely weak self-consistent periodic field. Combining this result with the Pauli exclusion principle, which applies to electrons with no regard to the interaction, we obtain the *quasifree electron model*. The quasifree electron moves with the effective mass m^* which is different from the gravitational mass m_e. In this model the quantum states for the electron in a crystal are characterized by wave vector (k vector: $\boldsymbol{k}$) and energy

$$\varepsilon = E(k) . \tag{1.6}$$

At 0 K, all of the lowest energy states are filled with electrons, and there exists a sharp Fermi surface represented by

$$E(k) = \varepsilon_F , \tag{1.7}$$

where ε_F is the *Fermi energy*. Experimentally, the electrons in alkali metals, which form body-centered cubic (bcc) lattices, including lithium (Li), sodium (Na), and potassium (K), behave like quasifree electrons.

1.2.8
"Electrons" and "Holes"

"Electrons" ("holes") in the text are defined as *quasiparticles* possessing charge e (magnitude) that circulate counterclockwise (clockwise) when viewed from the tip of the applied magnetic field vector $\boldsymbol{B}$. This definition is used routinely in semiconductor physics. We use the quotation-marked "electron" to distinguish it from the generic electron having the gravitational mass m_e. A "hole" can be regarded as a particle having positive charge, positive mass, and positive energy. The "hole" does not, however, have the same effective mass m^* (magnitude) as the "electron," so that "holes" are not true antiparticles like positrons. We will see that "electrons" and "holes" are thermally excited particles and they are closely related to the curvature of the Fermi surface (see Chapter 3).

1.2.9
The Gate Field Effect

Graphene and nanotubes are often subjected to the so-called *gate voltage* in experiments. We will show here that the gate voltage polarizes the conductor and hence the surface charges ("electrons," "holes") are induced. The actual conductor may have a shape and a particular Fermi surface. But in all cases surface charges are induced by electric fields. If a bias voltage is applied, then some charges can move and generate currents.

A.
Let us take a rectangular metallic plate and place it under an external electric field $\boldsymbol{E}$, see Figure 1.4.

When the upper and lower sides are parallel to the field $\boldsymbol{E}$, then the remaining two side surfaces are polarized so as to reduce the total electric field energy. If the plate is rotated, then all side surfaces are polarized.

B.
Let us now look at the electric field effect in k-space. Assume a quasifree electron system which has a spherical Fermi surface at zero field. Upon the application of a static field $\boldsymbol{E}$, the Fermi surface will be shifted towards the right by $qE\tau/m^*$, where τ is the mean free time and m^* the effective mass, as shown in Figure 1.5. There is a steady current since the sphere is off from the center O. We may assume

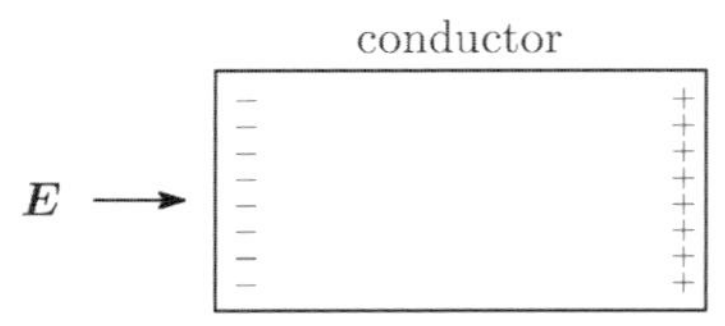

Figure 1.4 The surface charges are induced in the conductor under an external electric field $\boldsymbol{E}$.

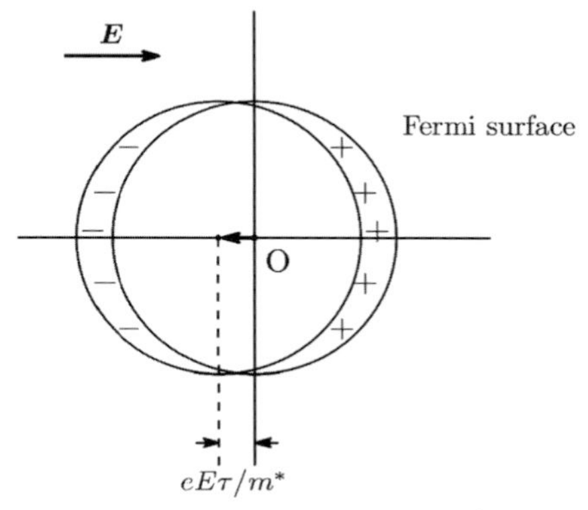

Figure 1.5 The Fermi surface is shifted by $eE\tau/m^*$ due to the electric field $\boldsymbol{E}$.

that the ionic lattice is stationary. Then, there is an unbalanced charge distribution as shown, where we assumed $q = -e < 0$. This effect will appear only on the surface of the metal. We used the fermionic nature of electrons in **B**.

1.3 Book Layout

In Chapters 2 and 3 kinetic theory and Bloch electron dynamics are developed, respectively. Phonon and electron–phonon interaction are discussed in Chapter 4. These chapters are preliminaries for the theory of the conductivity of carbon nanotubes, which is discussed in Chapters 5 and 6. Semiconducting SWNTs are discussed in Chapter 6. A quantum statistical theory of superconductivity is summarized in Chapter 7. Chapter 8 deals with the supercurrents in metallic SWNTs, starting with the BCS-like Hamiltonian and deriving expressions for a linear dispersion relation, and a critical (superconducting) temperature. Metallic SWNTs exhibit non-Ohmic behavior, and charged particles appear to run through the tube length with no scattering. We interpret this in terms of the condensed Cooper pairs (pairons).

An applied static magnetic field induces a profound change in the electron states. Pauli's paramagnetic and Landau's diamagnetism are described in Chapter 9. Landau states generate an oscillatory density of states that induces de Haas–van Alphen oscillation which is discussed in Chapter 10. The Quantum Hall Effect (QHE) in GaAs/AlGaAs is summarized in Chapter 11. The QHE in graphene observed at room temperature is discussed in Chapter 12. The QHE occurs where the "hole" ("electron") density becomes high near the neck Fermi surface, which develops by charging the graphene through the gate voltage. The different temperatures generate different carrier densities and the resulting carrier diffusion generates a thermal electromotive force. A new formula for the Seebeck coefficient is obtained and is applied to multiwalled carbon nanotubes in Chapter 13. In Chapter 14, we discuss miscellaneous topics.

1.4 Suggestions for Readers

Graphene and CNTs are composed entirely of carbons but their lattice structures are distinct from each other. The simple free electron model does not work. To describe the electrical conduction of graphene and CNTs it is necessary to understand a number of advanced topics including superconductivity and Fermi surfaces.

1.4.1 Second Quantization

Reading Chapter 7 Superconductivity requires a knowledge of second quantization. The authors suggest that the readers learn the second quantization in two steps.

1. Dirac solved the energy-eigenvalue problem for a simple harmonic oscillator in the Heisenberg picture, using creation and annihilation operators $(a^\dagger, a)$, see Chapter 4, Section 4.3. We follow Dirac [9] and obtain the eigenvalues, $(n' + 1/2)\hbar\omega$, where n' is the eigenvalues of $n = a^\dagger a$, $n' = 0, 1, 2, \ldots$
2. Read Appendix A.1, where a general theory for a quantum many-boson and fermion system is presented.

1.4.2 Semiclassical Theory of Electron Dynamics

Electrons and phonons are regarded as *waves packets* in solids. Dirac showed that the wave packets move, following classical equations of motion [9]. The conduction electron ("electron," "hole") size is equal to the orthogonal unit cell size. The phonon size is about two orders of magnitude greater at room temperature. The "electron" and "hole" move with effective masses m^* which are distinct from the gravitational effective mass m_e. Bloch electron dynamics are described in Chapter 3.

1.4.3 Fermi Surface

The time-honored WS cell model can be used for cubic lattice systems including a diamond lattice. For hexagonal systems including graphene and graphite an orthogonal unit cell model must be used to establish the k-space. Read Sections 5.2 and 5.4. The same orthogonal unit cell model must be used for the discussion of phonons.

In our quantum statistical theory we do not jump to conclusions. We make arguments backed up by step-by-step calculations. This is the surest way of doing and learning physics for ordinary men and women.

References

1 Iijima, S. (1991) *Nature*, **354**, 56.
2 Iijima, S. and Ichihashi, T. (1993) *Nature*, **363**, 603.
3 Bethune, D.S., Klang, C.H., de Vries, M.S., Gorman, G., Savoy, R., Vazquez, J., and Beyers, R. (1993) *Nature*, **363**, 605.
4 Saito, R., Dresselhaus, G., and Dresselhaus, M.S. (1998) *Physical Properties of Carbon Nanotubes*, Imperial College Press, London.
5 Dresselhaus, M.S. *et al.* (1955) *Carbon*, **33**, 883.
6 Saito, R., Fujita, M., Dresselhaus, G., and Dresselhaus, M.S. (1992) *Appl. Phys. Lett.*, **60**, 2204.
7 Moriyam, S., Toratani, K., Tsuya, D., Suzuki, M., Aoyagi, Y., and Ishibashi, K. (2004) *Physica E*, **24**, 46.
8 Kittel, C. (2005) *Introduction to Solid State Physics*, 8th edn, John Wiley & Sons, Inc., New York.
9 Dirac, P.A.M. (1958) *Principles of Quantum Mechanics*, 4th edn, Oxford University Press, London, pp. 121–125 and 136–139.
10 London, F. (1938) *Nature* **141**, 643.
11 London, F. (1964) *Superfluids*, vol. I and II, Dover, New York.
12 Pauli, W. (1940) *Phys. Rev.*, **58**, 716.
13 Ehrenfest, P. and Oppenheimer (1931) *J.R. Phys. Rev.*, **37**, 333.
14 Bethe, H.A. and Bacher, R.F. (1936) *Rev. Mod. Phys.*, **8**, 193.
15 Fujita, S. and Morabito, D.L. (1998) *Mod. Phys. Lett. B*, **12**, 753.

2
Kinetic Theory and the Boltzmann Equation

Elements of the kinetic theory of gas dynamics, the Boltzmann equation method, and electrical conduction are discussed in this chapter.

2.1
Diffusion and Thermal Conduction

In order to clearly understand diffusion let us look at the following simple situation. Imagine that four particles are in space a, and two particles are in space b as shown in Figure 2.1.

Assuming that both spaces a and b have the same volume, we may say that the particle density is higher in a than in b. We assume that half of the particles in each space will be heading toward the boundary CC′. It is then natural to expect that in due time two particles would cross the boundary CC′ from a to b, and one particle from b to a. This means that more particles would pass the boundary from a to b, that is, from the side of high density to that of low density. This is the cause of diffusion.

The essential points in the above arguments are the reasonable assumptions that

(i) the particles flow out from a given space in all directions with the same probability, and
(ii) the rate of this outflow is proportional to the number of particles contained in that space.

In the present case condition (i) will be assured by the fact that each particle collides with other particles frequently so that it may forget how it originally entered the space, and may leave with no preferred direction. From a more quantitative perspective it is found that the particle current j is proportional to the density gradient ∇n with n being the number density:

$$j = -D\nabla n \,, \tag{2.1}$$

where D is called the *diffusion coefficient*. This linear relation (2.1) is called *Fick's law* [1, 2].

Electrical Conduction in Graphene and Nanotubes, First Edition. S. Fujita and A. Suzuki.

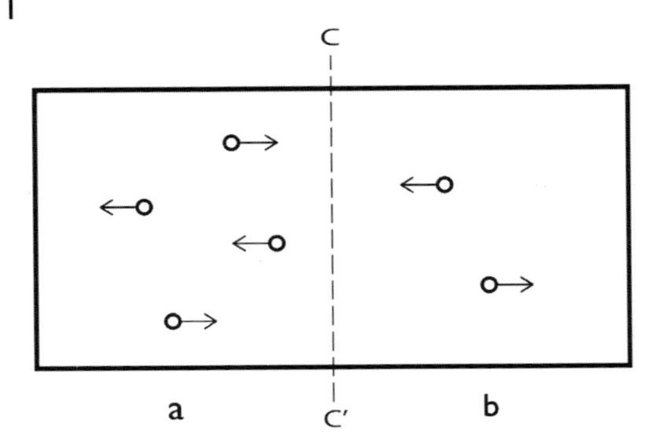

Figure 2.1 If the particles flow out in all directions with no preference, there will be more particles crossing the imaginary boundary CC′ in the a to b direction than in the opposite direction.

Consider next thermal conduction. Assume that the spaces a and b are occupied by the same number of the particles. Further assume that the temperature T is higher in b than in a. Then, the particle speed is higher in b than in a on the average. In due time a particle crosses the boundary CC′ from a to b and another crosses the boundary CC′ from b to a. Then, the energy is transferred through the boundary. In a more detailed study *Fourier's law* [1, 2] is observed:

$$q = -K\nabla T \,, \tag{2.2}$$

where $\boldsymbol{q}$ is the heat (energy) current and the proportionality constant K is called the *thermal conductivity*.

2.2
Collision Rate: Mean Free Path

Let us consider a particle moving through a medium containing n molecules per unit volume. If the particle proceeds with a speed v, it will sweep the volume $(v\mathrm{d}t) \times A = vA\mathrm{d}t$ during the time interval $\mathrm{d}t$, where A represents the cross section. See Figure 2.2.

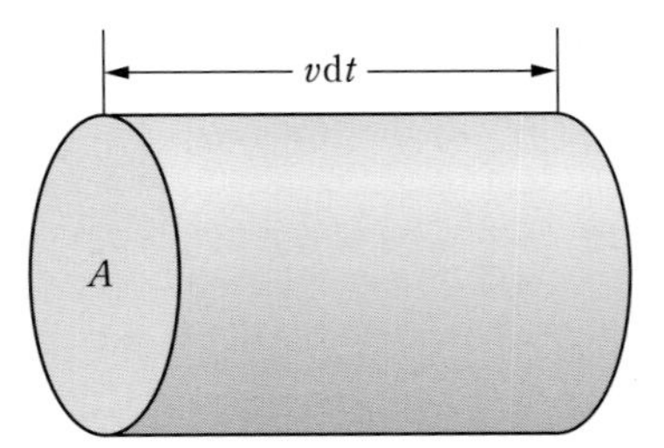

Figure 2.2 A particle moving with a speed v sweeps the volume $(v\mathrm{d}t)A$ during the time $\mathrm{d}t$, where A represents the cross section.

The particle would collide with any molecule if the latter lies within the cylinder. Now, the number of molecules in the cylindrical volume $v\mathrm{d}t$ is $n(vA\mathrm{d}t) = nvA\mathrm{d}t$. Dividing this number by $\mathrm{d}t$, we obtain the number of collisions per unit time:

$$\omega_c = nvA\,. \tag{2.3}$$

This frequency ω_c is called the *collision rate*. Note that the collision rate depends linearly on the speed v, the number density n, and the cross section A.

The above consideration may be applied to the molecular collision in a gas. In this case, the particle in question is simply one of the molecules. Let us estimate the collision rate for a typical gas. For neon gas, the interaction potential between two atoms has a range of a few ångstroms,

$$R(\text{range}) \approx 2\,\text{Å} = 2 \times 10^{-8}\,\text{cm}\,. \tag{2.4}$$

Therefore, the total cross section A has the following order of magnitude:

$$A = \pi R^2 \approx 3.14 \times (2 \times 10^{-8})^2\,\text{cm}^2 = 1.2 \times 10^{-15}\,\text{cm}^2\,. \tag{2.5}$$

A typical atom has a kinetic energy of the order $3/2k_\text{B}T$. Therefore, it has the thermal speed

$$v_\text{thermal} = \left\langle \left(\frac{2\varepsilon}{m}\right)^{1/2} \right\rangle = \left(2 \cdot \frac{3}{2}\frac{k_\text{B}T}{m}\right)^{1/2} = \left(\frac{3k_\text{B}T}{m}\right)^{1/2}\,. \tag{2.6}$$

Using the data $T = 273$ K and m (neon) $= 20\,\mathrm{m_p}$ (proton mass), we then obtain

$$v_\text{thermal} = 5.9 \times 10^4\,\text{cm}\,\text{s}^{-1} = 590\,\text{m}\,\text{s}^{-1}\,. \tag{2.7}$$

It is interesting to note that this molecular speed has the same order of magnitude as the speed of sound $340\,\text{m}\,\text{s}^{-1}$. At $0°$ C and 1 atmospheric pressure, the number of molecules per cm^3 is

$$n = 2.69 \times 10^{19}\,\text{cm}^{-3}\,. \tag{2.8}$$

If we substitute the values from (2.5)–(2.7) into (2.3), we obtain

$$\omega_c = 1.91 \times 10^9\,\text{s}^{-1}\,. \tag{2.9}$$

Here we see that atoms collide with each other an enormous number of times in a second.

Strictly speaking, we should choose the average speed of approach between atoms rather than the thermal speed when estimating the value of v when both molecules move. Such a choice, however, will not change the order of magnitude of ω_c. In a more quantitative treatment, this choice becomes important.

The inverse of the collision rate defined by

$$\tau_c \equiv \frac{1}{\omega_c} = \frac{1}{n v_\text{thermal} A} \tag{2.10}$$

is called the *collision time*, or the *average time between collisions*. Substituting the numerical value from (2.9), we get

$$\tau_c = \frac{1}{1.91 \times 10^9\ \mathrm{s}^{-1}} = 5.22 \times 10^{-10}\ \mathrm{s}\ . \tag{2.11}$$

Let us now compare this time τ_c with the average collision duration, the average time that the molecule spends within the force range R of another particle. The latter is defined by

$$\tau_d \equiv \frac{R}{v_{\text{thermal}}}\ . \tag{2.12}$$

Using the numerical values from (2.5) and (2.6), we have

$$\tau_d = 3.4 \times 10^{-13}\ \mathrm{s}\ . \tag{2.13}$$

Comparison between (2.11) and (2.13) shows that

$$\tau_c \gg \tau_d\ . \tag{2.14}$$

This means that in a typical gas the molecules move freely most of the time, and occasionally collide with each other.

By multiplying the thermal speed v_{thermal} by the average time between collision τ_c, we obtain

$$v_{\text{thermal}} \times \tau_c \equiv l \quad \text{(mean free path)}\ . \tag{2.15}$$

This quantity, called the *mean free path*, gives a measure of the distance that a typical molecule covers between successive collisions. From (2.10) and (2.15), we obtain

$$l = v_{\text{thermal}} \times \tau_c = v_{\text{thermal}} \times (nAv_{\text{thermal}})^{-1} = (nA)^{-1}\ . \tag{2.16}$$

Note that the mean free path does not depend on the speed of the particle, and therefore has a value independent of temperature. Introducing the numerical values from (2.7) and (2.11), we obtain

$$l \approx 3.1 \times 10^{-5}\ \mathrm{cm} = 3100\ \text{Å} \gg R(\sim 2\ \text{Å})\ . \tag{2.17}$$

We see that the mean free path is about three orders of magnitude greater than the force range.

Problem 2.2.1. Using the numerical values introduced in the present section, estimate

1. The probability of a particular particle being within the force range of another particle, and
2. The probability of a particular particle being within the force range of two particles simultaneously.

Problem 2.2.2. Assume that the probability of collision for a small distance Δx is given by $\Delta x/l$. Show that the probability of the particle proceeding without collision for a finite distance x is given by $\exp(-x/l)$. Hint: $\lim_{n\to\infty}(1-x/n)^n = \mathrm{e}^{-x}$.

2.3 Electrical Conductivity and Matthiessen's Rule

Let us consider a system of electrons moving independently in a potential field of impurities, which act as scatterers. The impurities are assumed to be distributed uniformly.

Under the action of an electric field $\boldsymbol{E}$ that points along the positive x-axis, a classical electron with mass m will move following Newton's equation of motion:

$$m\frac{\mathrm{d}v_x}{\mathrm{d}t} = -eE\,, \tag{2.18}$$

in the absence of an impurity. Solving this, we obtain

$$v_x = -\frac{e}{m}Et + v_x^0\,, \tag{2.19}$$

where v_x^0 is the x-component of the initial velocity. For a free electron the velocity v_x can increase indefinitely and leads to infinite conductivity.

In the presence of impurities, the uniform acceleration will be interrupted by scattering. When the electron hits a scatterer (impurity), the velocity will suffer an abrupt change in direction and grow again following (2.19) until the electron hits another scatterer. Let us denote the average time between successive scatterings or mean free time by τ_f. The average velocity $\langle v_x\rangle$ is then given by

$$\langle v_x\rangle = -\frac{e}{m}E\tau_\mathrm{f}\,, \tag{2.20}$$

where we assumed that the electron forgets its preceding motion every time it hits a scatterer, and the average initial velocity after collision is zero:

$$\langle v_x^0\rangle = 0\,. \tag{2.21}$$

The charge current density (average current per unit volume) j is given by

$$\begin{aligned} j &= (\text{charge} : -e) \times (\text{number density} : n) \times (\text{velocity} : \langle v_x\rangle) \\ &= -en\langle v_x\rangle = n\frac{e^2\tau_\mathrm{f}}{m}E\,, \end{aligned} \tag{2.22}$$

where we used (2.20). According to *Ohm's law*, the current density j is proportional to the applied electric field E when this field is small:

$$j = \sigma E\,. \tag{2.23}$$

The proportionality factor σ is called the *electrical conductivity* [1, 2]. It represents the facility with which the current flows in response to the electric field. Combining the last two equations, we obtain

$$\boxed{\sigma = n\frac{e^2\tau_f}{m}} \,. \tag{2.24}$$

This equation is used in the qualitative discussion of the electrical transport phenomenon. The inverse mass-dependence law means that the ion contribution to electrical transport in an ionized gas will be smaller by at least three orders of magnitude than the electron contribution. Also note that the conductivity is higher if the number density is greater and if the mean free time is greater.

The inverse of the mean free time τ_f,

$$\Gamma = \frac{1}{\tau_f} \tag{2.25}$$

is called the *scattering rate* or the *relaxation rate*. Roughly speaking this Γ represents the mean frequency with which the electron is scattered by impurities (scatterers). The scattering rate Γ is given by

$$\Gamma = n_I v A \,, \tag{2.26}$$

where n_I, v, and A are respectively the density of scatterers, the electron speed, and the scattering cross section.

If there is more than one kind of scatterer, the scattering rate may be computed by the addition law:

$$\Gamma = n_1 v_1 A_1 + n_2 v_2 A_2 + \ldots \equiv \Gamma_1 + \Gamma_2 + \ldots \tag{2.27}$$

This is called *Matthiessen's rule.*

Historically and still today, the analysis of resistance data for a conductor is performed as follows: if the electrons are scattered by impurities and again by phonons (quanta of lattice vibrations), the total resistance will be written as the sum of the resistances due to each separate cause of scattering:

$$R_{\text{total}} = R_{\text{impurity}} + R_{\text{phonon}} \,. \tag{2.28}$$

This is the original statement of Matthiessen's rule. In further detail, the electron–phonon scattering depends on temperature because of the changing phonon population while the effect of the electron–impurity scattering is temperature-independent. By separating the resistance into two parts, one temperature-dependent and the other temperature-independent, we may apply Matthiessen's rule.

Problem 2.3.1. Free electrons are confined within a long rectangular planer strip. Assume that each electron is diffusely scattered at the boundary so that it may move in all directions without preference after the scattering. Find the mean free path along the length of the strip. Calculate the conductivity.

Problem 2.3.2. Assume the same condition as in Problem 2.3.1 for the case in which electrons are confined within a long circular cylinder. Find the conductivity.

2.4
The Hall Effect: "Electrons" and "Holes"

In this section we discuss the *Hall effect*, which was discovered in 1879 by E.H. Hall and published in [3]. The "electron" ("hole") has a negative (positive) charge $-e$ $(+e)$, $e = 4.80 \times 10^{-10}$ esu $= 1.6 \times 10^{-19}$ C. As we see later, "electrons" and "holes" play very important roles in the microscopic theory of electrical transport and superconductivity. Let us consider a conducting wire connected to a battery. If a magnetic field $\boldsymbol{B}$ is applied, the field penetrates the wire. The *Lorentz force,*

$$\boldsymbol{F} = q\boldsymbol{v} \times \boldsymbol{B} , \tag{2.29}$$

where q is the charge of the carrier, may then affect the electron's classical motion. If so, the picture of the straight line motion of a free electron in kinetic theory has to be modified significantly. If the field $\boldsymbol{B}$ (magnitude) is not too high and the stationary state is considered, the actual physical situation turns out to be much simpler.

Take the case in which the field $\boldsymbol{B}$ is applied perpendicular to a wire of rectangular cross section as shown in Figure 2.3.

Experiments show that a voltage V_{H} is generated perpendicular to the field $\boldsymbol{B}$ *and* electric current $\boldsymbol{j}$ such that a steady current flows in the wire apparently unhindered. This is the essence of the Hall effect. We may interpret this condition as follows. Let us write the current density $\boldsymbol{j}$ as

$$\boldsymbol{j} = qn\boldsymbol{v}_{\mathrm{d}} , \tag{2.30}$$

where n is the density of conduction electrons and $\boldsymbol{v}_{\mathrm{d}}$ the *drift velocity*. A charge carrier having a velocity equal to the drift velocity $\boldsymbol{v}_{\mathrm{d}}$ is affected by the *Lorentz force*:

$$\boldsymbol{F} = q(\boldsymbol{E}_{\mathrm{H}} + \boldsymbol{v}_{\mathrm{d}} \times \boldsymbol{B}) , \tag{2.31}$$

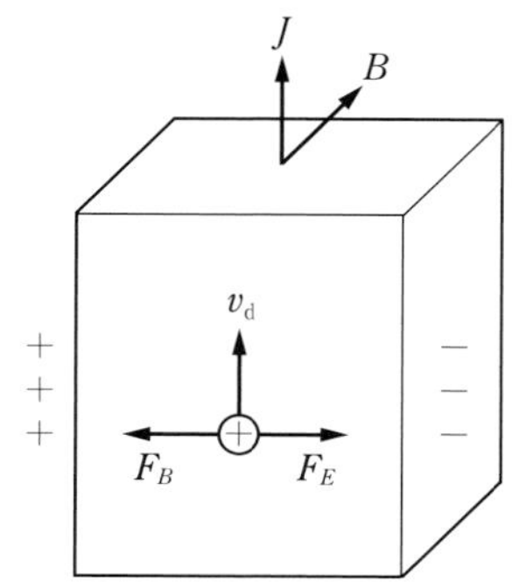

Figure 2.3 Schematic view of Hall's experiment. The magnetic and electric forces $(\boldsymbol{F}_B, \boldsymbol{F}_E)$ are balanced to zero in the Hall effect measurement.

where $\boldsymbol{E}_\mathrm{H}$ is the electric field due to the *Hall voltage* V_H. In the geometry shown in Figure 2.3, only the x-component of the force $\boldsymbol{F}$ is relevant. If the net force vanishes:

$$q(E_\mathrm{H} + v_\mathrm{d} B) = 0 \,, \tag{2.32}$$

then the carrier can proceed along the wire (z-direction) unhindered.

Let us check our model calculation. We define the *Hall coefficient* R_H by

$$R_\mathrm{H} \equiv \frac{E_\mathrm{H}}{j B} \,, \tag{2.33}$$

where the three qualities (E_H, j, B) on the right-hand side can be measured. Using (2.30) and (2.32), we obtain

$$\boxed{R_\mathrm{H} = \frac{1}{qn}} \,. \tag{2.34}$$

The Hall coefficient is a material parameter, because we will get different numbers for R_H if we do experiments with identical magnetic fields and current densities, but with different materials. The experimental values for $-1/(qnR_\mathrm{H})$ in some metals are given in Table 2.1.

For alkali metals the agreement between theory and experiment is nearly perfect. The measured Hall coefficient R_H is negative for most metals. This can be understood by assuming that the charge carriers are "*electrons*" having a negative charge $q = -e$. However, there are exceptions. As we see in Table 2.1, Al, Be and others exhibit positive Hall coefficients. This can be explained only by assuming that in these metals the main charge carriers are "*holes*" possessing a positive charge $q = +e$. This is a quantum many-body effect. As we shall see later, the existence of "electrons" and "holes" is closely connected to the curvature of the Fermi surface. Nonmagnetic metals which have "holes" tend to be superconductors, as will be explained later.

Table 2.1 Hall coefficients of selected metals.

Metal	Valence	$-1/(nqR_\mathrm{H})$
Li	1	−0.8
Na	1	−1.2
K	1	−1.1
Cu	1	−1.5
Ag	1	−1.3
Au	1	−1.5
Be	2	0.2
Mg	2	0.4
In	3	0.3
Al	3	0.3

2.5
The Boltzmann Equation

In the method of a Boltzmann equation the qualitative arguments in the simple kinetic theory will be formulated in more precise terms. In some simple cases, the description of the transport phenomena by this method is exact. In more complicated cases this method is an approximation, but it is widely used.

Let us consider the electron–impurity system, a system of free electrons with uniformly distributed impurities. We introduce a *momentum distribution function* $\varphi(\boldsymbol{p}, t)$ defined such that $\varphi(\boldsymbol{p}, t)\mathrm{d}^3 p$ gives the relative probability of finding an electron in the element $\mathrm{d}^3 p$ at time t. This function will be normalized such that

$$\frac{2}{(2\pi\hbar)^3}\int \mathrm{d}^3 p\varphi(\boldsymbol{p}, t) = \frac{N}{V} = n\ . \tag{2.35}$$

The electric current density $\boldsymbol{j}$ is given in terms of φ as follows:

$$\boldsymbol{j} = \frac{-e}{(2\pi\hbar)^3}\int \mathrm{d}^3 p\frac{\boldsymbol{P}}{m}\varphi(\boldsymbol{p}, t)\ . \tag{2.36}$$

The function φ can be obtained by solving the Boltzmann equation, which may be set up in the following manner.

The change in the distribution function φ will be caused by the force acting on the electrons in the element $\mathrm{d}^3 p$ *and* by collision. We may write this change in the form:

$$\frac{\partial\varphi}{\partial t} = \left.\frac{\mathrm{d}\varphi}{\mathrm{d}t}\right|_{\text{force}} + \left.\frac{\mathrm{d}\varphi}{\mathrm{d}t}\right|_{\text{collision}}\ . \tag{2.37}$$

The *force term* $\mathrm{d}\varphi/\mathrm{d}t|_{\text{force}}$, caused by the force $-e\boldsymbol{E}$ acting on the electrons can be expressed by

$$\left.\frac{\mathrm{d}\varphi}{\mathrm{d}t}\right|_{\text{force}} = -e\boldsymbol{E}\cdot\frac{\partial\varphi}{\partial\boldsymbol{p}}\ . \tag{2.38}$$

If the density of impurities, n_{I}, is low and the integration between electron and impurity has a short range, the electron will be scattered by one impurity at a time. We may then write the *collision term* in the following form:

$$\left.\frac{\mathrm{d}\varphi}{\mathrm{d}t}\right|_{\text{collision}} = \int \mathrm{d}\Omega\,\frac{p}{m}n_{\mathrm{I}}I(|\boldsymbol{p}|, \theta)[\varphi(\boldsymbol{p}', t) - \varphi(\boldsymbol{p}, t)]\ , \tag{2.39}$$

where $\mathrm{d}\Omega$ is the solid angle[1] in a spherical coordinate system and $I(|\boldsymbol{p}|, \theta)$ is the *differential cross section*. In fact, the rate of collision is given by (density of

1) In a spherical coordinate system, $\mathrm{d}\Omega \equiv \sin\theta\mathrm{d}\theta\mathrm{d}\phi = 2\pi\sin\theta\mathrm{d}\theta$, where θ is the scattering angle and ϕ is the azimuth angle of its orthogonal projection on a reference plane measured from the a fixed reference direction $\boldsymbol{p}$ on that plane. the angle between the initial and final momenta $\boldsymbol{p}$ and $\boldsymbol{p}'$.

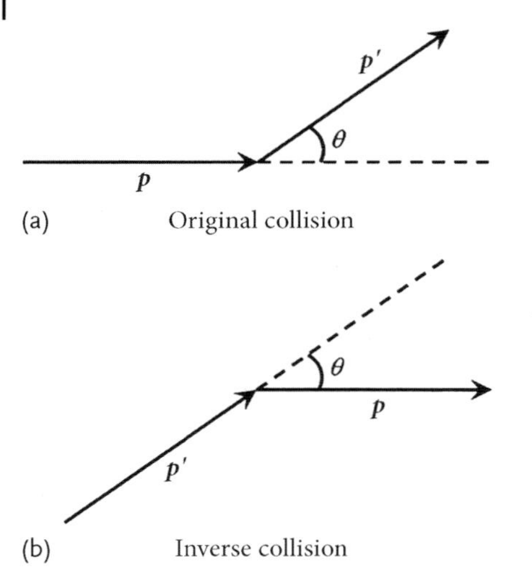

Figure 2.4 In (a), the electron suffers a change in momentum from $\boldsymbol{p}$ to $\boldsymbol{p}'$ after scattering. The inverse collision is shown in (b).

scatterers)×(speed)×(total cross section). If we apply this rule to the flux of particles with momentum $\boldsymbol{p}$, we can obtain the second integral of (2.39), the integral with the minus sign. The integral corresponds to the *loss* of the flux due to the collision (see Figure 2.4a). The flux of particles with momentum $\boldsymbol{p}$ can *gain* by inverse collision, which is shown in Figure 2.4b. The contribution of the *inverse collision* is represented by the first integral.

So far, we have neglected the fact that electrons are fermions and are therefore subject to the Pauli exclusion principle. We will now look at the effect of quantum statistics.

If the final momentum state $\boldsymbol{p}'$ was already occupied, then the scattering from state $\boldsymbol{p}$ to state $\boldsymbol{p}'$ should not have occurred. The probability of this scattering therefore should be reduced by the factor $1 - \varphi(\boldsymbol{p}', t)$, which represents the probability that the final state $\boldsymbol{p}'$ is unoccupied. Consideration of the exclusion principle thus modifies the Boltzmann collision term given in (2.39) to

$$\left.\frac{\mathrm{d}\varphi}{\mathrm{d}t}\right|_{\text{collision}} = \int \mathrm{d}\Omega \frac{p}{m} n_{\mathrm{I}} I(|\boldsymbol{p}|, \theta) \{\varphi(\boldsymbol{p}', t)[1 - \varphi(\boldsymbol{p}, t)] - \varphi(\boldsymbol{p}, t)[1 - \varphi(\boldsymbol{p}', t)]\} \,. \tag{2.40}$$

When (2.40) is expanded, the two terms proportional to $\varphi(\boldsymbol{p}', t)\varphi(\boldsymbol{p}, t)$ in the curly brackets cancel each other out. We then have the same collision term as given by (2.39).

Gathering the results from (2.37) through (2.39), we obtain

$$\frac{\partial \varphi(\boldsymbol{p}, t)}{\partial t} + e\boldsymbol{E} \cdot \frac{\partial \varphi(\boldsymbol{p}, t)}{\partial \boldsymbol{p}} = n_{\mathrm{I}} \int \mathrm{d}\Omega \frac{p}{m} I(|\boldsymbol{p}|, \theta)[\varphi(\boldsymbol{p}', t) - \varphi(\boldsymbol{p}, t)] \,. \tag{2.41}$$

This is the *Boltzmann equation* for the electron–impurity system. This equation is linear in φ, and much simpler than the Boltzmann equation for a dilute gas. In particular, we can solve (2.41) by elementary methods and calculate the conductivity σ. We will do this in the next section. For simple forms of the scattering cross section, we can also solve (2.41) as an initial value problem (Problems 2.5.1 and 2.5.2).

As mentioned at the beginning of this section, the Boltzmann equation is very important, but it is an approximate equation. If the impurity density is high or if the range of the interaction is not short, then we must consider simultaneous scatterings by two or more impurities. If we include the effect of the Coulomb interaction among electrons, which has hitherto been ignored, the collision term should be further modified. It is, however, very difficult to estimate appropriate corrections arising from these various effects.

Problem 2.5.1. Obtain the Boltzmann equation for an electron–impurity system in two dimensions, which may be deduced by inspection, from (2.41). Assume that all electrons have the same velocity at the initial time $t = 0$. Further, assuming no electric field ($\boldsymbol{E} = 0$) and isotropic scattering (I = constant), solve the Boltzmann equation.

Problem 2.5.2. Solve the Boltzmann equation (2.41) for three dimensions with the same condition as in Problem 2.5.1. Define the Boltzmann H-function by $H(t) \equiv (1/n) \int \mathrm{d}^3 p \varphi(\boldsymbol{p}, t) \ln(\varphi(\boldsymbol{p}, t)/n)$. Evaluate $\mathrm{d}H/\mathrm{d}t$. Plot $H(t)$ as a function of time.

2.6 The Current Relaxation Rate

Let us assume that a small constant electric field $\boldsymbol{E}$ is applied to the electron–impurity system and that a stationary homogeneous current is established. We take the positive x-axis along the field $\boldsymbol{E}$. In the stationary state, $\partial\varphi/\partial t = 0$: the distribution function φ depends on momentum $\boldsymbol{p}$ only. From (2.41) the Boltzmann equation for φ is then given by

$$-eE\frac{\partial\varphi(\boldsymbol{p})}{\partial p_x} = \frac{n_\mathrm{I}}{m}\int \mathrm{d}\Omega\, |\boldsymbol{p}| [\varphi(\boldsymbol{p}') - \varphi(\boldsymbol{p})] \,. \tag{2.42}$$

We wish to solve this equation and calculate the conductivity σ.

In the absence of the field $\boldsymbol{E}$, the system, by assumption, is characterized by the equilibrium distribution function, that is, the Fermi distribution function for free electrons:

$$\varphi_0(\boldsymbol{p}) = f(\varepsilon_p) \equiv \frac{1}{\mathrm{e}^{\beta(\varepsilon_p - \mu)} + 1} \,. \tag{2.43}$$

With the small field E, the function φ deviates from φ_0. Let us regard φ as a function of E and expand it in powers of E:

$$\begin{aligned}\varphi(\boldsymbol{p}) &= \varphi_0 + \varphi_1 + \cdots \\ &= f(\varepsilon_p) + \varphi_1(\boldsymbol{p}) + \cdots ,\end{aligned} \tag{2.44}$$

where the subscripts denote the order in E. For the determination of the conductivity σ we need φ_1 only. Let us introduce (2.44) in (2.42), and compare terms of the same order in E.

In the zeroth order we have

$$\frac{n_{\mathrm{I}}}{m}\int \mathrm{d}\Omega\,|\boldsymbol{p}|[\varphi(\boldsymbol{p}') - \varphi(\boldsymbol{p})] = 0 . \tag{2.45}$$

Since the energy is conserved in each scattering,

$$\varepsilon_{p'} = \varepsilon_p , \tag{2.46}$$

$\varphi_0(\boldsymbol{p}) \equiv f(\varepsilon_p)$ clearly satisfies (2.45). In the first order in E, we obtain

$$\begin{aligned}\text{lhs} &= -eE\frac{\partial\varphi_0}{\partial p_x} = -eE\frac{\partial f(\varepsilon_p)}{\partial p_x} \\ &= -eE\frac{\partial\varepsilon_p}{\partial p_x}\frac{\mathrm{d}f}{\mathrm{d}\varepsilon_p} = -eE\frac{p_x}{m}\frac{\mathrm{d}f}{\mathrm{d}\varepsilon_p} .\end{aligned} \tag{2.47}$$

Therefore, we obtain from (2.42)

$$-eE\frac{p_x}{m}\frac{\mathrm{d}f}{\mathrm{d}\varepsilon_p} = \frac{n_{\mathrm{I}}}{m}\int \mathrm{d}\Omega\,|\boldsymbol{p}|\,I[\varphi_1(\boldsymbol{p}') - \varphi_1(|\boldsymbol{p}|)] . \tag{2.48}$$

For the moment, let us neglect the first term on the rhs. In this case, the rhs equals $-(n_{\mathrm{I}}/m)|\boldsymbol{p}|\varphi_1(|\boldsymbol{p}|)\int \mathrm{d}\Omega\, I$. Then, the function $\varphi_1(|\boldsymbol{p}|)$ is proportional to p_x. Let us now try a solution of the form:

$$\varphi_1(|\boldsymbol{p}|) = p_x \Phi(\varepsilon_p) , \tag{2.49}$$

where $\Phi(\varepsilon_p)$ is a function of ε_p (no angular dependence). Substitution of (2.49) into (2.48) yields

$$-eEp_x\frac{\mathrm{d}f}{\mathrm{d}\varepsilon_p} = n_{\mathrm{I}}\int \mathrm{d}\Omega\,|\boldsymbol{p}|\,I\,\Phi(\varepsilon_p)(p'_x - p_x) . \tag{2.50}$$

Let us look at the integral on the rhs. We introduce a new frame of reference with the polar axis (the z-axis) pointing along the vector $\boldsymbol{p}$ as shown in Figure 2.5. The old positive x-axis, which is parallel to the electric field $\boldsymbol{E}$, can be specified by the angle (θ, φ). From the diagram we have

$$p_x = |\boldsymbol{p}|\cos\theta = p\cos\theta . \tag{2.51}$$

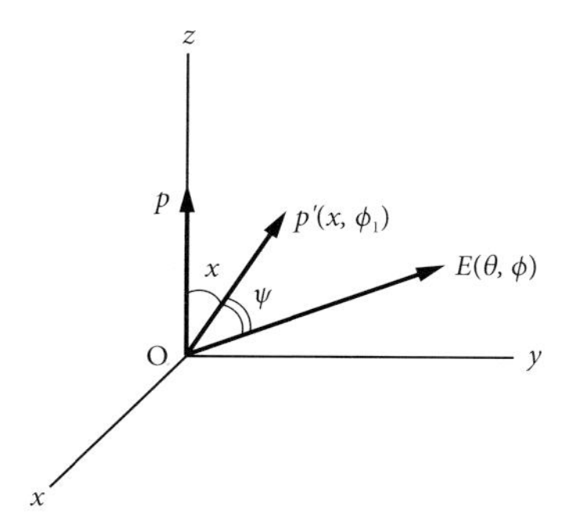

Figure 2.5 A new frame of reference in which the positive z-axis points in the direction of the fixed vector $\boldsymbol{p}$. In this frame, the direction of the electric field $\boldsymbol{E}$ is specified by (θ, ϕ) and that of the momentum $\boldsymbol{p}'$ by (χ, ϕ_1).

The vector $\boldsymbol{p}'$ can be represented by (p, χ, ϕ_1). If we denote the angle between $\boldsymbol{p}'$ and $\boldsymbol{E}$ by ψ, we have

$$p'_x = p \cos\psi \ . \tag{2.52}$$

To express $\cos\phi$ in terms of the angles (θ, ϕ) and (χ, ϕ_1), we use the vector decomposition property and obtain

$$\begin{aligned} \cos\psi &= \frac{\boldsymbol{p}'}{p} \cdot \frac{\boldsymbol{E}}{E} \\ &= (\boldsymbol{i} \sin\chi \cos\phi_1 + \boldsymbol{j} \sin\chi \sin\phi_1 + \boldsymbol{k} \cos\chi) \\ &\quad \cdot (\boldsymbol{i} \sin\theta \cos\phi + \boldsymbol{j} \sin\theta \sin\phi + \boldsymbol{k} \cos\theta) \\ &= \sin\chi \sin\theta [\cos\phi_1 \cos\phi + \sin\phi_1 \sin\phi] + \cos\chi \cos\theta \end{aligned}$$

or

$$\cos\psi = \sin\chi \sin\theta \cos(\phi - \phi_1) + \cos\chi \cos\theta \ . \tag{2.53}$$

Let us now consider the first integral in (2.50):

$$\begin{aligned} A \equiv n_{\mathrm{I}} &\int \mathrm{d}\Omega \, p \, I(p, \chi) \Phi(\varepsilon_p) (p \cos\psi) \\ = n_{\mathrm{I}} &\int_0^{2\pi} \mathrm{d}\phi_1 \int_0^{\pi} \mathrm{d}\chi \sin\chi \, p^2 I(p, \chi) \Phi(\varepsilon_p) [\sin\chi \sin\theta \cos(\phi - \phi_1) \\ &+ \cos\chi \cos\theta] \ . \end{aligned} \tag{2.54}$$

Since

$$\int_0^{2\pi} \mathrm{d}\phi_1 \cos(\phi - \phi_1) = 0 \ , \tag{2.55}$$

the first integral can be dropped. We then obtain

$$A = n_{\mathrm{I}} \int_0^{2\pi} \mathrm{d}\varphi_1 \int_0^{\pi} \mathrm{d}\chi \sin\chi\, p\, I(p,\chi) \cos\chi [p \cos\theta\, \Phi(\varepsilon_p)]$$

$$= \varphi_1(p) n_{\mathrm{I}} \int \mathrm{d}\Omega\, p\, I(p,\chi) \cos\chi \,. \tag{2.56}$$

The second integral in (2.50), therefore, is proportional to $\varphi(p)$. We thus obtain the solution:

$$\varphi_1(p) = eE \frac{p_x}{m} \frac{\mathrm{d}f}{\mathrm{d}\varepsilon_p} \frac{1}{\Gamma(p)} \,, \tag{2.57}$$

where $\Gamma(p)$ is given by

$$\Gamma(p) \equiv \frac{n_{\mathrm{I}}}{m} \int \mathrm{d}\Omega\, p\, I(p,\chi) |1 - \cos\chi| > 0 \,. \tag{2.58}$$

The Γ here is positive and depends only on $p \equiv |\boldsymbol{p}|$ (or equivalently on the energy ε_p); it has the dimension of frequency and is called the *energy-dependent current relaxation rate* or simply the *relaxation rate*. Its inverse is called the *relaxation time*.

The electric current density j_x can be calculated from

$$j_x = \frac{-2e}{(2\pi\hbar)^3 m} \int \mathrm{d}^3 p\, p_x \varphi(\boldsymbol{p}) \,. \tag{2.59}$$

We introduce $\varphi = \varphi_0 + \varphi_1 + \ldots$ in this expression. The first term gives a vanishing contribution (no current in equilibrium). The second term yields, using (2.55),

$$j_x = -\frac{2e^2 E}{(2\pi\hbar)^3 m^2} \int \mathrm{d}^3 p \frac{p_x^2}{\Gamma(p)} \frac{\mathrm{d}f}{\mathrm{d}\varepsilon_p} \,. \tag{2.60}$$

Comparing this with Ohm's law, $j_x = \sigma E$, we obtain the following expression for the conductivity:

$$\sigma = \frac{2e^2}{(2\pi\hbar)^3 m^2} \int \mathrm{d}^3 p \frac{p_x^2}{\Gamma(p)} \left(-\frac{\mathrm{d}f}{\mathrm{d}\varepsilon_p} \right) = \frac{4}{3} \frac{e^2}{(2\pi\hbar)^3} \int \mathrm{d}^3 p \frac{1}{\Gamma(p)} \varepsilon \frac{\mathrm{d}f}{\mathrm{d}\varepsilon_p} \,, \tag{2.61}$$

where we used $\varepsilon = (p_x^2 + p_y^2 + p_z^2)/(2m)$. A few applications of this formula will be discussed later.

Problem 2.6.1. For a classical hard-sphere interaction, the scattering cross section I is given by $1/2a^2$, where a is the radius of the sphere. Evaluate the relaxation rate $\Gamma(p)$ given by (2.58). Using this result, calculate the conductivity σ through (2.61). Verify that the conductivity calculated is temperature-independent.

Problem 2.6.2. Formula (2.61) was obtained with the assumption that the equilibrium distribution function in the absence of the field is given by the Fermi distribution function (2.43).

1. Verify that the same formula applies when we assume that the equilibrium distribution function is given by the Boltzmann distribution function.
2. Show that the conductivity calculated by this formula does not depend on the temperature.

References

1 Kittel, C. (2005) *Introduction to Solid State Physics*, 8th edn, John Wiley & Sons, Inc., New York, p. 142 and p. 518.

2 Ashcroft, N.W. and Mermin, N.D. (1976) *Solid State Physics*, Saunders College, Philadelphia, p. 7, p. 20, and p. 602.

3 Hall, E.H. (1879) *Am. J. Math.*, **2**, 287.

3
Bloch Electron Dynamics

To properly develop a microscopic theory of the conduction a deeper understanding of the properties of normal metals than what is provided by the free-electron model is required. On the basis of the Bloch theorem, the Fermi liquid model is derived. At 0 K, the normal metal is shown to have a sharp Fermi surface, which is experimentally supported by the fact that the heat capacity is linear in temperature at the lowest temperatures. Electrons and holes, which appear in the Hall effect measurements, are defined in terms of the curvature of the Fermi surface. Newtonian equations of motion for a Bloch electron (wavepacket) are derived and discussed.

3.1
Bloch Theorem in One Dimension

Let us consider a periodic potential $\mathcal{V}(x)$ in one dimension, see Figure 3.1, that satisfies

$$\mathcal{V}(x + na) = \mathcal{V}(x), \quad -\infty < x < \infty \tag{3.1}$$

where a is the lattice constant and n is an integer.

The Schrödinger energy-eigenvalue equation for an electron is

$$\left[-\frac{\hbar^2}{2m}\frac{\mathrm{d}^2}{\mathrm{d}x^2} + \mathcal{V}(x)\right]\psi_E(x) = E\psi_E(x) . \tag{3.2}$$

Clearly the wavefunction $\psi_E(x + na)$ also satisfies the same equation. Therefore, $\psi_E(x + na)$ is likely to be different from $\psi_E(x)$ only by an x-independent phase:

$$\psi_E(x + na) = \mathrm{e}^{\mathrm{i}kna}\psi_E(x) , \tag{3.3}$$

where k is real, see below. Equation (3.3) represents a form of the *Bloch theorem* [1]. It generates far-reaching consequences in the theory of conduction electrons. Let us prove (3.3). Since $\psi(x)$ and $\psi(x + na)$ satisfy the same equation, they are linearly dependent:

$$\psi(x + na) = c(na)\psi(x) . \tag{3.4}$$

Electrical Conduction in Graphene and Nanotubes, First Edition. S. Fujita and A. Suzuki.

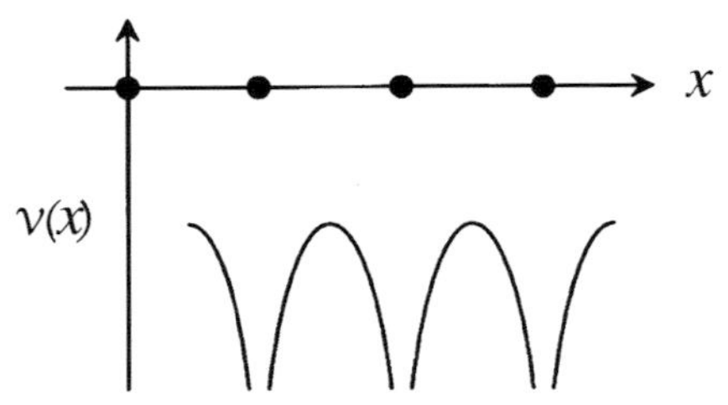

Figure 3.1 A periodic potential in one dimension. The blackened circles (•) indicate lattice ions.

Using (3.4) twice, we obtain

$$\psi(x+na+ma) = c(na)\psi(x+ma) = c(na)c(ma)\psi(x) = c(na+ma)\psi(x) .$$

Since the wavefunction $\psi(x)$ does not vanish in general, we obtain

$$c(na + ma) = c(na)c(ma) \quad \text{or} \quad c(x + y) = c(x)c(y) . \tag{3.5}$$

Solving this functional equation, we obtain (Problem 3.1.1)

$$c(y) = \exp(\lambda y) , \tag{3.6}$$

where λ is a constant. Because the wavefunction ψ in (3.4) must be finite for all ranges, constant λ must be a pure imaginary number:

$$\lambda = \mathrm{i}k , \tag{3.7}$$

where k is real. Combining (3.4), (3.6), and (3.7) we obtain (3.3). □

Let us discuss a few physical properties of the Bloch wavefunction ψ. By taking the absolute square of (3.3), we obtain

$$\boxed{|\psi(x + na)|^2 = |\psi(x)|^2} . \tag{3.8}$$

The following three main properties are observed:

(a) The probability distribution function $P(x) \equiv |\psi(x)|^2$ is *lattice periodic*:

$$P(x) \equiv |\psi(x)|^2 = P(x + na) , \quad \text{for any } n . \tag{3.9}$$

(b) The exponential function of a complex number $\exp(\mathrm{i}y)$ (y real) is periodic: $\exp(\mathrm{i}(y + 2\pi m)) = \exp(\mathrm{i}y)$, where m is an integer. We may choose the real number k in (3.3), called the k-number (2π times the wavenumber), to have a fundamental range:

$$-\frac{\pi}{a} \le k \le \frac{\pi}{a} ; \tag{3.10}$$

the two end points $(-\pi/a, \pi/a)$ are called the *Brillouin boundary* (points). This fundamental range for k is called the *first Brillouin zone.*

(c) In general, there are a number of energy gaps (forbidden regions of energy) in which no solutions of (3.2) exist (see Figure 3.4 and Section 3.2). The energy eigenvalues E are characterized by the k-number and the *zone number* (or *band index*) j, which enumerates the *energy bands*:

$$E = E_j(k) . \tag{3.11}$$

This property (c) is not obvious, and it will be illustrated by examples in Section 3.2.

To further explore the nature of the Bloch wavefunction ψ, let us write

$$\psi_E(x) = u_{j,k}(x) \exp(\mathrm{i}kx) , \tag{3.12}$$

and substitute it into (3.3). If the function $u_{j,k}(x)$ is lattice periodic,

$$u_{j,k}(x + na) = u_{j,k}(x) , \tag{3.13}$$

then (3.3) is satisfied (Problem 3.1.2). Equation (3.12) represents a second form of the Bloch theorem. The Bloch wavefunction $\psi(x) = u_{j,k}(x) \exp(\mathrm{i}kx)$ has great similarity with the free-particle wavefunction:

$$\psi_{\text{free}}(x) = c \exp(\mathrm{i}kx) , \tag{3.14}$$

where c is a constant. The connection may be illustrated as shown in Figure 3.2. For the free particle, the k-number can range from $-\infty$ to ∞, and the energy is

$$E_{\text{free}} = \frac{p^2}{2m} \equiv \frac{\hbar^2 k^2}{2m} \tag{3.15}$$

with *no* gaps. This feature is different from the properties (b) and (c) in the list above.

An important similarity arises when we write the time-dependent wavefunction $\psi(x, t)$ in the running wave form:

$$\psi_E(x, t) = U(x) \exp(\mathrm{i}(kx - \omega t)) , \tag{3.16}$$

where the frequency ω is defined by

$$\omega = \begin{cases} \hbar^{-1} E_j(k) & \text{for the Bloch electron,} \\ \hbar^{-1} E_{\text{free}} & \text{for the free electron ,} \end{cases} \tag{3.17}$$

and the amplitude $U(x)$ is defined by

$$U(x) = \begin{cases} u_{j,k}(x) & \text{for the Bloch electron,} \\ c & \text{for the free electron .} \end{cases} \tag{3.18}$$

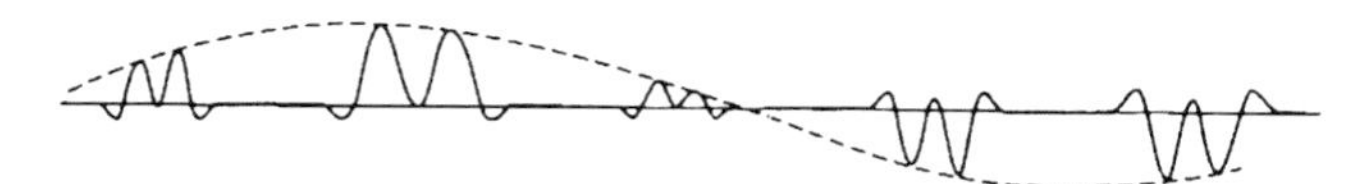

Figure 3.2 Variation of the real (or imaginary) part of the wavefunction $\psi_E(x)$.

Equation (3.16) shows that the Bloch wavefunction $\psi_E(x)$ represents a *running wave* characterized by k-number k, angular frequency ω, and *wave train* $u_{j,k}$.

The *group velocity* v of the Bloch wavepacket is given by

$$v \equiv \frac{\partial \omega}{\partial k} \equiv \hbar^{-1} \frac{\partial E}{\partial k} . \tag{3.19}$$

By applying the (quantum) principle of the wave-particle duality, we say that the Bloch electron moves with the *dispersion* (energy-momentum) *relation*:

$$E = \varepsilon_j(\hbar k) \equiv \varepsilon(p) . \tag{3.20}$$

The velocity v is given by (3.19). This gives a picture of great familiarity. We generalize our theory to the three-dimensional case in Section 3.3.

Problem 3.1.1. Solve the following functional equations:

$$\begin{aligned} &\text{(i)} \quad f(x+y) = f(x) + f(y) \\ &\text{(ii)} \quad f(x+y) = f(x) f(y) \end{aligned}$$

Hints: Differentiate (i) with respect to x, and convert the result into an ordinary differential equation. Take the logarithm of (ii), and use (i). Answer: (i) $f(x) = cx$, (ii) $f(x) = \mathrm{e}^{\lambda x}$, where c and λ are constants.

Problem 3.1.2.

1. Show that Bloch's wavefunction (3.12) satisfies (3.3).
2. Assuming (3.3), derive (3.12).

3.2 The Kronig–Penney Model

The Bloch energy-eigenvalues in general have bands and gaps. We show this by taking the *Kronig–Penney* (KP) *model* [8]. Let us consider a periodic square-well potential $\mathcal{V}(x)$ with depth $V_0(< 0)$ and well width $\beta(\equiv a - \alpha)$ as shown in Figure 3.3:

$$\mathcal{V}(x) = \begin{cases} 0 & \text{if} \quad na < x < na + \alpha , \\ V_0 & \text{if} \quad na - \beta \le x \le na , \end{cases} \tag{3.21}$$

where $a = \alpha + \beta$ and the n are integers.

The Schrödinger energy-eigenvalue equation for an electron can be written as in (3.2). Since this is a linear homogeneous differential equation with constant coefficients, the wavefunction $\psi(x)$ should have the form

$$\psi(x) = c \exp(\gamma x) , \quad (c, \gamma \text{ are constants}) . \tag{3.22}$$

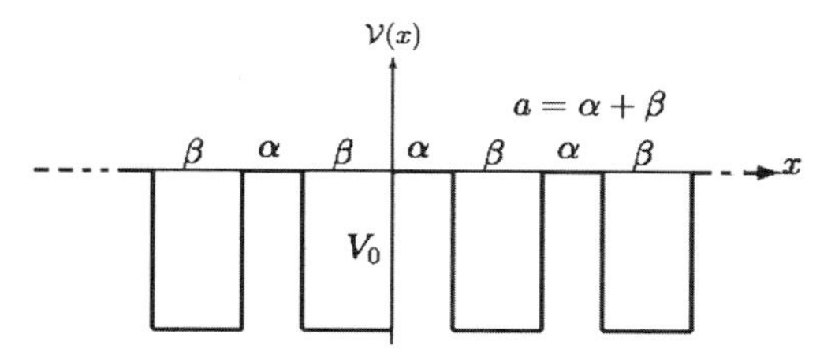

Figure 3.3 A Kronig–Penney potential has a square-well with depth V_0 and width β periodically arranged with period $a = \alpha + \beta$.

According to the Bloch theorem in (3.12), this function $\psi(x)$ can be written as

$$\psi_k(x) = u_k(x)\exp(\mathrm{i}kx) , \tag{3.23}$$

$$u_k(x) = u_k(x + a) . \tag{3.24}$$

The condition that the function $\psi(x)$ be *continuous* and *analytic* at the well boundary yields the following relationships (Problem 3.2.1):

$$\cos(ka) = \cosh(K\alpha)\cos(\mu\beta) + \frac{K^2 - \mu^2}{2\,K\mu}\sinh(K\alpha)\sin(\mu\beta) \equiv f(E) \tag{3.25}$$

$$E \equiv \begin{cases} -\dfrac{\hbar^2 K^2}{2m} , & na < x < na + \alpha , \\ V_0 + \dfrac{\hbar^2\mu^2}{2m} , & na - \beta \leq x \leq na , \quad V_0 < 0 . \end{cases} \tag{3.26}$$

By solving (3.25) with (3.26), we obtain the eigenvalue E as a function of k. The *band edges* are obtained from

$$f(E) = \pm 1 , \tag{3.27}$$

which corresponds to the limits of $\cos(ka)$. Numerical studies of (3.25) indicate that (i) there are, in general, a number of negative- and positive-energy bands; (ii) at each band edge, an *effective mass* m^* can be defined, whose value can be positive or negative and whose absolute value can be greater or less than the electron mass m; and (iii) the effective mass is positive at the lower edge of each band, and it is negative at the upper edge. A typical dispersion relation for the model, showing energy bands and energy gaps, is shown in Figure 3.4. At the lowest band edge ε_0 we have

$$f(\varepsilon_0) = 1 . \tag{3.28}$$

Near this edge the dispersion (energy-k) relation calculated from (3.25) is (Problem 3.2.2)

$$\varepsilon = \varepsilon_0 + \frac{\hbar^2 k^2}{2m^*} \quad (\varepsilon_0 < 0) , \tag{3.29}$$

$$m^* \equiv -\hbar^2 a^{-2} f'(\varepsilon_0) . \tag{3.30}$$

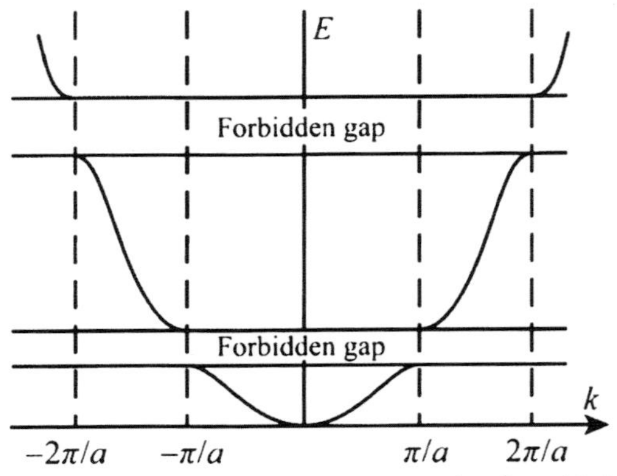

Figure 3.4 E–k diagram showing energy bands and forbidden gaps.

This one-dimensional KP model can be used to study a simple three-dimensional system. Let us take an orthorhombic (orc) lattice of unit lengths (a_1, a_2, a_3), with each lattice point representing a short-range attractive potential center (ion). The Schrödinger equation (3.2) for this system is hard to solve.

Let us now construct a model potential $\mathcal{V}$ defined by

$$\mathcal{V}(x, y, z) = \mathcal{V}_1(x) + \mathcal{V}_2(y) + \mathcal{V}_3(z) , \tag{3.31}$$

$$\mathcal{V}_j(u) = \begin{cases} V_0 \ (< 0) & \text{if } n a_j - \beta \leq u \leq n a_j , \\ 0 & \text{otherwise.} \end{cases} \tag{3.32}$$

Here the n are integers. A similar two-dimensional model is shown in Figure 3.5. The domains in which $\mathcal{V} \neq 0$ are parallel plates of thickness β $(< a_j)$ separated by a_j in the direction x_j, $(x_1, x_2, x_3) = (x, y, z)$. The intersection of any two plates are straight beams of cross section β^2, where the potential $\mathcal{V}$ has the value $2V_0$. The intersections of three plates, where the potential $\mathcal{V}$ has the value $3V_0$, are cubes of side length β. The set of these cubes form an orc lattice, a configuration similar to that of the commercially available molecular lattice model made up of balls and sticks. Note: Each square-well potential $\mathcal{V}_j$ has three parameters (V_0, β, a_j), and this model represents the true potential fairly well [12]. The Schrödinger equation for the 3D model Hamiltonian

$$\begin{aligned} \mathcal{H} &\equiv \frac{p_x^2 + p_y^2 + p_z^2}{2m} + \mathcal{V}_1(x) + \mathcal{V}_2(y) + \mathcal{V}_3(z) \\ &= \frac{p_x^2}{2m} + \mathcal{V}_1(x) + \frac{p_y^2}{2m} + \mathcal{V}_2(y) + \frac{p_z^2}{2m} + \mathcal{V}_3(z) \end{aligned} \tag{3.33}$$

can now be reduced to three one-dimensional KP equations. We can then write an expression for the energy of our model system near the lowest band edge as

$$E = \frac{\hbar^2}{2m_1} k_x^2 + \frac{\hbar^2}{2m_2} k_y^2 + \frac{\hbar^2}{2m_3} k_z^2 + \text{constant} , \tag{3.34}$$

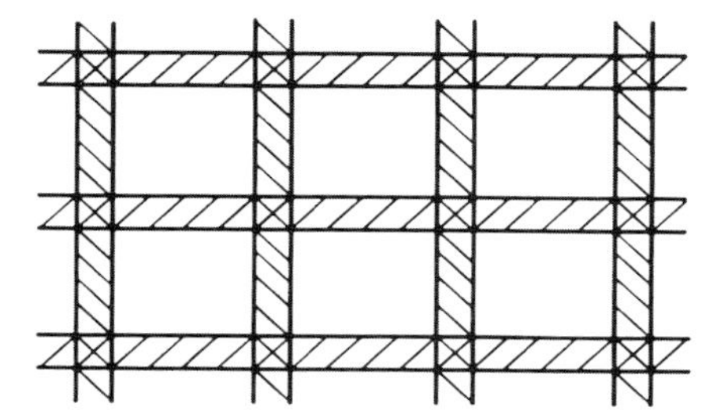

Figure 3.5 A 2D model potential. Each singly shaded stripe has a potential energy (depth) V_0. Each cross-shaded square has a potential energy $2V_0$.

where $\{m_j\}$ are effective masses defined by (3.30) with $a = a_j$.

Equation (3.34) is identical to what is intuitively expected of the energy-k relation for the electron in the orc lattice. It is stressed that we derived it from first principles, assuming a three-dimensional model Hamiltonian $\mathcal{H}$. Our study demonstrates qualitatively how electron energy bands and gaps are generated from the Schrödinger equation for an electron moving in a three-dimensional lattice space.

Problem 3.2.1. Derive (3.25). Hint: The wavefunction ψ and the energy eigenvalue E are given by

$$\psi = \begin{cases} A\mathrm{e}^{\mathrm{i}Kx} + B\mathrm{e}^{-\mathrm{i}Kx}\,, & E = -\hbar^2 K^2/(2m) \quad \text{if } na < x < na + a \\ C\mathrm{e}^{\mu x} + D\mathrm{e}^{-\mu x}\,, & E = V_0 + \hbar^2\mu^2/(2m) \quad \text{if } na - \beta \le x \le na\,. \end{cases}$$

Use ψ and $\mathrm{d}\psi/\mathrm{d}x$ are continuous at the potential boundaries, say, $x = 0$ and $x = a$

Problem 3.2.2. Derive (3.30).

3.3 Bloch Theorem in Three Dimensions

Let us take a monovalent metal such as sodium (Na). The Hamiltonian $\mathcal{H}$ of the system may be represented by

$$\mathcal{H} = \sum_{j=1}^{N} \frac{p_j^2}{2m} + \sum_{j>k}\sum \frac{k_0 e^2}{|\boldsymbol{r}_j - \boldsymbol{r}_k|} + \sum_{\alpha=1}^{N} \frac{P_\alpha^2}{2M} + \sum_{\alpha>\gamma}\sum \frac{k_0 e^2}{|\boldsymbol{R}_\alpha - \boldsymbol{R}_\gamma|} - \sum_j \sum_\alpha \frac{k_0 e^2}{|\boldsymbol{r}_j - \boldsymbol{R}_\alpha|}\,, \tag{3.35}$$

where $k_0 \equiv (4\pi\varepsilon_0)^{-1}$. The sums on the rhs represent, respectively, the kinetic energy of electrons, the interaction energy among electrons, the kinetic energy of ions, the interaction energy among ions, and the interaction energy between electrons and ions. The metal as a whole is electrically neutral, and hence the number of electrons equals the number of ions. Both numbers are denoted by N.

At the lowest temperatures the ions are almost stationary near the equilibrium lattice points. Because of quantum zero-point motion, the ions are not at rest even at 0 K. But this fact does not affect the following argument. The system then can be viewed as the one in which the electrons move in a periodic lattice potential $\mathcal{V}$. The Hamiltonian of this idealized system that depends on the electron variables only can be written as

$$\mathcal{H} = \sum_{j=1}^{N} \frac{p_j^2}{2m} + \sum_{j>k}\sum \frac{k_0 e^2}{|\boldsymbol{r}_j - \boldsymbol{r}_k|} + \sum_j \mathcal{V}(\boldsymbol{r}_j) + C \,, \tag{3.36}$$

where $\mathcal{V}(\boldsymbol{r})$ represents the lattice potential, and the constant energy C depends on the lattice configuration.

Let us drop the Coulomb interaction energy from (3.36). We then have

$$\mathcal{H} = \sum_j \frac{p_j^2}{2m} + \sum_j \mathcal{V}(\boldsymbol{r}_j) + C \,. \tag{3.37}$$

For definiteness, we consider an infinite orc lattice. We choose a Cartesian frame of coordinates (x, y, z) along the lattice axes. The potential $\mathcal{V}(x, y, z) = \mathcal{V}(\boldsymbol{r})$ is lattice periodic:

$$\mathcal{V}(\boldsymbol{r}) = \mathcal{V}(\boldsymbol{r} + \boldsymbol{R}) \tag{3.38}$$

$$\boldsymbol{R} \equiv n_1 a \boldsymbol{i} + n_2 b \boldsymbol{j} + n_3 c \boldsymbol{k} \,, \quad (n_j\text{: integers}) \tag{3.39}$$

where (a, b, c) are lattice constants and the vector $\boldsymbol{R}$ is called the *Bravais lattice vector.*[1] The Schrödinger equation for an electron is

$$\left[-\frac{\hbar^2}{2m}\nabla^2 + \mathcal{V}(\boldsymbol{r})\right]\psi_{\boldsymbol{k}}(\boldsymbol{r}) = \varepsilon\psi_{\boldsymbol{k}}(\boldsymbol{r}) \,, \tag{3.40}$$

where ε is the energy. Clearly $\psi_{\boldsymbol{k}}(\boldsymbol{r} + \boldsymbol{R})$ also satisfies the same equation. The values of the wavefunction $\psi_{\boldsymbol{k}}$ at $\boldsymbol{r}$ and $\boldsymbol{r} + \boldsymbol{R}$ may therefore be different only by a $\boldsymbol{r}$-independent phase:

$$\boxed{\psi_{\boldsymbol{k}}(\boldsymbol{r} + \boldsymbol{R}) = \mathrm{e}^{\mathrm{i}\boldsymbol{k}\cdot\boldsymbol{R}}\psi_{\boldsymbol{k}}(\boldsymbol{r})} \,, \tag{3.41}$$

where $\boldsymbol{k} \equiv (k_x, k_y, k_z)$ is called a *k-vector.* Equation (3.41) represents a form of *Bloch's theorem* [1, 2]. It generates far-reaching consequences in the theory of conduction electrons.

1) The vector $\boldsymbol{R}$ is a translational vector, which expresses a lattice periodicity.

The three principal properties of the *Bloch wavefunctions* are:

1. The probability distribution $P(\boldsymbol{r})$ is lattice periodic:

$$P(\boldsymbol{r}) \equiv |\psi_{\boldsymbol{k}}(\boldsymbol{r})|^2 = P(\boldsymbol{r} + \boldsymbol{R}) \,. \tag{3.42}$$

2. The k-vector $\boldsymbol{k} = (k_x, k_y, k_z)$ in (3.41) has the fundamental range:

$$-\frac{\pi}{a} \leq k_x \leq \frac{\pi}{a} \,, \quad -\frac{\pi}{b} \leq k_y \leq \frac{\pi}{b} \,, \quad -\frac{\pi}{c} \leq k_z \leq \frac{\pi}{c} \,. \tag{3.43}$$

   The end points that form a rectangular box, are called the *Brillouin boundary*.
3. The energy eigenvalues ε have energy gaps, and the allowed energies ε can be characterized by the *zone number* j and the *k-vector* $\boldsymbol{k}$:

$$\varepsilon = \varepsilon_j(\hbar \boldsymbol{k}) \equiv \varepsilon_j(\boldsymbol{p}) \,. \tag{3.44}$$

Using (3.41), we can express the *Bloch wavefunction* ψ in the form:

$$\boxed{\psi_{\boldsymbol{k}}(\boldsymbol{r}) \equiv \psi_{j,\boldsymbol{k}}(\boldsymbol{r}) = u_{j,\boldsymbol{k}}(\boldsymbol{r}) \exp(\mathrm{i}\boldsymbol{k} \cdot \boldsymbol{r})} \,, \tag{3.45}$$

where $u(\boldsymbol{r})$ is a *periodic* function:

$$\boxed{u_{j,\boldsymbol{k}}(\boldsymbol{r} + \boldsymbol{R}) = u_{j,\boldsymbol{k}}(\boldsymbol{r})} \,. \tag{3.46}$$

Formula (3.45) along with (3.46) is known as the *Bloch theorem* [1, 2]. The wavefunction $\psi_{\boldsymbol{k}}(\boldsymbol{r})$ in the form (3.45) resembles a plane wave describing the motion of a free particle, but here the wave is modulated by a periodic function $u_{\boldsymbol{k}}(\boldsymbol{r})$. The k-vector $\boldsymbol{k}$ is connected with the momentum $\boldsymbol{p}$ by $\boldsymbol{p} = \hbar \boldsymbol{k}$.

Equations (3.44)–(3.46) indicate that the Bloch wavefunction $\psi_{\boldsymbol{k}}(\boldsymbol{r})$, associated with quantum numbers $(j, \boldsymbol{k})$, is a plane wave characterized by k-vector $\boldsymbol{k}$, angular frequency $\omega \equiv \varepsilon_j(\hbar \boldsymbol{k})/\hbar$ and wave train $u_{j,\boldsymbol{k}}(\boldsymbol{r})$. It is clear from (3.45) that

$$|\psi_{j,\boldsymbol{k}}(\boldsymbol{r})|^2 = |u_{j,\boldsymbol{k}}(\boldsymbol{r})|^2 = |\psi_{j,\boldsymbol{k}}(\boldsymbol{r} + \boldsymbol{R})|^2 \,. \tag{3.47}$$

Hence, the probability is lattice periodic. The Bloch theorem holds in 3D or 2D only if the Bravais vector is expressed in the Cartesian frame of reference. This is relevant to graphene physics, and will be further discussed in Chapter 5, Section 5.2.

Problem 3.3.1.

1. Show that the Bloch wavefunction (3.41) satisfies (3.40).
2. Assuming (3.41), derive Equations (3.45) and (3.46).

3.4
Fermi Liquid Model

We consider a monovalent metal, whose Hamiltonian $\mathcal{H}_A$ is given by (3.35):

$$\mathcal{H}_A = \sum_{j=1}^{N} \frac{p_j^2}{2m} + \sum_{j>k}\sum \frac{k_0 e^2}{|\boldsymbol{r}_j - \boldsymbol{r}_k|} + \sum_{\alpha=1}^{N} \frac{P_\alpha^2}{2M} + \sum_{\alpha>\gamma}\sum \frac{k_0 e^2}{|\boldsymbol{R}_\alpha - \boldsymbol{R}_\gamma|} - \sum_j \sum_\alpha \frac{k_0 e^2}{|\boldsymbol{r}_j - \boldsymbol{R}_\alpha|} . \tag{3.48}$$

The motion of the set of N electrons is correlated because of the interelectronic interaction. If we omit the ionic kinetic energy, and the interionic and interelectronic Coulomb interaction from (3.48), we obtain

$$\mathcal{H}_B = \sum_j \frac{p_j^2}{2m} + \sum_j \mathcal{V}(\boldsymbol{r}_j) + C , \tag{3.49}$$

which characterizes a system of electrons moving in the bare lattice potential.

Since the metal as a whole is neutral, the Coulomb interaction among the electrons, among the ions, and between electrons and ions, all have the same order of magnitude, and hence they are equally important. We now pick out one electron in the system. This electron is interacting with the system of N ions and $N - 1$ electrons, the system (medium) having the net charge $+e$. These other $N - 1$ electrons should, in accordance with Bloch's theorem, be distributed with the lattice periodicity and all over the crystal in equilibrium. The charge per lattice ion is greatly reduced from e to e/N because the net charge e of the medium is shared equally by N ions. Since N is a large number, the selected electron moves in an extremely weak effective lattice potential $\mathcal{V}_{\text{eff}}$ as characterized by the model Hamiltonian:

$$h_C = \frac{p^2}{2m} + \mathcal{V}_{\text{eff}}(\boldsymbol{r}) , \quad \mathcal{V}_{\text{eff}}(\boldsymbol{r} + \boldsymbol{R}) = \mathcal{V}_{\text{eff}}(\boldsymbol{r}) . \tag{3.50}$$

In other words any chosen electron moves in an environment far different from what is represented by the bare lattice potential $\mathcal{V}$. It moves almost freely in an extremely weak effective lattice potential $\mathcal{V}_{\text{eff}}$. This picture was obtained with the aide of Bloch's theorem, and hence it is a result of quantum theory. To illustrate let us examine the same system from the classical point of view. In equilibrium the classical electron distribution is lattice periodic, so there is one electron near each ion. The electron will not move in the greatly reduced field.

We now assume that electrons move independently in the effective potential field $\mathcal{V}_{\text{eff}}$. The total Hamiltonian for the idealized system may then be represented by

$$\mathcal{H}_C = \sum_j h_C(\boldsymbol{r}_j, \boldsymbol{p}_j) \equiv \sum_j \frac{p_j^2}{2m} + \sum_j \mathcal{V}_{\text{eff}}(\boldsymbol{r}_j) . \tag{3.51}$$

This Hamiltonian $\mathcal{H}_C$ is a far better approximation to the original Hamiltonian $\mathcal{H}_A$ than the Hamiltonian $\mathcal{H}_B$. In $\mathcal{H}_C$ both interelectronic and interionic Coulomb repulsion are not neglected but are taken into consideration self-consistently. This theoretical model is a *one-electron-picture approximation*, but it is hard to improve on by any simple method. The model in fact forms the basis for the band theory of electrons.

We now apply Bloch's theorem to the Hamiltonian $\mathcal{H}_C$ composed of the kinetic energy and the interaction energy $\mathcal{V}_{\text{eff}}$. We then obtain the *Bloch energy bands* $\varepsilon_j(\hbar \boldsymbol{k})$ and the *Bloch states* characterized by band index j and k-vector $\boldsymbol{k}$. The Fermi–Dirac statistics obeyed by the electrons can be applied to the Bloch electrons with no regard to interaction. This means that there is a certain Fermi energy ε_F for the ground state of the system. Thus, there is a sharp Fermi surface represented by

$$\varepsilon_j(\hbar \boldsymbol{k}) = \varepsilon_F \,, \tag{3.52}$$

which separates the electron-filled k-space (low-energy side) from the empty k-space (high-energy side). The Fermi surface for a real metal in general is complicated in contrast to the free-electron Fermi sphere represented by

$$\frac{p^2}{2m} \equiv \frac{p_x^2 + p_y^2 + p_z^2}{2m} = \varepsilon_F \,. \tag{3.53}$$

The independent electron model with a sharp Fermi surface at 0 K is called the *Fermi liquid model* of Landau [3]. As we show later, many thermal properties of conductors are dominated by those electrons near the Fermi surface. The shape of the Fermi surface is very important for the conduction behavior. In the following section, we shall examine the Fermi surfaces of some metals.

The Fermi liquid model was obtained in the static lattice approximation in which the motion of the ions is neglected. If the effect of moving ions (phonons: quanta of lattice vibrations) is taken into account, a new model is required. The electron–phonon interaction turns out to be very important in the theory of superconductivity, which will be discussed in Chapter 7.

3.5 The Fermi Surface

Why does a particular metal exist in a particular crystalline state? This is a good question. The answer must involve the composition and nature of the atoms constituting the metal and the interaction between the component particles. To illustrate let us take Na, which forms a bcc lattice. This monovalent metal may be thought of as an ideal composite system of electrons and ions. The system Hamiltonian may be approximated by $\mathcal{H}_A$ in (3.48), which consists of the kinetic energies of electrons and ions and the Coulomb interaction energies among and between electrons and ions. This is an approximation since the interaction between electron and ion deviates significantly from the ideal Coulomb law at short distances because each ion has core electrons. At any rate the study of the ground state energy of the ideal model favors a fcc lattice structure, which is not observed for this metal. If multivalent metals like Pb and Sn are considered, the condition becomes even more complicated, since the core electrons forming part of the ions have anisotropic charge distribution. Because of this complexity it is customary in solid state physics to assume the experimentally known lattice structures first, then proceed to study the Fermi surface.

Once a lattice is selected, the Brillouin zone is fixed. For an orthorhombic (orc) lattice the Brillouin zone is a rectangular box defined by (3.43). We now assume a large periodic box of volume:

$$V = (N_1 a)(N_2 b)(N_3 c) , \quad N_1 N_2 N_3 \gg 1 . \tag{3.54}$$

Let us find the number N of the quantum states in the first Brillouin zone. With the neglect of the spin degeneracy, the number N is equal to the total k-space volume enclosed by the Brillouin boundary divided by unit k-cell volume:

$$\left(\frac{2\pi}{a}\right)\left(\frac{2\pi}{b}\right)\left(\frac{2\pi}{c}\right) \div \left(\frac{2\pi}{N_1 a}\right)\left(\frac{2\pi}{N_2 b}\right)\left(\frac{2\pi}{N_3 c}\right) = N_1 N_2 N_3 , \tag{3.55}$$

which equals the number of ions in the normalization volume. It is also equal to the number of conduction electrons in a monovalent metal. Thus, the first Brillouin zone can contain twice (because of spin degeneracy) the number of conduction electrons for the monovalent metal. This means that at 0 K, half of the Brillouin zone may be filled by electrons. Something similar to this actually happens to alkali metals including Li, Na, K. These metals form bcc lattices. All experiments indicate that the Fermi surface is nearly spherical and entirely within the first Brillouin zone. The Fermi surface of Na is shown in Figure 3.6.

The *nearly free electron model* (NFEM) developed by Harrison [5] can predict a Fermi surface for any metal in the first approximation. This model is obtained by applying Heisenberg's uncertainty principle and Pauli's exclusion principle to a solid. Hence, it has a general applicability unhindered by the complications due to particle–particle interaction. Briefly in the NFEM, the first Brillouin zone is drawn for a chosen metal. Electrons are filled, starting from the center of the zone, with the assumption of a free electron dispersion relation. If we apply the NFEM to alkali metals, we simply obtain the Fermi sphere as shown in Figure 3.6.

Noble metals, including copper (Cu), silver (Ag), and gold (Au) are monovalent fcc metals. The Brillouin zone and Fermi surface of Cu are shown in Figure 3.7. The Fermi surface is far from spherical. Notice that the Fermi surface approaches the Brillouin boundary at *right* angles. This arises from the mirror symmetry possessed by the fcc lattice.

For a divalent metal like calcium (Ca) (fcc), the first Brillouin zone can in principle contain all of the conduction electrons. However, the Fermi surface must approach the zone boundary at right angles, which distorts the ideal configuration considerably. Thus, the real Fermi surface for Ca has a set of unfilled corners in the first zone, and the overflow electrons are in the second zone. As a result Ca is

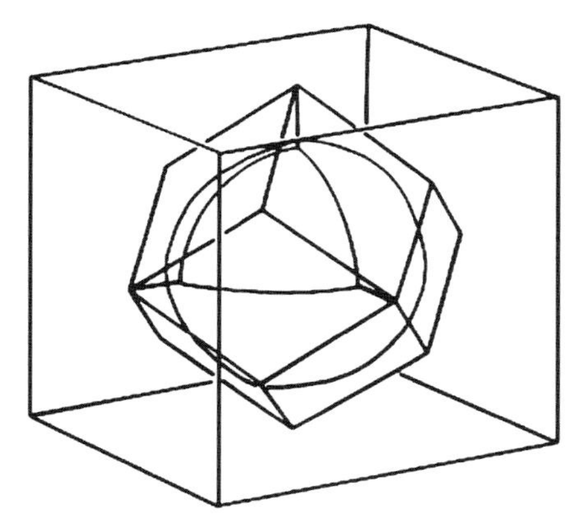

Figure 3.6 The Fermi surface of Na (bcc) is spherical within the first Brillouin zone.

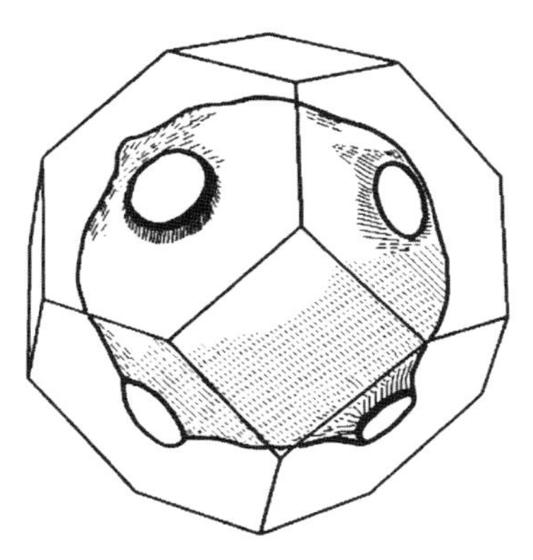

Figure 3.7 The Fermi surface of Cu (fcc) bulges out in the ⟨111⟩ direction to make contact with the hexagonal zone faces.

a metal, and not an insulator. Besides Ca has electrons and holes. Divalent beryllium (Be) forms a hexagonal closed packed (hcp) crystal. The Fermi surfaces in the second zone constructed in the NFEM and observed [7], are shown in Figure 3.8a

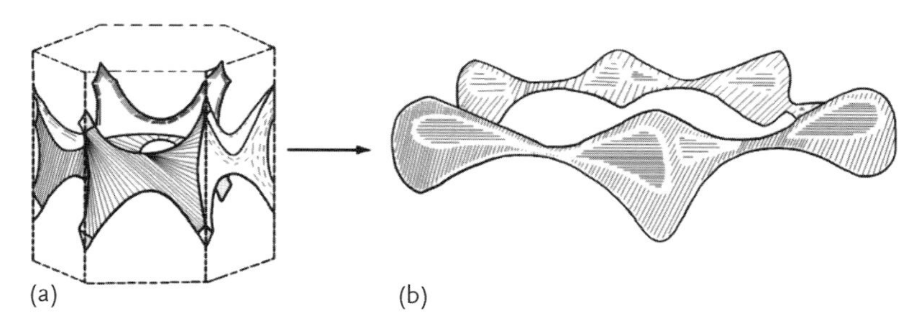

Figure 3.8 The Fermi surfaces in the second zone for Be. (a) NFEM "monster," (b) measured "coronet." The coronet encloses unoccupied states.

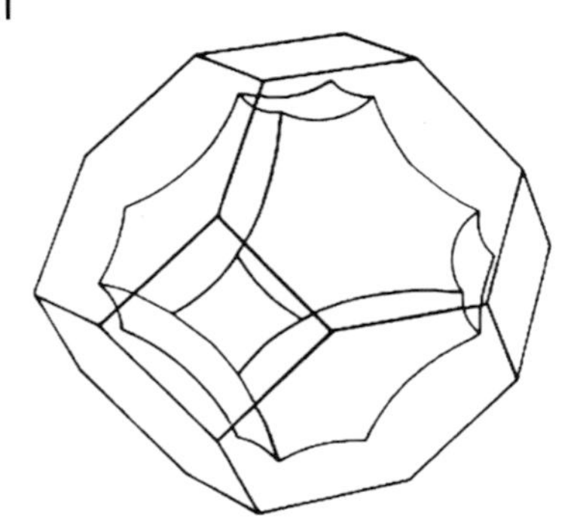

Figure 3.9 The Fermi surface constructed by Harrison's model (NFEM) in the second zone for Al. The convex surface encloses vacant states.

and b, respectively. Let us now consider trivalent aluminum (Al), which forms a fcc lattice. The first Brillouin zone is entirely filled with electrons. The second zone is half filled with electrons, starting with the zone boundary as shown in Figure 3.9. For a more detailed description of the Fermi surface of metals, see standard texts on solid state physics [2, 6, 8]. Al and Be are superconductors, while Na and Cu are not.

3.6
Heat Capacity and Density of States

The band structures of conduction electrons are quite different from metal to metal. In spite of this, the electronic heat capacities at very low temperatures are all similar, which is shown in this section. We first show that any normal metal having a sharp Fermi surface has a *T-linear heat capacity*. We draw the density of states $\mathcal{D}(\varepsilon)$, and the Fermi distribution function $f(\varepsilon)$ as a function of the kinetic energy ε in Figure 3.10. The change in $f(\varepsilon)$ is appreciable only near the Fermi energy ε_{F}. The number of excited electrons, ΔN, is estimated by

$$\Delta N = \mathcal{D}(\varepsilon_{\mathrm{F}}) k_{\mathrm{B}} T \,. \tag{3.56}$$

Each thermally excited electron will move up with an extra energy of the order $k_{\mathrm{B}} T$. The approximate change in the total energy ΔE is given by multiplying these two factors:

$$\Delta E = \Delta N k_{\mathrm{B}} T = \mathcal{D}(\varepsilon_{\mathrm{F}})(k_{\mathrm{B}} T)^2 \,. \tag{3.57}$$

Differentiating this with respect to T, we obtain an expression for the heat capacity:

$$C_V \simeq \frac{\partial}{\partial T} \Delta E = 2 k_{\mathrm{B}}^2 \mathcal{D}(\varepsilon_{\mathrm{F}}) T \,, \tag{3.58}$$

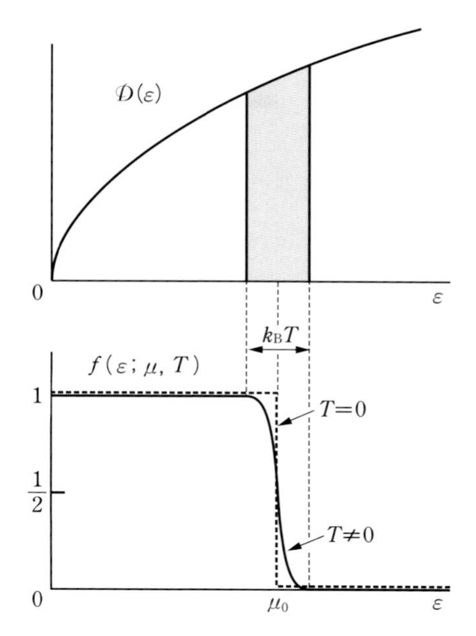

Figure 3.10 The density of states in energy, $\mathcal{D}(\varepsilon)$, and the Fermi distribution function $f(\varepsilon)$ are drawn as a function of the kinetic energy ε. The change in f is appreciable only near the Fermi energy ε_F if $k_B T \ll \varepsilon_F$. The shaded area represents approximately the number of thermally excited electrons.

which indicates the T-linear dependence. This T-dependence comes from the Fermi distribution function. Using the more rigorous calculation, we obtain

$$C_V = \frac{1}{3}\pi^2 k_B^2 \mathcal{D}(\varepsilon_F) T \, . \tag{3.59}$$

The density of states, $\mathcal{D}(\varepsilon_F)$, for any 3D normal metal can be evaluated from

$$\mathcal{D}(\varepsilon) = \sum_j \frac{2}{(2\pi\hbar)^3} \int dS \frac{1}{|\nabla \varepsilon_j(\hbar \boldsymbol{k})|} \, , \tag{3.60}$$

where the factor 2 is due to the spin degeneracy, and the surface integration is carried out over the Fermi surface represented by

$$\varepsilon_j(\hbar \boldsymbol{k}) = \varepsilon_F \, . \tag{3.61}$$

As an illustration, consider a free electron system having the Fermi sphere:

$$\varepsilon = \frac{p_x^2 + p_y^2 + p_z^2}{2m} \equiv \varepsilon_F \, . \tag{3.62}$$

The gradient $\nabla \varepsilon(\boldsymbol{p})$ at any point of the surface has a constant magnitude p_F/m, and the surface integral is equal to $4\pi p_F^2$. Equation (3.60), then, yields

$$\mathcal{D}(\varepsilon_F) = \frac{2}{(2\pi\hbar)^3} \frac{4\pi p_F^2}{p_F/m} = \frac{2^{1/2} m^{3/2}}{\pi^2 \hbar^3} \varepsilon_F^{1/2} \, . \tag{3.63}$$

As a second example, consider the ellipsoidal surface represented by

$$\varepsilon = \frac{p_x^2}{2m_1} + \frac{p_y^2}{2m_2} + \frac{p_z^2}{2m_3} \,. \tag{3.64}$$

After elementary calculations, we obtain (Problem 3.6.1)

$$\mathcal{D}(\varepsilon) = \frac{2^{1/2}}{\pi^2 \hbar^3} (m_1 m_2 m_3)^{1/2} \varepsilon^{1/2} \,, \tag{3.65}$$

which shows that the density of states still grows like $\varepsilon^{1/2}$, but the coefficient depends on the three effective masses (m_1, m_2, m_3).

Problem 3.6.1.

1. Compute the momentum-space volume between the surfaces represented by $\varepsilon = p_x^2/2m_1 + p_y^2/2m_2 + p_z^2/2m_3$ and $\varepsilon + d\varepsilon = p_x'^2/2m_1 + p_y'^2/2m_2 + p_z'^2/2m_3$. By counting the number of momentum states in this volume in the bulk limit, obtain (3.65).
2. Derive (3.65), starting from the general formula (3.60). Hint: Convert the integral over the ellipsoidal surface to one over a spherical surface.

3.7
The Density of State in the Momentum Space

In many applications of quantum statistical mechanics we meet the need for converting the sum over quantum states into an integral. This conversion becomes necessary when we first find *discrete* quantum states for a finite box, and then seek the sum over states in the *bulk limit*. The necessity of such a conversion does not arise in the spin problem. This conversion is a welcome procedure because the resulting integral, in general, is easier to handle than the sum. The conversion is of a purely mathematical nature, but it is an important step in carrying out statistical mechanical computations.

Let us first examine a sum over momentum states $p_k \equiv 2\pi\hbar k/L$ corresponding to a one-dimensional motion. Let us take the sum

$$\sum_k A(p_k) \,, \tag{3.66}$$

where $A(p_k)$ is an arbitrary function of p. The discrete momentum states $p_k \equiv 2\pi\hbar k/L$ are equally spaced, as shown by the short bars in Figure 3.11. As the normalization length L is made greater, the spacing (distance) between two successive states, $2\pi\hbar/L$, becomes smaller. This means that the number of states par unit momentum interval increases as L increases. We denote the number of states within

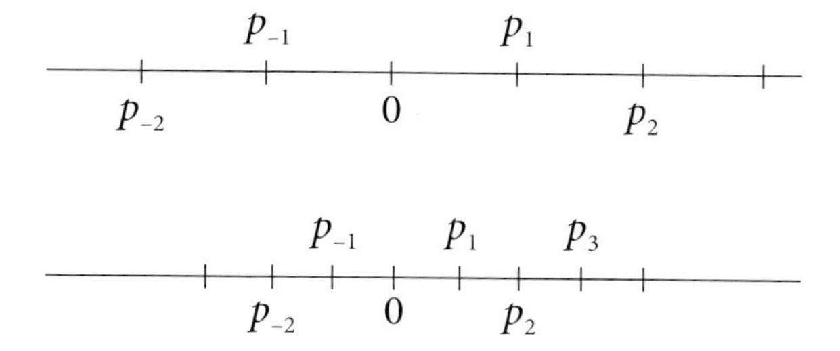

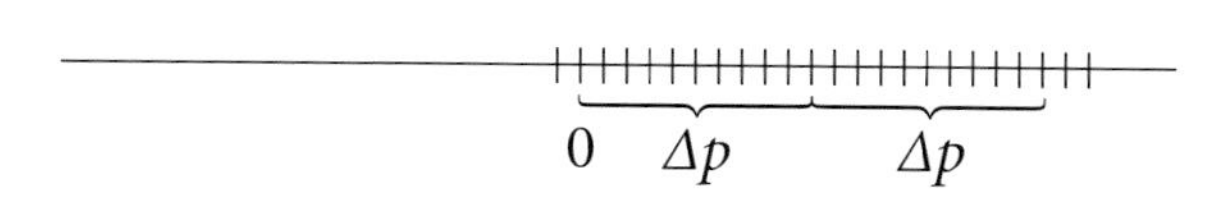

Figure 3.11 The linear momentum states are represented by short bars forming a linear lattice with unit spacing equal to $2\pi\hbar/L$. As the normalization length L is made greater, the spacing becomes smaller.

a small momentum interval Δp by Δn. We now take the ratio,

$$\frac{\Delta n}{\Delta p} \equiv \frac{\text{number of states in } \Delta p}{\Delta p} . \tag{3.67}$$

Dividing both the numerator and denominator by the number of states, we get

$$\frac{\Delta n}{\Delta p} \equiv \frac{1}{\text{momentum spacing per state}} = \frac{1}{2\pi\hbar/L} = \frac{L}{2\pi\hbar} . \tag{3.68}$$

Note that this ratio $\Delta n/\Delta p$ increases linearly with the normalization length L.

Let us now consider a quantity

$$\sum_l A(p_l) \frac{\Delta n}{\Delta_l p} \Delta_l p , \tag{3.69}$$

where $\Delta_l p$ is the l-th interval and p_l represents a typical value of p within the interval $\Delta_l p$, say the p-value at the midpoint of $\Delta_l p$. Since

$$\left[\frac{\Delta n}{\Delta_l p} \Delta_l p\right] = [\Delta n] = (\text{number}) ,$$

the two sums (3.66) and (3.69) have the same dimension. Furthermore, their numerical values will be close if (a) the function $A(p)$ is a smooth function of p, and (b) there exist many states in $\Delta_l p$ so that $\Delta n/\Delta_l p$ can be regarded as the density of states. Condition (b) is satisfied for the momentum states $\{p_k\}$ when the length L is made sufficiently large. We can then expect that in the bulk limit, by choosing infinitesimally small intervals : $\Delta_l p \to \mathrm{d}p$, expressions (3.66) and (3.69) will have the same value. That is,

$$\sum_{k(\text{states})} A(p_k) \to \sum_{\Delta_l p} A(p) \frac{\Delta n}{\Delta_l p} \Delta_l p \tag{3.70}$$

as $L \to \infty$. But by definition, the sum on the rhs becomes the integral $\int \mathrm{d}p\, A(p)\mathrm{d}n/\mathrm{d}p$, where

$$\frac{\mathrm{d}n}{\mathrm{d}p} \equiv \lim_{\Delta p \to 0} \frac{\Delta n}{\Delta p} = \frac{L}{2\pi\hbar} . \tag{3.71}$$

In summary, we therefore have

$$\sum_k A(p_k) \to \int_{-\infty}^{\infty} \mathrm{d}p\, A(p) \frac{\mathrm{d}n}{\mathrm{d}p} \Delta_l p \quad \text{as} \quad L \to \infty . \tag{3.72}$$

It is stressed that condition (a) depends on the nature of the function A. Therefore, if $A(p)$ is singular at some point, condition (a) is not satisfied and this may invalidate the limit (3.72). Such exceptional cases do occur. We further note that the *density of state* $\mathrm{d}n/\mathrm{d}p = L/(2\pi\hbar)$ does not depend on the momentum.

The sum-to-integral conversion, which we have discussed, can easily be generalized for a multidimensional case. For example, in three dimensions, we have

$$\sum_{\boldsymbol{p}_k} A(\boldsymbol{p}_k) \to \int \mathrm{d}^3 p\, A(\boldsymbol{p}) \mathcal{D}(\boldsymbol{p}) \quad \text{as} \quad V \equiv L^3 \to \infty , \tag{3.73}$$

where

$$\mathcal{D}(\boldsymbol{p}) \equiv \frac{\mathrm{d}n}{\mathrm{d}^3 p} \tag{3.74}$$

is called the *density of states per unit volume* in momentum space.

Let us choose a periodic cubic box of side length L for the normalization. The density of states $\mathrm{d}n/\mathrm{d}^3 p$ can then be calculated by extending the arguments leading to (3.68). The result is given by

$$\mathcal{D}(\boldsymbol{p}) = \frac{1}{(2\pi\hbar/L)^3} = \frac{L^3}{(2\pi\hbar)^3} . \tag{3.75}$$

For electrons, the *spin degeneracy* doubles this density. We therefore obtain

$$\mathcal{D}(\boldsymbol{p}) \equiv \frac{\mathrm{d}n}{\mathrm{d}^3 p} = \frac{2L^3}{(2\pi\hbar)^3} \tag{3.76}$$

with spin degeneracy.

Let us now use this result and simplify the normalization condition, which relates the chemical potential μ with the average density n:

$$n \equiv \frac{N}{V} = \mathrm{Lim} \frac{1}{V} \sum_{\kappa} \langle n_\kappa \rangle_{\mathrm{gc}} = \mathrm{Lim} \frac{1}{V} \sum_{\kappa} f(\varepsilon_\kappa) , \tag{3.77}$$

where $\langle n_\kappa \rangle_{\mathrm{gc}}$ denotes a grand canonical ensemble average of the number operator

n_κ, resulting in the Fermi distribution function $f(\varepsilon_\kappa)$.[2] Note that κ denotes the set of states and ε_κ the single-electron energy associated with state κ and the symbol "Lim" indicates the bulk limit. By choosing $A(\boldsymbol{p}) = f(p^2/2m)$, we obtain

$$\begin{aligned} n &= \mathrm{Lim}\frac{1}{V}\int \mathrm{d}^3 \frac{p^2}{2m}\frac{\mathrm{d}n}{\mathrm{d}^3 p} = \mathrm{Lim}\frac{1}{V}\int \mathrm{d}^3 \frac{p^2}{2m}\frac{2L^3}{(2\pi\hbar)^3} \\ &= \frac{2}{(2\pi\hbar)^3}\int \mathrm{d}^3 \frac{p^2}{2m} \, . \end{aligned} \tag{3.78}$$

As a second example, we take the energy density of the system. The total energy of the many-particle system is given by

$$E \equiv \sum_\kappa \varepsilon_\kappa \langle n_\kappa \rangle_{\mathrm{gc}} = \sum_\kappa \varepsilon_\kappa f(\varepsilon_\kappa) \, , \tag{3.79}$$

where $\varepsilon_\kappa \equiv p_\kappa^2/(2m)$ is the kinetic energy of the electron with momentum $\boldsymbol{p}_\kappa$. From (3.79), we obtain the energy density $e \equiv E/V$:

$$\begin{aligned} e &\equiv \frac{E}{V} = \mathrm{Lim}\frac{1}{V}\langle \mathcal{H} \rangle_{\mathrm{gc}} = \mathrm{Lim}\frac{1}{V}\sum_\kappa \varepsilon_\kappa f(\varepsilon_\kappa) \\ &= \mathrm{Lim}\frac{1}{V}\int \mathrm{d}^3 p \frac{p^2}{2m} f\left(\frac{p^2}{2m}\right)\frac{\mathrm{d}n}{\mathrm{d}^3 p} = \frac{2}{(2\pi\hbar)^3}\int \mathrm{d}^3 p \frac{p^2}{2m} f\left(\frac{p^2}{2m}\right) \, . \end{aligned} \tag{3.80}$$

Equations (3.79) and (3.78) were obtained starting with the momentum eigenvalues corresponding to the periodic boundary condition. The results in the bulk limit, however, do not depend on the choice of the boundary condition.

The Fermi energy ε_F, by definition, is the chemical potential μ_0 at absolute zero. We may look at this connection as follows. For a box of finite volume V, the momentum states are quantized. As the volume V is increased, the unit cell volume in momentum space, $(2\pi\hbar)^3/V$ decreases like V^{-1}. However, in the process of the bulk limit we must increase the number of electrons, N, in proportion to V. Therefore, the radius of the *Fermi sphere* within which all momentum states are filled with electrons neither grows nor shrinks. Obviously, this configuration corresponds to the lowest energy state for the system. The Fermi energy $\mu_0 \equiv p_\mathrm{F}^2/(2m)$ represents the energy of an electron with the momentum magnitude p_F at the surface of the Fermi sphere. If we attempt to add an extra electron to the Fermi sphere, we must bring in an electron with an energy equal to μ_0.

2) We show briefly the derivation of the Fermi distribution function $f(\varepsilon_\kappa)$. A detailed derivation is given in for example [13]. When the Hamiltonian of a system, $\mathcal{H}$, is given by the sum of a *single-particle* Hamiltonian h, that is, $\mathcal{H} = \sum_j h_j$, the grand canonical ensemble average of a number operator n_κ then gives the Fermi distribution function $f(\varepsilon_\kappa)$:

$$\begin{aligned} \langle n_\kappa \rangle_{\mathrm{gc}} &\equiv \frac{\mathrm{TR}\{n_\kappa \mathrm{e}^{\alpha\mathcal{N}-\beta\mathcal{H}}\}}{\mathrm{TR}\{\mathrm{e}^{\alpha\mathcal{N}-\beta\mathcal{H}}\}} \\ &= \frac{1}{\mathrm{e}^{\beta\varepsilon_\kappa-\alpha}+1} \equiv f(\varepsilon_\kappa) \, , \end{aligned}$$

where the symbol "TR" denotes a many-body trace, $\mathcal{N} \equiv \sum_\kappa n_\kappa$, ε_κ is the eigenvalue of h, $\beta \equiv 1/(k_\mathrm{B} T)$, and $\alpha \equiv \beta\mu$ (μ: the chemical potential).

Problem 3.7.1. The momentum eigenvalues for a particle in a rectangular box with sides of unequal length (L_x, L_y, L_z) are given by

$$p_{x,j} \equiv \frac{2\pi\hbar}{L_x} j \,, \quad p_{y,k} \equiv \frac{2\pi\hbar}{L_y} k \,, \quad p_{z,l} \equiv \frac{2\pi\hbar}{L_z} l \,.$$

Assuming that the particles are fermions, verify that the Fermi energy ε_F is the same, independent of the box shape, in the bulk limit.

3.8 Equations of Motion for a Bloch Electron

In this section we discuss how conduction electrons respond to the applied electromagnetic fields. Let us recall that in the Fermi liquid model each electron in a crystal moves independently in an extremely weak lattice periodic effective potential $\mathcal{V}_{\text{eff}}(\boldsymbol{r})$:

$$\mathcal{V}_{\text{eff}}(\boldsymbol{r} + \boldsymbol{R}) = \mathcal{V}_{\text{eff}}(\boldsymbol{r}) \equiv \mathcal{V}(\boldsymbol{r}) \,. \tag{3.81}$$

We write down the Schrödinger equation:

$$\left[-\frac{\hbar^2}{2m} \nabla^2 + \mathcal{V}(\boldsymbol{r}) \right] \psi(\boldsymbol{r}) = \varepsilon \psi(\boldsymbol{r}) \,. \tag{3.82}$$

According to Bloch's theorem, the wavefunction ψ satisfies

$$\psi_{j,\boldsymbol{k}}(\boldsymbol{r} + \boldsymbol{R}) = \mathrm{e}^{\mathrm{i}\boldsymbol{k}\cdot\boldsymbol{R}} \psi_{j,\boldsymbol{k}}(\boldsymbol{r}) \,. \tag{3.83}$$

The Bravais vector $\boldsymbol{R}$ can take on only discrete values, and its minimum length can equal the lattice constant a_0. This generates a limitation on the domain in $\boldsymbol{k}$. For example the values for each $k_\alpha (\alpha = x, y, z)$ for a sc lattice are limited to $[-\pi/a_0, \pi/a_0]$. This means that the Bloch electron's wavelength $\lambda \equiv 2\pi/k$ has a lower bound:

$$\lambda \geq 2a_0 \,. \tag{3.84}$$

The Bloch electron state is characterized by k-vector $\boldsymbol{k}$, band index j and energy

$$\varepsilon = \varepsilon_j(\hbar\boldsymbol{k}) \equiv \varepsilon_j(\boldsymbol{p}) \,. \tag{3.85}$$

Here we defined the lattice momentum by $\boldsymbol{p} \equiv \hbar\boldsymbol{k}$. The *energy-momentum* (or *dispersion*) *relation* represented by (3.85) can be probed by transport measurements. A metal is perturbed from the equilibrium condition by an applied electric field; the deviations of the electron distribution from the equilibrium move in the crystal to reach and maintain a stationary state. The deviations, that is, the *localized Bloch wavepackets*, should extend over one unit cell or more. This is so because no wavepackets constructed from waves whose wavelengths have the lower bounds $2a_0$ can be localized within distances less than a_0.

Dirac demonstrated [9] that for any p-dependence of the kinetic energy ($\varepsilon = \varepsilon_j(\boldsymbol{p})$) the center of a quantum wavepacket, identified as the position of the corresponding particle, moves in accordance with Hamilton's equations of motion. Hence, the Bloch electron representing the wavepacket moves in the classical mechanical way under the action of the force averaged over the lattice constants. The lattice force $-\partial\mathcal{V}_{\text{eff}}/\partial x$ averaged over a unit cell vanishes:

$$\left\langle -\frac{\partial}{\partial x}\mathcal{V}_{\text{eff}}(\boldsymbol{r})\right\rangle_{\text{unit cell}} \equiv -a_0^{-3}\iint \mathrm{d}y\mathrm{d}z\int_0^{a_0}\mathrm{d}x\frac{\partial}{\partial x}\mathcal{V}_{\text{eff}}(x,y,z) = 0\,. \tag{3.86}$$

Thus, the only important forces acting on the Bloch electron are the electromagnetic forces.

We now formulate dynamics for the Bloch electron as follows. First, from the quantum principle of wave-particle duality, we postulate that

$$(\hbar k_x, \hbar k_y, \hbar k_z) = (p_x, p_y, p_z) \equiv (p_1, p_2, p_3) \equiv \boldsymbol{p}\,. \tag{3.87}$$

Second, we introduce a model Hamiltonian,

$$\mathcal{H}_0(p_1, p_2, p_3) \equiv \varepsilon_j(\hbar k_1, \hbar k_2, \hbar k_3)\,. \tag{3.88}$$

Third, we generalize our Hamiltonian $\mathcal{H}$ to include the electromagnetic interaction energy:

$$\mathcal{H} = \mathcal{H}_0(\boldsymbol{p} - q\boldsymbol{A}) + q\phi\,, \tag{3.89}$$

where $(\boldsymbol{A}, \phi)$ are vector and scalar potentials generating electromagnetic interaction energy:

$$\boldsymbol{E} = -\nabla\phi(\boldsymbol{r},t) - \frac{\partial}{\partial t}\boldsymbol{A}(\boldsymbol{r},t)\,, \quad \boldsymbol{B} = \nabla\times\boldsymbol{A}(\boldsymbol{r},t)\,, \tag{3.90}$$

where $\boldsymbol{r} \equiv (x_1, x_2, x_3)$. By using the standard procedures, we then obtain Hamilton's equations of motion:

$$\dot{\boldsymbol{r}} \equiv \frac{\mathrm{d}\boldsymbol{r}}{\mathrm{d}t} = \boldsymbol{v} = \frac{\partial\mathcal{H}}{\partial\boldsymbol{p}} = \frac{\partial\mathcal{H}_0}{\partial\boldsymbol{p}}\,, \quad \dot{\boldsymbol{p}} \equiv \frac{\mathrm{d}\boldsymbol{p}}{\mathrm{d}t} = -\frac{\partial\mathcal{H}}{\partial\boldsymbol{r}} = -\frac{\partial\mathcal{H}_0}{\partial\boldsymbol{r}} - q\frac{\partial\phi}{\partial\boldsymbol{r}}\,. \tag{3.91}$$

The first equation defines the velocity $\boldsymbol{v} \equiv (v_1, v_2, v_3)$. Notice that in the zero-field limit these equations are in agreement with the general definition of a group velocity:

$$v_{g,i} \equiv \frac{\partial\omega(\boldsymbol{k})}{\partial k_i}\,, \quad \omega(\boldsymbol{k}) \equiv \frac{\varepsilon(\boldsymbol{p})}{\hbar} \quad \text{(wave picture)}\,, \tag{3.92}$$

$$v_i \equiv \frac{\partial\varepsilon(\boldsymbol{p})}{\partial p_i} \quad \text{(particle picture)}\,. \tag{3.93}$$

The first of (3.91) gives the velocity $\boldsymbol{v}$ as a function of $\boldsymbol{p} - q\boldsymbol{A}$. Inverting this functional relation, we have

$$\boldsymbol{p} - q\boldsymbol{A} = \boldsymbol{f}(\boldsymbol{v}) . \tag{3.94}$$

Using (3.91)–(3.94), we obtain

$$\frac{\mathrm{d}\boldsymbol{f}}{\mathrm{d}t} = q(\boldsymbol{E} + \boldsymbol{v} \times \boldsymbol{B}) . \tag{3.95}$$

Since the vector $\boldsymbol{f}$ is a function of the velocity $\boldsymbol{v}$, (3.95) describes how $\boldsymbol{v}$ changes by the action of the Lorentz force (the right-hand term).

To see the nature of (3.95), we take a quadratic dispersion relation represented by

$$\varepsilon = \frac{p_1^2}{2m_1^*} + \frac{p_2^2}{2m_2^*} + \frac{p_3^2}{2m_3^*} + \varepsilon_0 , \tag{3.96}$$

where $\{m_i^*\}$ are *effective masses* and ε_0 is a constant. The effective masses $\{m_i^*\}$ may be positive or negative. Depending on their values, the energy surface represented by (3.96) is ellipsoidal or hyperboloidal. See Figure 3.12. If the Cartesian axes are taken along the major axes of the ellipsoid, (3.95) can be written as (Problem 3.8.1)

$$m_i^* \frac{\mathrm{d}v_i}{\mathrm{d}t} = q(\boldsymbol{E} + \boldsymbol{v} \times \boldsymbol{B})_i . \tag{3.97}$$

These are *Newtonian equations of motion*: mass × acceleration = force. Only a set of three effective masses $\{m_i^*\}$ are introduced. The Bloch electron moves in an anisotropic environment if the effective masses are different.

Let us now go back to the general case. The function $\boldsymbol{f}$ may be determined from the dispersion relation as follows: take a point A at the constant-energy surface represented by (3.85) in the k-space. We choose to point the positive normal vector in the direction in which energy increases. A *normal curvature* κ is defined as the inverse of the radius of the contact circle at A (in the plane containing the normal vector) times the curvature sign δ_{A}:

$$\kappa \equiv \delta_{\mathrm{A}} R_{\mathrm{A}}^{-1} , \tag{3.98}$$

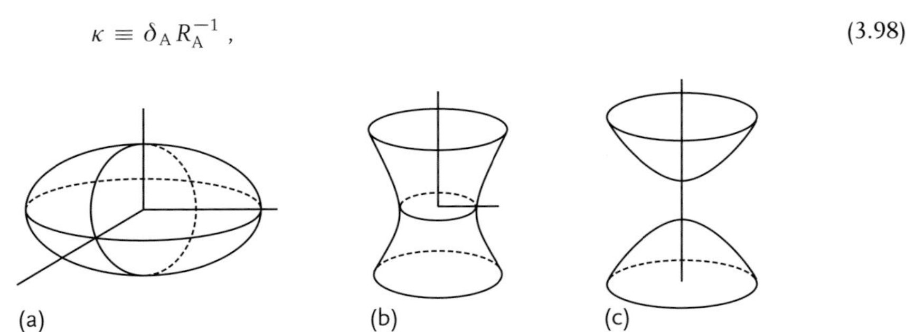

Figure 3.12 (a) Ellipsoid, (b) hyperboloid of one sheet (neck), (c) hyperboloid of two sheets (inverted double caps).

where δ_A is $+1$ or -1 according to whether the center of the contact circle is on the positive side (which contains the positive normal) or not. In space-surface theory [11], the two planes that contain the greatest and smallest normal curvatures are known to be mutually orthogonal. They are, by construction, orthogonal to the contact plane at A. Therefore, the intersections of these two planes and the contact plane can form a Cartesian set of orthogonal axes with the origin at A, called the *principal axes of curvatures*. By using this property, we define *principal masses* m_i by

$$\frac{1}{m_i} \equiv \frac{\partial^2 \varepsilon}{\partial p_i^2}\,, \tag{3.99}$$

where ∂p_i is the differential along the principal axis i. If we choose a Cartesian coordinate system along the principal axes, (3.95) can be written as (Problem 3.8.2)

$$m_i \frac{\mathrm{d}v_i}{\mathrm{d}t} = q(\boldsymbol{E} + \boldsymbol{v} \times \boldsymbol{B})_i\,. \tag{3.100}$$

Note that these equations are similar to those of (3.97). The principal masses $\{m_i\}$, however, are defined at each point on the constant-energy surface, and hence they depend on $\boldsymbol{p}$ and $\varepsilon_j(\boldsymbol{p})$. Let us take a simple example: an ellipsoidal constant-energy surface represented by (3.96) with all positive m_i^*. At extremal points, for example, $(p_{1,\max}, 0, 0) = (\sqrt{2(\varepsilon - \varepsilon_0) m_1^*}, 0, 0)$, the principal exes of curvatures match the major axes of the ellipsoid. Then the principal masses $\{m_i\}$ can simply be expressed in terms of the constant effective masses $\{m_j^*\}$ (Problem 3.8.3).

The proof of the equivalence between (3.95) and (3.100) is carried out as follows. Since f_i are functions of (v_1, v_2, v_3), we obtain

$$\frac{\mathrm{d}f_i}{\mathrm{d}t} = \sum_j \frac{\partial f_i}{\partial v_j}\frac{\mathrm{d}v_j}{\mathrm{d}t} = \sum_j \left(\frac{\partial v_j}{\partial f_i}\right)^{-1} \frac{\mathrm{d}v_j}{\mathrm{d}t}\,. \tag{3.101}$$

The velocities v_i from (3.91) can be expressed in terms of the first p-derivatives. Thus, in the zero-field limit:

$$\frac{\partial v_j}{\partial f_i} \to \frac{\partial^2 \varepsilon}{\partial p_i \partial p_j} \equiv \frac{1}{m_{ij}}\,, \tag{3.102}$$

which defines the *symmetric mass tensor elements* $\{m_{ij}\}$. By using (3.101) and (3.102), we can re-express (3.95) as

$$\sum_j m_{ij} \frac{\mathrm{d}v_j}{\mathrm{d}t} = q(\boldsymbol{E} + \boldsymbol{v} \times \boldsymbol{B})_i\,, \quad i = 1, 2, 3\,, \tag{3.103}$$

which is valid in *any* Cartesian frame of reference. The mass tensor $\{m_{ij}\}$ is real and symmetric, and hence can always be diagonalized by a principal-axes transformation [11]. The principal masses $\{m_i\}$ are given by (3.99) and the principal axes are given by the principal axes of curvature. □

In (3.99) the third principal mass m_3 is defined in terms of the second derivative, $\partial^2 \varepsilon / \partial p_3^2$, in the energy-increasing (p_3-)direction. The first and second principal masses (m_1, m_2) can be connected with the two principal radii of curvature, (P_1, P_2), which by definition equal the inverse principal curvatures (κ_1, κ_2) (Problem 3.8.4):

$$\frac{1}{m_j} = -\kappa_j v \equiv -\frac{v}{P_j} , \quad v \equiv |\boldsymbol{v}| , \quad \frac{1}{P_j} \equiv \kappa_j . \tag{3.104}$$

The equations in (3.104) are very useful relations. The signs (definitely) and magnitudes (qualitatively) of the first two principal masses (m_1, m_2) can be obtained by a visual inspection of the constant-energy surface, an example of which is the Fermi surface. The sign of the third principal mass m_3 can also be obtained by inspection: the mass m_3 is positive or negative according to whether the center of the contact circle is on the negative or the positive side. For example, the system of free electrons has a spherical constant-energy surface represented by $\varepsilon = p^2/(2m)$ with the normal vector pointing outward. By inspection the principal radii of curvatures at every point of the surface are negative, and the principal masses (m_1, m_2) are positive and equal to m. The third principal mass m_3 is also positive and equal to m. Equation (3.100) was derived from the energy-k relation (3.85) without referring to the Fermi energy. It is valid for all wave vectors $\boldsymbol{k}$ and all band indices j.

Problem 3.8.1. Assume a quadratic dispersion relation (3.96) and derive (3.97).

Problem 3.8.2. Assume a general dispersion relation (3.88) and derive (3.100).

Problem 3.8.3. Consider the ellipsoidal constant-energy surface represented by (3.96) with all $m_i^* > 0$. At the six extremal points, the principal axes of the curvatures match the major axes of the ellipsoidal. Demonstrate that the principal masses $\{m_i\}$ at one of these points can be expressed simply in terms of the effective masses $\{m_j^*\}$.

Problem 3.8.4. Verify (3.104). Hint: Use a Taylor expansion.

References

1. Bloch, F. (1928) *Z. Phys.*, **555**, 52.
2. Ashcroft, N.W. and Mermin, N.D. (1976) *Solid State Physics*, Saunders College, Philadelphia, Chaps. 8 and 15.
3. Landau, L.D. (1956) *Zh. Eksp.Teor. Fiz.*, **30**, 1058 [English transl.: (1957) *Sov. Phys. JETP*, **3**, 920].
4. Suzuki, A. and Fujita, S. (2008) *Foundation of Statistical Thermodynamics*, Kyoritsu Pub., Tokyo, Chap. 18 (in Japanese).
5. Harrison, W.A. (1960) Phys. Rev., **118**, 1190.
6. Schönberg, D. and Gold, A.V. (1969) Physics of Metals 1: *Electrons*, ed. Ziman, J.M., Cambridge University Press, Cambridge, UK, p. 112.
7. Harrison, W.A. (1979) *Solid State Theory*, Dover, New York.
8. Kittel, C. (1986) *Introduction to Solid State Physics*, 6th edn, John Wiley & Sons, Inc., New York, Chap. 9.
9. Dirac, P.A.M. (1958) *Principles of Quantum Mechanics*, 4th edn, Oxford University Press, London, pp. 121–125.
10. Eisenhart, L.P (1940) *Introduction to Differential Geometry*, Princeton University Press, Princeton.
11. Kibble, T.W.B. (1966) *Classical Mechanics*, McGraw-Hill, Maidenhead, England, pp. 166–171.
12. Shukla, K. (1990) *Kronig–Penney models and their applications to solids*, PhD, State University of New York at Buffalo.
13. Balescu, R. *Equilibrium and Nonequilibrium Statistical Mechanics*, John Wiley & Sons, Inc., New York, 1975), pp. 167–169.

4
Phonons and Electron–Phonon Interaction

Phonons, electron–phonon interaction, and the phonon-exchange attraction are discussed in this chapter.

4.1
Phonons and Lattice Dynamics

In the present section, we review a general theory of heat capacity based on lattice dynamics.

Let us take a crystal composed of N atoms. The potential energy $\mathcal{V}$ depends on the configuration of N atoms located at $(\boldsymbol{r}_1, \boldsymbol{r}_2, \cdots, \boldsymbol{r}_N)$. We regard this energy $\mathcal{V}$ as a function of the *displacements* of the atoms,

$$\boldsymbol{u}_j \equiv \boldsymbol{r}_j - \boldsymbol{r}_j^{(0)} , \tag{4.1}$$

measured from the equilibrium positions $\boldsymbol{r}_j^{(0)}$.

Let us Taylor expand the potential

$$\mathcal{V} = \mathcal{V}(\boldsymbol{u}_1, \boldsymbol{u}_2, \ldots, \boldsymbol{u}_N) \equiv \mathcal{V}(u_{1x}, u_{1y}, u_{1z}, u_{2x}, \ldots)$$

in terms of small displacements $\{u_{j\mu}\}$:

$$\mathcal{V} = \mathcal{V}_0 + \sum_j \sum_{\mu=x,y,z} u_{j\mu} \left[\frac{\partial \mathcal{V}}{\partial u_{j\mu}}\right]_0 + \frac{1}{2} \sum_j \sum_\mu \sum_k \sum_\nu u_{j\mu} u_{k\nu} \left[\frac{\partial^2 \mathcal{V}}{\partial u_{j\mu} \partial u_{k\nu}}\right]_0 + \cdots , \tag{4.2}$$

where all partial derivatives are evaluated at $\boldsymbol{u}_1 = \boldsymbol{u}_2 = \cdots = 0$, which is indicated by subscript 0. We may set the constant $\mathcal{V}_0$ equal to zero with no loss of rigor. By assumption, the lattice is stable at the equilibrium configuration. Then the potential $\mathcal{V}$ must be at a minimum, requiring that the first-order derivatives vanish:

$$\left[\frac{\partial \mathcal{V}}{\partial u_{j\mu}}\right]_0 = 0 . \tag{4.3}$$

Electrical Conduction in Graphene and Nanotubes, First Edition. S. Fujita and A. Suzuki.

For small oscillations we may keep terms of the second order in $u_{j\mu}$ only. We then have

$$\mathcal{V} \simeq \mathcal{V}' \equiv \sum_j \sum_\mu \sum_k \sum_\nu \frac{1}{2} A_{j\mu k\nu} u_{j\mu} u_{k\nu} , \tag{4.4}$$

where

$$A_{j\mu k\nu} \equiv \left[\frac{\partial^2 \mathcal{V}}{\partial u_{j\mu} \partial u_{k\nu}} \right]_0 . \tag{4.5}$$

The prime (′) on $\mathcal{V}$ indicating the *harmonic approximation*, will be dropped hereafter. The kinetic energy of the system is given by

$$\mathcal{T} \equiv \sum_j \frac{1}{2} m \dot{r}_j^2 = \sum_j \frac{1}{2} m \dot{u}_j^2 \equiv \sum_j \sum_\mu \frac{1}{2} m \dot{u}_{j\mu}^2 . \tag{4.6}$$

The kinetic energy of all the atoms is the sum of their individual kinetic energies.

We can now write down the Lagrangian $\mathcal{L} \equiv \mathcal{T} - \mathcal{V}$ as

$$\mathcal{L} = \sum_j \sum_\mu \frac{1}{2} m \dot{u}_{j\mu}^2 - \sum_j \sum_\mu \sum_k \sum_\nu \frac{1}{2} A_{j\mu k\nu} u_{j\mu} u_{k\nu} . \tag{4.7}$$

The Lagrangian $\mathcal{L}$ in the harmonic approximation is *quadratic* in $u_{j\mu}$ and $\dot{u}_{j\mu}$. According to the theory of *principal-axis transformation* [1], we can in principle transform the Hamiltonian (total energy) $\mathcal{H} = \mathcal{T} + \mathcal{V}$ for the system into the sum of the energies of the normal modes of oscillations:

$$\mathcal{H} = \sum_{\kappa=1}^{3N} \frac{1}{2} \left(P_\kappa^2 + \omega_\kappa^2 Q_\kappa^2 \right) , \tag{4.8}$$

where $\{Q_\kappa, P_\kappa\}$ are the normal coordinates and momenta, and ω_κ are characteristic frequencies. We note that there are exactly $3N$ normal modes.

Let us first calculate the heat capacity by means of classical statistical mechanics. This Hamiltonian $\mathcal{H}$ is quadratic in canonical variables (Q_κ, P_κ). Hence, the equipartition theorem holds. We multiply the average thermal energy $k_B T$ for each mode by the number of modes $3N$ and obtain $3N k_B T$ for the average energy $\langle \mathcal{H} \rangle$. Differentiating this with respect to T, we obtain $3N k_B$ for the heat capacity, in agreement with *Dulong–Petit's law*. It is interesting to observe that we obtained this result without knowing the actual distribution of normal-mode frequencies. The fact that there are $3N$ normal modes played an important role.

Let us now use quantum theory and calculate the heat capacity based on formula (4.8). The energy eigenvalues of the Hamiltonian $\mathcal{H}$ are given by

$$E(\{n_\kappa\}) = \sum_\kappa \hbar \omega_\kappa \left(\frac{1}{2} + n_\kappa \right) , \quad n_\kappa = 0, 1, 2, \ldots . \tag{4.9}$$

We can interpret (4.9) in terms of *phonons* as follows: the energy of the lattice vibrations is characterized by the set of numbers of phonons $\{n_\kappa\}$ in the normal modes $\{\kappa\}$. Taking the canonical-ensemble average of (4.9), we obtain

$$\langle E(\{n\})\rangle = \sum_\kappa \hbar\omega_\kappa \left(\frac{1}{2} + \langle n_\kappa\rangle\right) = \sum_\kappa \hbar\omega_\kappa \left(\frac{1}{2} + f_0(\hbar\omega_\kappa)\right) \equiv E(T) , \tag{4.10}$$

where

$$f_0(\varepsilon) \equiv \frac{1}{\exp(\varepsilon/k_B T) - 1} \tag{4.11}$$

is the *Planck distribution function*.

The normal-mode frequencies $\{\omega_\kappa\}$ depend on the normalization volume V, and they are densely populated for large V. In the bulk limit, we may convert the sum over the normal modes into a frequency integral and obtain

$$E(T) = E_0 + \int_0^\infty d\omega \mathcal{D}(\omega)\hbar\omega\, f_0(\hbar\omega) , \tag{4.12}$$

$$E_0 \equiv \frac{1}{2}\int_0^\infty d\omega \mathcal{D}(\omega)\hbar\omega , \tag{4.13}$$

where $\mathcal{D}(\omega)$ is the *density of states* (DOS) in angular frequency defined such that the

$$\text{number of modes in the interval } (\omega, \omega + d\omega) \equiv \mathcal{D}(\omega) . \tag{4.14}$$

The constant E_0 represents a temperature-independent zero-point energy.

Differentiating $E(T)$ with respect to T, we obtain for the heat capacity:

$$C_V \equiv \left(\frac{\partial E}{\partial T}\right)_V = \int_0^\infty d\omega \mathcal{D}(\omega)\hbar\omega \frac{\partial f_0(\hbar\omega)}{\partial T} . \tag{4.15}$$

This expression was obtained under the harmonic approximation only, which is good at very low temperatures.

To proceed further, we have to know the density of normal modes, $\mathcal{D}(\omega)$. To find the set of characteristic frequencies $\{\omega_\kappa\}$ requires solving an algebraic equation of $3N$th order, and we need the frequency distribution for large N. This is not a simple matter. In fact, a branch of mathematical physics whose principal aim is to find the frequency distribution, is called *lattice dynamics*. Figure 4.1 represents a result obtained by Walker [2] after an analysis of the X-ray scattering data for aluminum (Al), based on lattice dynamics. Some remarkable features of

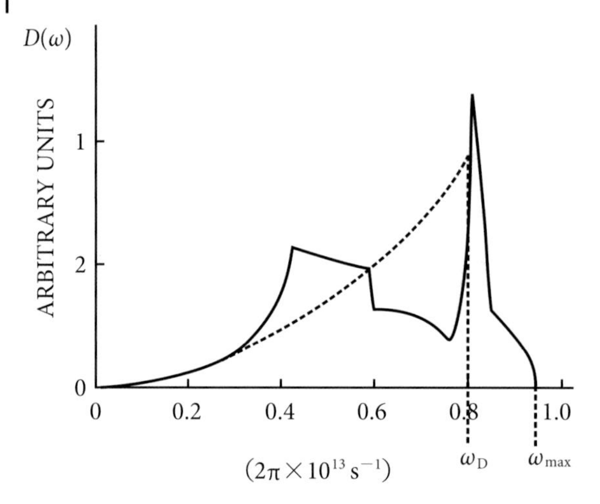

Figure 4.1 The density of normal modes $\mathcal{D}(\omega)$ in the angular frequency ω for aluminum. The solid curve represents the data deduced from X-ray scattering measurements due to Walker [2]. The broken lines indicate the Debye distribution with $\Theta_D = 328$ K.

the curve are:

1. At low frequencies,

$$\mathcal{D}(\omega) \propto \omega^2 . \tag{4.16}$$

2. There exists a maximum frequency ω_{max} such that

$$\mathcal{D}(\omega) = 0 \quad \text{for} \quad \omega \geq \omega_{max} . \tag{4.17}$$

3. A few sharp peaks exist below ω_{max}.

Feature (1) is common to all crystals. The low-frequency modes can be described adequately in terms of longitudinal and transverse elastic waves. This region can be represented very well by the Debye continuum model [3], indicated by the broken line. Feature (2) is connected with the lattice structure. Briefly, no normal modes of extreme short wavelengths (extreme high frequencies) exist. There is a limit frequency ω_{max}. Sharp peaks were first predicted by van Hove [4] on topological grounds. These peaks are often referred to as *van Hove singularities* [5, 6].

The van Hove singularities, the jumps in the derivative of the density of states, occur in two and three dimensions when the constant-frequency plane touches the Brillouin boundary *and* undergoes a curvature inversion. Similar singularities occur for conduction electrons, too. We give a full account of the van Hove singularities in the following Section 4.2.

As we will see later, the cause of superconductivity lies in the electron–phonon interaction [7]. The microscopic theory, however, can be formulated in terms of the generalized BCS Hamiltonian [7], see Section 7.3, where all phonon variables are eliminated. In this sense the details of lattice dynamics are secondary to our main concern. The following point, however, is noteworthy. All lattice dynamical calculations start with the assumption of a real crystal lattice. For example, to treat aluminum, we start with a face-centered cubic (fcc) lattice having empirically known lattice constants. The equations of motion for a set of ions are solved under the assumption of a periodic lattice-box boundary condition. Thus, the k-vectors used in both lattice dynamics and Bloch electron dynamics are the same. The domain of the k-vectors can be restricted to the same first Brillouin zone. Colloquially speaking, phonons (bosons) and electrons (fermions) live together in the same Brillouin zone, which is equivalent to saying that electrons and phonons share the same house (crystal lattice). This affinity between electrons and phonons makes the conservation of momentum in the electron–phonon interaction physically meaningful. Thus, the fact that the electron–phonon interaction is the cause of superconductivity is not accidental.

4.2 Van Hove Singularities

We define the *van Hove singularities* as jumps in the derivative of the density of states. The property arising from the reflection symmetry that the constant-frequency plane touches the Brillouin boundary at right angles [15] must be used to complete discussion of the van Hove singularities in 2D and 3D.

In Section 4.2.1, the normal modes of N coupled harmonic oscillators are obtained and discussed. Low-frequency phonons in simple-cubic (sc), body-centered cubic (bcc), tetragonal (tet), and fcc lattices are analyzed in Section 4.2.2. Van Hove singularities and others are discussed in Section 4.2.3.

4.2.1 Particles on a Stretched String (Coupled Harmonic Oscillators)

Let us take a system of N particles, each with mass m, separated by the equilibrium distance l on a stretched string with a tension τ. We consider small vertical displacements $y_j(t)$. We assume a *periodic boundary condition*:

$$y_{j+N}(t) = y_j(t) \,, \quad j = 1, 2, \ldots, N \,. \tag{4.18}$$

The kinetic energy $\mathcal{T}$ is given by

$$\mathcal{T} = \frac{m}{2}\left(\dot{y}_1^2 + \dot{y}_2^2 + \cdots + \dot{y}_N^2\right) \,, \quad \dot{y} \equiv \frac{\mathrm{d}y}{\mathrm{d}t} \,. \tag{4.19}$$

The potential energy $\mathcal{V}$ is given by

$$\mathcal{V} = \frac{\tau}{2l}\left[(y_2 - y_1)^2 + (y_3 - y_2)^2 + \cdots + (y_1 - y_N)^2\right] \,, \tag{4.20}$$

which is invariant under cyclic permutation. The Lagrangian $\mathcal{L}$ is then

$$\begin{aligned}\mathcal{L} &= \mathcal{T} - \mathcal{V} \\ &= \sum_{j=1}^{N} \left[\frac{m}{2} \dot{y}_j^2 - \frac{\tau}{2l} (y_{j+1} - y_j)^2 \right] . \end{aligned} \tag{4.21}$$

The equations of motion are

$$m \ddot{y}_j = \frac{\tau}{l} \left[y_{j+1} - 2 y_j + y_{j-1} \right] . \tag{4.22}$$

We assume solutions of the form:

$$y_j(t) = \exp[-\mathrm{i}(\omega t - j k l)] , \tag{4.23}$$

where k is a k-number. From the periodic boundary condition (4.18) this k-number satisfies

$$\exp(\mathrm{i}kjl) = \exp[\mathrm{i}k(j + N)l] \quad \text{for any } j (\leq N) , \tag{4.24}$$

or

$$\exp(\mathrm{i}kNl) = 1 , \tag{4.25}$$

whose solution is

$$k = \left(\frac{2\pi}{Nl} \right) n \equiv k_n , \quad n = 0, \pm 1, \pm 2, \ldots . \tag{4.26}$$

Note that k has a dimension of an inverse distance. Substituting (4.23) into (4.22), we obtain

$$\begin{aligned}\omega^2 \exp[-\mathrm{i}(\omega t - k j l)] = &-\frac{\tau}{ml} [\exp(\mathrm{i}kl) + \exp(-\mathrm{i}kl) - 2] \\ &\times \exp[-\mathrm{i}(\omega t - k j l)] , \end{aligned} \tag{4.27}$$

or

$$\omega^2 = \frac{2\tau}{ml} \left[1 - \cos(kl) \right] = 4 \omega_0^2 \sin^2 \left(\frac{kl}{2} \right) , \tag{4.28}$$

where $\omega_0 \equiv \sqrt{\tau / ml}$ is the (positive) frequency of a single harmonic oscillator. By convention, we choose a positive angular frequency ω:

$$\omega > 0 . \tag{4.29}$$

We then obtain a dispersion relation:

$$\omega = 2 \omega_0 \left| \sin \left(\frac{kl}{2} \right) \right| . \tag{4.30}$$

A general solution is composed of

$$\exp(-\mathrm{i}\omega_n t)\exp(\mathrm{i}k_n x) , \tag{4.31}$$

where k_n and ω_n are respectively given by

$$k_n = \left(\frac{2\pi}{Nl}\right) n , \quad \omega_n = 2\omega_0 \left|\sin\left(\frac{k_n l}{2}\right)\right| . \tag{4.32}$$

These solutions characterized by (k_n, ω_n) represent waves propagating along the line with the phase velocity

$$c_n = \frac{\omega_n}{k_n} . \tag{4.33}$$

We take first the case of an even number of particles, N. Then, $N/2$ is an integer. The number of degrees of freedom for the system is N. Then, there exist only $N k_n$. The frequency ω_n is an increasing function of $|k_n|$. We may choose an equal number, $N/2$, of positive $+k_n$ states and negative $-k_n$ states.

The Lagrangian $\mathcal{L}$ can be written as

$$\mathcal{L} = \frac{1}{2}\sum_{\{k_n\}} \left(\dot{Q}_{k_n}^2 - \omega_n^2 Q_{k_n}^2\right) , \tag{4.34}$$

where Q_{k_n} are the normal coordinates and ω_n normal mode frequencies.

We introduce canonical momenta

$$P_{k_n} \equiv \frac{\partial \mathcal{L}}{\partial \dot{Q}_{k_n}} = \dot{Q}_{k_n} \tag{4.35}$$

and construct a Hamiltonian $\mathcal{H}$

$$\mathcal{H} = \frac{1}{2}\sum_{\{k_n\}} \left(P_{k_n}^2 + \omega_n^2 Q_{k_n}^2\right) . \tag{4.36}$$

The frequencies $\{\omega_n\}$ are distributed, following the dispersion relation (4.32).

In the low-frequency (energy) region we obtain from (4.30)

$$\omega_n = \omega_0 l |k_n| \quad \text{for small } |k_n| . \tag{4.37}$$

This relation is relevant to the low-energy (acoustic) phonons in a solid.

4.2.2 Low-Frequency Phonons

1. Simple Cubic Lattice: Let us consider small oscillations for a system of atoms forming a sc lattice. We note that polonium (Po) forms a sc lattice [16]. We use a harmonic approximation. Assume a longitudinal traveling wave along a cubic

axis (x-axis). Imagine hypothetical planes perpendicular to the x-axis containing atoms forming a square lattice. This plane has a mass per unit square of side-length a (the lattice constant), equal to the atomic mass m. The plane is subjected to a restoring force per cm^2 equal to Young modulus Y. The dynamics of a set of the parallel planes are similar to that of the coupled harmonic oscillators discussed in Section 4.2.1.

Assume next a transverse wave traveling along the x-axis. The hypothetical planes containing the atoms are subjected to a restoring stress equal to the rigidity (shear) modulus S. The dynamics is also similar to the coupled harmonic oscillators in one dimension (1D).

Low-frequency phonons are those to which Debye's continuum solid model applies. The wave equations are

$$\frac{\partial^2 u_i}{\partial t^2} = c_i^2 \nabla^2 u_i \,, \tag{4.38}$$

where $i = \mathrm{l}$ (longitudinal) or t (transverse). The longitudinal-wave phase velocity c_l is

$$c_\mathrm{l} \equiv \sqrt{\frac{Y}{\rho}} \,, \tag{4.39}$$

where ρ is the mass density. The transverse-wave phase velocity c_t is

$$c_\mathrm{t} \equiv \sqrt{\frac{S}{\rho}} \,, \tag{4.40}$$

where S is the shear modulus. The waves are superposable. Hence, the movement of phonons is not restricted to the crystal's cubic directions. In short there is a k-vector, $\boldsymbol{k}$:

$$\boldsymbol{k} = k_x \hat{\boldsymbol{e}}_x + k_y \hat{\boldsymbol{e}}_y + k_z \hat{\boldsymbol{e}}_z \,, \tag{4.41}$$

where $\hat{\boldsymbol{e}}_x$, $\hat{\boldsymbol{e}}_y$, $\hat{\boldsymbol{e}}_z$ are the unit vectors in the Cartesian coordinates. The wave propagation is isotropic.

For the transverse waves the polarization vector points in a direction in the 2D polarization plane perpendicular to the k-vector. Hence, there are twice as many normal modes as those modes for the longitudinal waves.

2. Body-Centered Cubic Lattice: Alkali metals like sodium (Na) and potassium (K) form bcc lattices. The bcc lattice can be decomposed into two sublattices, one containing the corner atoms forming a sc lattice and the other containing body-center atoms also forming a sc lattice. These sc sublattices are similar with the same lattice constant a and the same crystal directions. The low-frequency phonons in a bcc lattice can travel in the same manner as the low-frequency phonons in the associated sc sublattice. In short they travel with a linear dispersion relation and isotropically.

3. Tetragonal Lattice: Indium (In) forms a tet lattice ($a_1 = a_2 \neq a_3, \alpha = \beta = \gamma = 90°$). Low-energy phonons may travel along one of the crystal axes (x-axis). We may introduce a set of hypothetical planes containing atoms. The plane has a mass density per cm^2 and it is subjected to a restoring stress. The dynamics of the plane motion is similar to the coupled harmonic oscillators discussed earlier. The mass density and the stress depend on the crystal axis along which the acoustic (low-energy) phonons travel. The elastic waves travel with linear dispersion relations. The wave propagation is angle-dependent or *anisotropic*, however, because the mass density and the restoring stress depend on the directions.
4. Face-Centered Cubic Lattice: Copper (Cu) and Al form fcc lattices. The fcc lattice can be decomposed into three sublattices, one containing the eight corner atoms forming a sc lattice with the side length a, a second containing the top and bottom face-center atoms forming a sc lattice with the side length a, a third containing the side face-center atoms forming a tetragonal lattice with the side-lengths $(a/\sqrt{2}, a/\sqrt{2}, a)$. The unit cell of a tetragonal sublattice contains two atoms. The acoustic longitudinal, transverse phonons associated with the sc-sublattices travel isotropically and the phonons associated with the tetragonal sublattice travel anisotropically. They have different phase speeds.

4.2.3 Discussion

Let us first take a sc lattice with the lattice constant a. We choose Cartesian coordinates (x, y, z) along the cubic axes with the center O at a unit cell center. The Brillouin boundary is cubic with the side-length $2\pi/a$. The original lattice has a reflection symmetry with respect to the plane: $x = 0$. It also has the same symmetry with respect to the $y = 0$-plane and the $z = 0$-plane. The low-frequency longitudinal and transverse phonons have linear dispersion relations. Hence, the density of states has a quadratic-ω dependence at low frequencies.

Second, we consider a bcc lattice. Since the bcc lattice can be decomposed into two similar sc sublattices with the lattice constant a and with the same cubic lattice directions, the density of states has a quadratic-ω dependence. The phonons for both sc and bcc lattices run isotropically. Hence, the constant-frequency surface at low frequencies is a sphere.

Third, we consider a tet lattice. The low-frequency dispersion relation is linear. The density of states has a ω-quadratic behavior. The phonon proceeds anisotropically, however.

Fourth, a fcc lattice is considered. A fcc lattice can be decomposed into two sc sublattices of the lattice constant a and a tetragonal sublattice of side-lengths $(a/\sqrt{2}, a/\sqrt{2}, a)$. The low-frequency dispersion relations for the longitudinal and transverse phonons are all linear. The density of states has a ω-quadratic behavior. The phonons associated with the tetragonal sublattice are anisotropic while the phonons associated with the sc lattice is isotropic. We note that such detailed phonon-mode behaviors are not available in Debye's continuum model theory [3].

In a historic paper [17] Einstein resolved the mystery concerning the near-absence of the lattice heat capacity of a diamond at room temperature. In his model each atom oscillates harmonically with the angular frequency ω_0. There is no frequency distribution. In our lattice-plane model developed in Section 4.2.2, the single-atom frequency ω_0 may be identified as

$$\omega_0 = \text{Einstein's model frequency} . \tag{4.42}$$

This ω_0 can be regarded as the *average* frequency:

$$\omega_0 = \langle \omega \rangle . \tag{4.43}$$

This view is supported by the fact that the Einstein temperature T_{E} is a little lower than the Debye temperature T_{D}:

$$k_{\text{B}} T_{\text{E}} = \hbar\omega_0 < k_{\text{B}} T_{\text{D}} . \tag{4.44}$$

For diamond, experiments indicate that

$$T_{\text{E}} = 1320\,\text{K} , \quad T_{\text{D}} = 1860\,\text{K} . \tag{4.45}$$

According to (4.28), the angular frequency ω approaches $2\omega_0$:

$$\omega = 2\omega_0 \sin\left(\frac{|k|\,l}{2}\right) \to 2\omega_0 \tag{4.46}$$

at highest $|k_{\text{N}}|l/2 = \pi/2$.

The density of states in 1D drops to zero at the maximum frequency. Aluminum forms a fcc lattice. The fcc lattice can be decomposed into two sc sublattices and one tet sublattice. Since the shear modulus S is smaller than the Young modulus Y, the transverse phonons have lower energies than the longitudinal phonons for the same k-values. The phonons associated with the sc sublattices are more easily excited than those associated with the tet sublattice since the restoring stresses are greater in the latter (more compact unit cell). Thus, the lowest-energy normal modes come from the transverse phonons associated with the sc sublattices. We may then conclude that the first peak in Figure 4.1 arises from the transverse sc phonons. The rise is greater than the Debye continuum model curve (dotted line). This is in line with the ω–k (dispersion) relation in (4.32), which has a negative second derivative for increasing k. The sharp peak arises as follows: for very small k, the dispersion relation is linear and the constant-frequency plane is a sphere. As the frequency (or energy) increases, the constant-frequency plane grows in size and, must approach and touch the six face-centers (singular points) of the cubic Brillouin zone boundary. The plane must approach the zone boundary at right angles because of the reflection symmetry possessed by the fcc lattice [15]. As the frequency increases, the constant-frequency plane grows in size, touches the boundary,

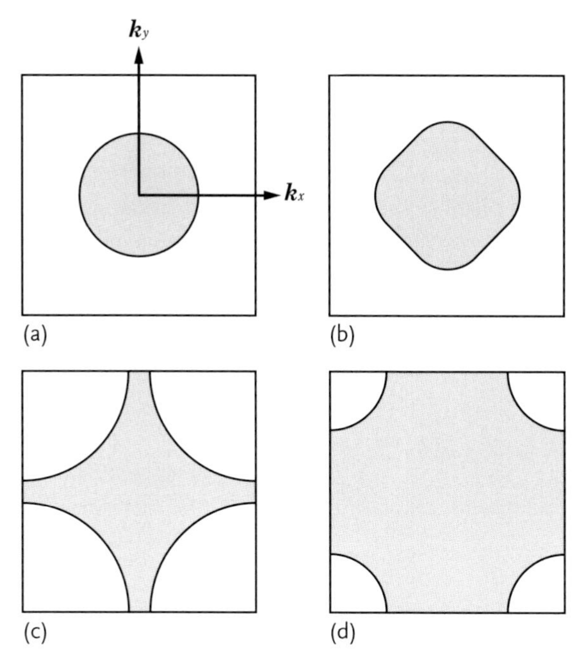

Figure 4.2 2D intersections at $k_z = 0$ of the constant frequency planes for a simple cubic lattice in the ω-increasing order: (a–d). The shaded areas correspond to the lowest-frequency (energy) sides. There is a singular point between (b,c) where the curvature sign changes. After [6].

changes its curvature sign, and eventually fills the whole k-space. The 2D cut of the surface development at $z = 0$ is illustrated in Figure 4.2. The small circle in Figure 4.2a represents the cut of a constant-frequency plane for small k. Between Figure 4.2b and 4.2c the curvature sign changes from convex to concave. We will show that the change point is the point of the van Hove singularity [6].

The density of phonon states in frequency per unit phase space, $\mathcal{D}(\omega)$, can be written as

$$\begin{aligned}\mathcal{D}(\omega) &= \frac{V}{(2\pi)^3}\int \mathrm{d}^3k\,\delta[\omega - \omega(\boldsymbol{k})] \\ &= \frac{V}{(2\pi)^3}\int_S \mathrm{d}S\frac{1}{|\nabla\omega(\boldsymbol{k})|} = \frac{V}{(2\pi)^3}\int_S \mathrm{d}S\frac{1}{|\partial\omega(\boldsymbol{k})/\partial\boldsymbol{k}|}\,,\end{aligned} \tag{4.47}$$

where the surface integral is over the surface S on which

$$\omega(\boldsymbol{k}) = \omega\ (= \text{constant}) \tag{4.48}$$

and in three dimensions:

$$\left|\frac{\partial\omega(\boldsymbol{k})}{\partial\boldsymbol{k}}\right| = \sqrt{\left(\frac{\partial\omega}{\partial k_x}\right)^2 + \left(\frac{\partial\omega}{\partial k_y}\right)^2 + \left(\frac{\partial\omega}{\partial k_z}\right)^2}\,.$$

The density is high at the contact where

$$\left|\frac{\partial \omega}{\partial \boldsymbol{k}}\right| = 0 , \quad \frac{1}{|\partial \omega / \partial \boldsymbol{k}|} = \infty . \tag{4.49}$$

But there are only six contact points, which have zero measure in the integral. They contribute nothing to the integral (4.47). The constant-frequency plane near the contact (singular points) is far from spherical. A further increase in ω generates a collection of singularities forming small circles on the zone boundary, which grows in size. The normal modes eventually fill up all the k-space and the density of states vanishes thereafter. In the range up to the curvature inversion point ω_{I} the density of the states $\mathcal{D}(\omega)$ increases with positive first and second derivatives:

$$\frac{\partial \mathcal{D}(\omega)}{\partial \omega} > 0 , \quad \frac{\partial^2 \mathcal{D}(\omega)}{\partial \omega^2} > 0 , \quad \omega < \omega_{\mathrm{I}} . \tag{4.50}$$

After the inversion point ω_{I}, we have

$$\frac{\partial \mathcal{D}(\omega)}{\partial \omega} < 0 , \quad \frac{\partial^2 \mathcal{D}(\omega)}{\partial \omega^2} < 0 , \quad \omega > \omega_{\mathrm{I}} . \tag{4.51}$$

The first region $\omega < \omega_{\mathrm{I}}$ in (4.50) corresponds to the case of the growing normal-mode-filled k-space (shaded): Figure 4.2a $\rightarrow$ b while the second case $\omega > \omega_{\mathrm{I}}$ in (4.51) to the shrinking empty k-space: Figure 4.2c $\rightarrow$ d. The density of states $D(\omega)$ has a maximum, and is continuous at the curvature-inversion point ω_{I} but its first derivative has a jump. The curvature inversion must exist since the total number of normal modes is precisely equal to the number of degrees of the system, and it must be contained within the Brillouin zone boundary. The curvature-inversion point is unstable, and can be reached smoothly neither from the low-frequency side nor the high-frequency side. There are unavoidable fluctuations and dispersions at this point, which may be probed by dynamic transport measurements.

For a 1D (linear) crystal, the frequency ω is given by $2\omega_0|\sin(kl/2)|$. By direct calculation we obtain

$$\mathcal{D}(\omega) = \frac{L}{2\pi} \int_{-L/2}^{L/2} \mathrm{d}k \delta \left[\omega - 2\omega_0 \sin\left(\frac{|k|\,l}{2}\right) \right] = \frac{2}{\pi} \frac{1}{\sqrt{4\omega_0^2 - \omega^2}} . \tag{4.52}$$

Because of the inversion symmetry, the constant-frequency lines approach the Brillouin zone boundary at right angles. There are no curvatures and no curvature inversions. The density of states diverges at the end points ($\omega = \pm 2\omega_0$). This singularity can be regarded as a precursor to the van Hove singularities occurring in 2D and 3D.

Essentially the same thing happens for the phonons associated with the tet sublattice. This explains the second peak in Figure 4.1. There are side-shoulders for the first and second peaks on the high-frequency sides. These are due to the longitudinal phonons.

In summary the van Hove singularities (peaks) occur when the constant-frequency plane touches the Brillouin zone boundary *and* undergoes a curvature inversion. There are two prominent peaks in Al (fcc) due to the transverse phonons associated with the sc sublattices *and* the tet sublattice.

4.3 Electron–Phonon Interaction

A crystal lattice is composed of a regular array of ions. If the ions move, then the electrons must move in a changing potential field. Fröhlich proposed an interaction Hamiltonian, which is especially suitable for transport and superconductivity problems. In the present section we derive the Fröhlich Hamiltonian [8, 9].

For simplicity let us takes a sc lattice. The normal modes of oscillations for a solid are longitudinal and transverse running waves characterized by wave vector $\boldsymbol{q}$ and frequency ω_q. First, consider the case of a longitudinal wave proceeding in the crystal axis x, which is represented by

$$u_q \exp(-\mathrm{i}\omega_q t + \mathrm{i}\boldsymbol{q} \cdot \boldsymbol{r}) = u_q \exp(-\mathrm{i}\omega_q t + \mathrm{i}qx) \,, \tag{4.53}$$

where u_q is the displacement in the x-direction. The wavelength $\lambda \equiv 2\pi/q (> 2a_0)$ is greater than twice the lattice constant a_0. The case, $\lambda = 12a_0$, is shown in Figure 4.3.

If we imagine a set of parallel plates containing a great number of ions fixed in each plate, we have a realistic picture of the lattice vibration mode. From Figure 4.3 we see that the density of ions changes in the x-direction. Hence, the longitudinal wave modes are also called the *density-wave* modes. The transverse wave modes can also be pictured from Figure 4.3 by imagining a set of parallel plates containing a great number of ions fixed in each plate and assuming the transverse displacements of the plates. Notice that this mode generates no charge-density variation.

Now, the Fermi velocity v_F in a typical metal is of the order $10^6\ \mathrm{m\,s^{-1}}$ while the speed of sound is of the order $10^3\ \mathrm{m\,s^{-1}}$. The electrons are then likely to move quickly to negate any electric field generated by the density variations associated with the lattice wave. Hence, the electrons may follow the lattice waves quite eas-

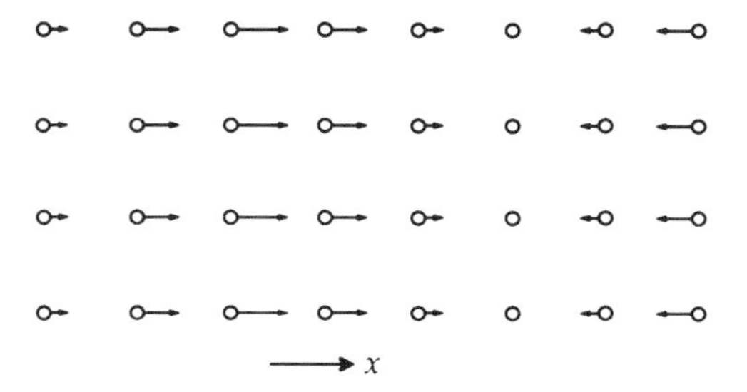

Figure 4.3 Longitudinal waves proceeding in the x-direction: $\lambda = 12a_0$. Circle (o) indicates a lattice ion.

ily. Given a traveling normal mode in (4.53), we may assume an electron density deviation of the form:

$$C_q \exp(-\mathrm{i}\omega_q t + \mathrm{i}\boldsymbol{q} \cdot \boldsymbol{r}) \,. \tag{4.54}$$

Since electrons follow phonons immediately for all ω_q, the factor C_q can be regarded as independent of ω_q. We further assume that the deviation is linear in $\boldsymbol{q} \cdot \boldsymbol{u}_q = q u_q$ and again in the electron density $n(\boldsymbol{r})$. Thus,

$$C_q = A_q q u_q n(\boldsymbol{r}) \,. \tag{4.55}$$

This is called the *deformation potential approximation* [18]. The dynamic response factor A_q is necessarily complex since there is a time delay between the field (cause) and the density variation (result). The traveling wave is represented by the exponential form (4.54). Complex conjugation of this equation yields $C_q^* \exp(\mathrm{i}\omega_q t - \mathrm{i}\boldsymbol{q} \cdot \boldsymbol{r})$. Using this form we can reformulate the electron's response, but the physics must be the same. From this consideration we obtain (Problem 4.3.1)

$$A_q = A_{-q}^* \,. \tag{4.56}$$

The classical displacement u_q changes, following the harmonic equation of motion:

$$\ddot{u}_q + \omega_q^2 u_q = 0 \,. \tag{4.57}$$

Let us write the corresponding Hamiltonian for each mode as

$$\mathcal{H} = \frac{1}{2}(p^2 + \omega^2 q^2) \,, \quad q \equiv u \,, \quad p \equiv \dot{q} \,, \quad \omega_q \equiv \omega \,, \tag{4.58}$$

where we dropped the mode index $\boldsymbol{q}$. If we assume the same quantum Hamiltonian $\mathcal{H}$ and the basic commutation relations:

$$[q, p] = \mathrm{i}\hbar \,, \quad [q, q] = [p, p] = 0 \,, \tag{4.59}$$

then the quantum description of a harmonic oscillator is complete. The equations of motion are (Problem 4.3.2)

$$\dot{q} = \frac{1}{\mathrm{i}\hbar}[q, \mathcal{H}] = p \,, \quad \dot{p} = \frac{1}{\mathrm{i}\hbar}[p, \mathcal{H}] = -\omega^2 q \,. \tag{4.60}$$

We introduce the dimensionless complex dynamical variables:

$$a^\dagger \equiv (2\hbar\omega)^{-1/2}(p + \mathrm{i}\omega q) \,, \quad a \equiv (2\hbar\omega)^{-1/2}(p - \mathrm{i}\omega q) \,. \tag{4.61}$$

Using (4.60) we obtain

$$\dot{a}^\dagger \equiv (2\hbar\omega)^{-1/2}(-\omega^2 q + \mathrm{i}\omega p) = \mathrm{i}\omega a^\dagger \,, \quad \dot{a} = -\mathrm{i}\omega a \,. \tag{4.62}$$

We can express (q, p) in terms of $(a, a^\dagger)$:

$$q = -\mathrm{i}\left(\frac{\hbar}{2\omega}\right)^{1/2}(a^\dagger - a)\,, \quad p = \left(\frac{\hbar\omega}{2}\right)^{1/2}(a^\dagger + a)\,. \tag{4.63}$$

Thus, we may work entirely in terms of $(a, a^\dagger)$. After straightforward calculations we obtain (Problem 4.3.3):

$$\begin{aligned}\hbar\omega a^\dagger a &= \frac{1}{2}(p + \mathrm{i}\omega q)(p - \mathrm{i}\omega q)\\ &= \frac{1}{2}[p^2 + \omega^2 q^2 + \mathrm{i}\omega(qp - pq)] = \mathcal{H} - \frac{1}{2}\hbar\omega\,.\end{aligned} \tag{4.64}$$

$$\hbar\omega a a^\dagger = \mathcal{H} + \frac{1}{2}\hbar\omega\,, \tag{4.65}$$

$$a a^\dagger - a^\dagger a \equiv [a, a^\dagger] = 1\,, \tag{4.66}$$

and we finally obtain from (4.64)–(4.66)

$$\begin{aligned}\mathcal{H} = \frac{1}{2}\hbar\omega(a^\dagger a + a a^\dagger) &= \hbar\omega\left(a^\dagger a + \frac{1}{2}\right)\\ &\equiv \hbar\omega\left(n + \frac{1}{2}\right)\,, \quad n \equiv a^\dagger a\,.\end{aligned} \tag{4.67}$$

The operators $(a^\dagger, a)$ satisfy the *Bose commutation rules* (4.66). The energy-eigenvalue equation for $\mathcal{H}$ is now reduced to finding the eigenvalues and eigenstates of the Hermitian operator

$$n = a^\dagger a = n^\dagger\,. \tag{4.68}$$

The eigenvalue equation is

$$n|n'\rangle = n'|n'\rangle\,, \tag{4.69}$$

where n' and $|n'\rangle$ are eigenvalues and eigenkets, respectively.

We shall show that n has as eigenvalues all nonnegative integers. Multiplying (4.69) by the eigenbra $\langle n'|$ from the left, we obtain

$$\langle n'|a^\dagger a|n'\rangle = n'\langle n'|n'\rangle\,. \tag{4.70}$$

Now, $\langle n'|a^\dagger a|n'\rangle$ is the squared length of the ket $a|n'\rangle$ and hence

$$\langle n'|a^\dagger a|n'\rangle \geq 0\,. \tag{4.71}$$

Also, by definition $\langle n'|n'\rangle > 0$. Hence from (4.70) and (4.71), we obtain

$$n' \geq 0\,, \tag{4.72}$$

the case of equality occurring only if

$$a|n'\rangle = 0\,. \tag{4.73}$$

Consider now $[a, n] = [a, a^\dagger a]$. We may use the following identities:

$$[A, BC] = B[A, C] + [A, B]C\ , \quad [AB, C] = A[B, C] + [A, C]B \tag{4.74}$$

and obtain

$$[a, a^\dagger a] = a^\dagger[a, a] + [a, a^\dagger]a = a \quad \text{or} \quad an - na = a\ . \tag{4.75}$$

Rearranging terms in (4.75) and multiplying it by the eigenket $|n'\rangle$ from the right, we obtain

$$na|n'\rangle = (an - a)|n'\rangle = (n' - 1)a|n'\rangle\ . \tag{4.76}$$

Now, if $a|n'\rangle \neq 0$, then $a|n'\rangle$ is, according to (4.76), an eigenket of n belonging to the eigenvalue $n' - 1$. Hence, for nonzero n', $n' - 1$ is another eigenvalue. We can repeat the argument and deduce that, if $n' - 1 \neq 0$, $n' - 2$ is another eigenvalue of n. Continuing in this way, we obtain a series of eigenvalues $n', n' - 1, n' - 2, \cdots$ that can terminate only with the value 0 because of inequality (4.72). By a similar process, we can show from the Hermitian conjugate of (4.75):

$$na^\dagger - a^\dagger n = a^\dagger \tag{4.77}$$

that the eigenvalue of n has *no* upper limit (Problem 4.3.4). Hence, the eigenvalues of n are nonnegative integers:

$$n' = 0, 1, 2, \ldots \tag{4.78}$$

□

The energy-eigenvalue equation for the harmonic oscillator is

$$\mathcal{H}|n\rangle = E|n\rangle\ , \tag{4.79}$$

where the energy E is given by

$$E = \hbar\omega\left(n + \frac{1}{2}\right) \equiv E_n\ , \quad n = 0, 1, 2, \ldots\ . \tag{4.80}$$

Here we omitted the primes for the eigenvalues since they are self-evident. In summary we obtain

- Eigenvalues of $n \equiv a^\dagger a$: $n' = 0, 1, 2, \ldots$
- Vacuum ket $|\phi\rangle$: $a|\phi\rangle = 0$
- Eigenkets of n: $|\phi\rangle, a^\dagger|\phi\rangle, (a^\dagger)^2|\phi\rangle, \ldots$ having the eigenvalues $0, 1, 2, \ldots$
- Eigenvalues of $\mathcal{H}$: $E = \left(n + \frac{1}{2}\right)\hbar\omega$

The main result (4.80) can be obtained by solving the Schrödinger equation:

$$\frac{1}{2}\left(-\hbar^2\frac{\partial^2}{\partial q^2} + \omega^2 q^2\right)\psi(q) = E\psi(q)\ , \tag{4.81}$$

which requires considerable mathematical skills.

The ket $|n\rangle$ can be expressed in terms of (a, a'). Let $|\phi_0\rangle$ be a normalized eigenket of n belonging to the eigenvalue 0 so that

$$n|\phi_0\rangle = a^\dagger a|\phi_0\rangle = 0 . \tag{4.82}$$

This ket is called the *vacuum ket*. It has the following property:

$$a|\phi_0\rangle = 0 . \tag{4.83}$$

Using the Bose commutation rules (4.66) we obtain a relation (Problem 4.3.5):

$$a(a^\dagger)^{n'} - (a^\dagger)^{n'} a = n'(a^\dagger)^{n'-1} , \tag{4.84}$$

which may be proved by induction. Multiply (4.84) by $a^\dagger$ from the left and operate the result to $|\Phi_0\rangle$. Using (4.83) we obtain

$$n(a^\dagger)^{n'}|\phi_0\rangle = n'(a^\dagger)^{n'}|\phi_0\rangle , \tag{4.85}$$

indicating that $(a^\dagger)^{n'}|\phi_0\rangle$ is an eigenket belonging to the eigenvalue n'. The square length of $(a^\dagger)^{n'}|\phi_0\rangle$ is

$$\langle\phi_0|a^{n'}(a^\dagger)^{n'}|\phi_0\rangle = n'\langle\phi_0|a^{n'-1}(a^\dagger)^{n'-1}|\phi_0\rangle = \cdots = n'! . \tag{4.86}$$

We see from (4.85) that $a|n'\rangle$ is an eigenket of n belonging to the eigenvalue $n' - 1$. Similarly, we can show from $[a, a^\dagger] = a^\dagger$ that $a^\dagger|n'\rangle$ is an eigenket of n belonging to the eigenvalue $n' + 1$. Thus, operator a, acting on the number eigenket, annihilates a particle, while operator $a^\dagger$ creates a particle. Therefore, a and $a^\dagger$ are called *annihilation* and *creation* operators, respectively.

In summary, the quantum Hamiltonian and the quantum states of a harmonic oscillator can be simply described in terms of the bosonic second-quantized operators $(a, a^\dagger)$.

We now go back to the case of the lattice normal modes. Each normal mode corresponds to a harmonic oscillator characterized by (q, ω_q). The displacements u_q can be expressed as

$$u_q = \mathrm{i}\left(\frac{\hbar}{2\omega_q}\right)^{1/2}\left(a_q - a_q^\dagger\right) , \tag{4.87}$$

where $(a_q, a_q^\dagger)$ are annihilation and creation operators satisfying the *Bose commutation rules*:

$$\left[a_q, a_p^\dagger\right] \equiv a_q a_p^\dagger - a_p^\dagger a_q = \delta_{p,q} , \quad [a_q, a_p] = \left[a_q^\dagger, a_p^\dagger\right] = 0 . \tag{4.88}$$

We can express the electron density (field) by

$$n(\boldsymbol{r}) = \psi^\dagger(\boldsymbol{r})\psi(\boldsymbol{r}) , \tag{4.89}$$

where $\psi(\boldsymbol{r})$ and $\psi^\dagger(\boldsymbol{r})$ are *annihilation* and *creation electron field operators*, respectively, satisfying the following *Fermi anticommutation rules*:

$$\{\psi(\boldsymbol{r}), \psi^\dagger(\boldsymbol{r}')\} \equiv \psi(\boldsymbol{r})\psi^\dagger(\boldsymbol{r}') + \psi^\dagger(\boldsymbol{r}')\psi(\boldsymbol{r}) = \delta^{(3)}(\boldsymbol{r} - \boldsymbol{r}') ,$$
$$\{\psi(\boldsymbol{r}), \psi(\boldsymbol{r}')\} = \{\psi^\dagger(\boldsymbol{r}), \psi^\dagger(\boldsymbol{r}')\} = 0 . \tag{4.90}$$

The field operators $\psi(\psi^\dagger)$ can be expanded in terms of the *momentum-state electron operators* $c_{\boldsymbol{k}}(c_{\boldsymbol{k}}^\dagger)$:

$$\psi(\boldsymbol{r}) = \frac{1}{\sqrt{V}} \sum_{\boldsymbol{k}} \exp(\mathrm{i}\boldsymbol{k} \cdot \boldsymbol{r}) c_{\boldsymbol{k}} , \quad \psi^\dagger(\boldsymbol{r}) = \frac{1}{\sqrt{V}} \sum_{\boldsymbol{k}'} \exp(-\mathrm{i}\boldsymbol{k}' \cdot \boldsymbol{r}) c_{\boldsymbol{k}'}^\dagger , \tag{4.91}$$

where operators $(c, c^\dagger)$ satisfy the *Fermi anticommutation rules*:

$$\left\{c_{\boldsymbol{k}}, c_{\boldsymbol{k}'}^\dagger\right\} \equiv c_{\boldsymbol{k}} c_{\boldsymbol{k}'}^\dagger + c_{\boldsymbol{k}'}^\dagger c_{\boldsymbol{k}} = \delta_{\boldsymbol{k},\boldsymbol{k}'}^{(3)} , \quad \{c_{\boldsymbol{k}}, c_{\boldsymbol{k}'}\} = \left\{c_{\boldsymbol{k}}^\dagger, c_{\boldsymbol{k}'}^\dagger\right\} = 0 . \tag{4.92}$$

Let us now construct an interaction Hamiltonian $\mathcal{H}_\mathrm{F}$, which has the dimension of an energy and which is Hermitian. We propose

$$\mathcal{H}_\mathrm{F} = \int \mathrm{d}^3 r \sum_{q} \left[A_q q u_q \exp(\mathrm{i}\boldsymbol{q} \cdot \boldsymbol{r}) \psi^\dagger(\boldsymbol{r}) \psi(\boldsymbol{r}) + \text{h.c.}\right] , \tag{4.93}$$

where h.c. denotes the Hermitian conjugate. Using (4.55), (4.87), and (4.91) we can express (4.93) in a second-quantized form (Problem 4.3.6):

$$\mathcal{H}_\mathrm{F} = \sum_{\boldsymbol{k}} \sum_{q} \left(V_q c_{\boldsymbol{k}+q}^\dagger c_{\boldsymbol{k}} a_q + V_q^* c_{\boldsymbol{k}}^\dagger c_{\boldsymbol{k}+q} a_q^\dagger\right) , \quad V_q \equiv \mathrm{i} A_q \left(\frac{\hbar}{2\omega_q}\right)^{1/2} q . \tag{4.94}$$

This is the *Fröhlich Hamiltonian*. Electrons describable in terms of $c_{\boldsymbol{k}}$'s are now coupled with phonons describable in terms of a_q's. The term

$$V_q c_{\boldsymbol{k}+q}^\dagger c_{\boldsymbol{k}} a_q \quad \left(V_q^* c_{\boldsymbol{k}}^\dagger c_{\boldsymbol{k}+q} a_q^\dagger\right)$$

can be pictured as an interaction process in which a phonon is absorbed (emitted) by an electron as represented by the Feynman diagram [10, 11] in Figure 4.4a(b). Note: at each vertex the momentum is conserved. The Fröhlich Hamiltonian $\mathcal{H}_\mathrm{F}$ is applicable for longitudinal phonons only. As noted earlier, the transverse lattice normal modes generate no charge density variations, making the electron–transverse phonon interaction negligible.

Problem 4.3.1. Prove (4.56).

Problem 4.3.2. Verify (4.60).

Problem 4.3.3. Verify (4.64)–(4.66).

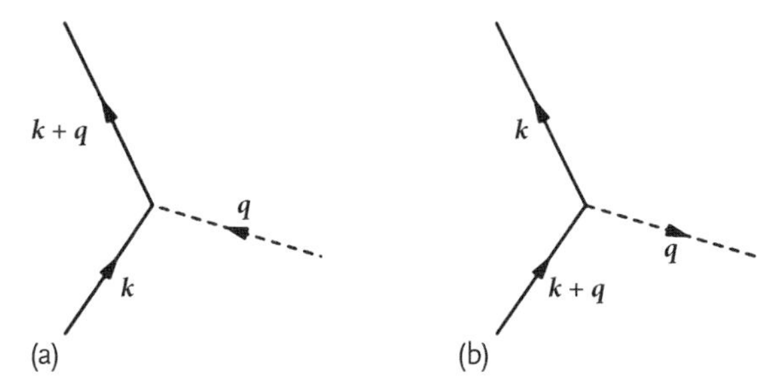

Figure 4.4 Feynman diagrams representing (a) absorption and (b) emission of a phonon (indicated by the dashed line) by an electron. They correspond to the interaction terms in (4.94).

Problem 4.3.4. Verify that the eigenvalues of n have no upper limit. As shown in the text the eigenvalues are separated by unity and the lower bound is zero. Hence the eigenvalues n' are $0, 1, 2, \cdots$.

Problem 4.3.5. Prove (4.84) by mathematical induction.

Problem 4.3.6. Show that the second-quantized form of $\mathcal{H}_\text{F}$ is given by (4.94).

4.4 Phonon-Exchange Attraction

By exchanging a phonon, two electrons can gain an attraction under a certain condition. In this section we treat this effect by using the many-body perturbation method [12, 13].

Let us consider an *electron–phonon system* characterized by

$$\begin{aligned}\mathcal{H} &= \sum_k \sum_s \varepsilon_k c_{ks}^\dagger c_{ks} + \sum_q \hbar\omega_q \left(a_q^\dagger a_q + \frac{1}{2}\right) \\ &\quad + \lambda \sum_k \sum_s \sum_q \left(V_q c_{k+q s}^\dagger c_{ks} a_q + \text{h.c.}\right) \\ &\equiv \mathcal{H}_\text{el} + \mathcal{H}_\text{ph} + \lambda \mathcal{H}_\text{F} \equiv \mathcal{H}_0 + \lambda \mathcal{V}\,, \quad (\mathcal{V} \equiv \mathcal{H}_\text{F})\end{aligned} \tag{4.95}$$

where the three sums represent respectively the total electron kinetic energy ($\mathcal{H}_\text{el}$), the total phonon energy ($\mathcal{H}_\text{ph}$), and the Fröhlich interaction Hamiltonian ($\mathcal{H}_\text{F}$) (see (4.94)).

For comparison we consider an *electron gas system* characterized by the Hamiltonian

$$\mathcal{H}_\text{C} = \sum_k \sum_s \varepsilon_k c_{ks}^\dagger c_{ks} + \frac{1}{2} \sum_{k_1 s_1} \cdots \sum_{k_4 s_4} \langle 1, 2|\mathcal{V}_\text{C}|3, 4\rangle c_2^\dagger c_1^\dagger c_3 c_4 \equiv \mathcal{H}_\text{el} + \mathcal{V}_\text{C}\,,$$

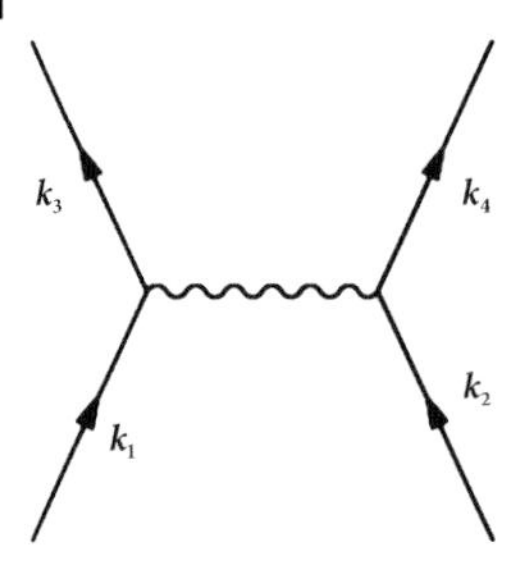

Figure 4.5 The Coulomb interaction represented by the horizontal wavy line generates the change in the momenta of two electrons.

(4.96)

where

$$\langle 1,2|\mathcal{V}_C|3,4\rangle \equiv \langle \boldsymbol{k}_1 s_1, \boldsymbol{k}_2 s_2|\mathcal{V}_C|\boldsymbol{k}_3 s_3, \boldsymbol{k}_4 s_4\rangle$$
$$= \frac{4\pi e^2 k_0}{V}\frac{1}{q^2}\delta_{\boldsymbol{k}_3+\boldsymbol{k}_4,\boldsymbol{k}_1+\boldsymbol{k}_2}\delta_{\boldsymbol{k}_3-\boldsymbol{k}_1,\boldsymbol{q}}\delta_{s_1,s_3}\delta_{s_2,s_4} \,. \tag{4.97}$$

The elementary interaction process can be represented by a diagram as in Figure 4.5. The wavy horizontal line represents the instantaneous Coulomb interaction $\mathcal{V}_C$. The net momentum of a pair of electrons is conserved:

$$\boldsymbol{k}_1 + \boldsymbol{k}_2 = \boldsymbol{k}_3 + \boldsymbol{k}_4 \,, \tag{4.98}$$

as seen by the appearance of the Kronecker delta in (4.97). Physically, the Coulomb force between a pair of electrons is an internal force, and hence it cannot change the net momentum.

We wish to find an *effective* Hamiltonian $\mathcal{V}_{\text{eff}}$ between a pair of electrons generated by a phonon exchange. If we look for this $\mathcal{V}_{\text{eff}}$ in the second order in the coupling constant λ, the likely candidates may be represented by the two Feynman diagrams[1] in Figure 4.6. Here, the time is measured upward. In the diagrams in Figure 4.6a,b, we follow the motion of two electrons. We may therefore consider a system of two electrons and obtain the effective Hamiltonian $\mathcal{V}_{\text{eff}}$ through a study of the evolution of *two-body density operator* ρ_2. For brevity we shall hereafter drop the subscript 2 on ρ indicating the two-body system.

The system-density operator $\rho(t)$ changes in time, following the *quantum Liouville equation*:

$$i\hbar\frac{\partial\rho(t)}{\partial t} = [\mathcal{H}, \rho] \equiv \mathcal{H}^{\times}\rho \,, \tag{4.99}$$

where the superscript $(\times)$ on $\mathcal{H}$ indicates the *Liouville operator*.[2] We assume the Hamiltonian $\mathcal{H}$ in (4.96), and study the time evolution of $\rho(t)$, using quantum

1) Historically, Feynman represented the elementary interaction processes in the Dirac picture (DP) by diagram. Such a diagram representation is very popular and widely used in quantum field theory [10, 11].
2) A Liouville operator is a superoperator which generates a commutator upon acting an operator.

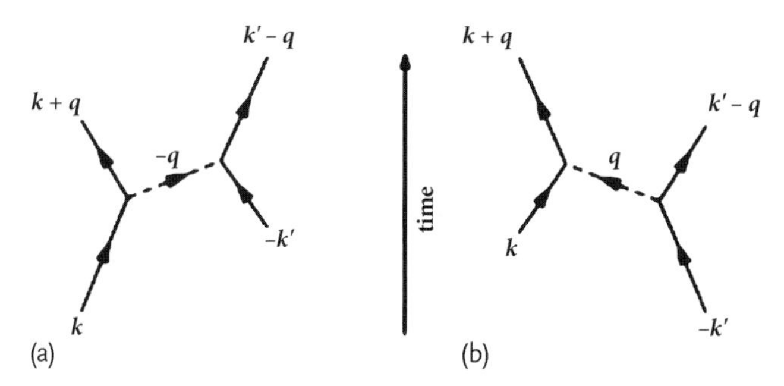

Figure 4.6 (a,b) A one-phonon exchange process generates the change in the momenta of two electrons similar to that caused by the Coulomb interaction.

many-body perturbation theory. We sketch only the important steps; more detailed calculations are given in Fujita and Godoy's book [12, 13].

Let us introduce *quantum Liouville operators*[3],

$$\mathcal{H}^{\times} \equiv \mathcal{H}_0^{\times} + \lambda \mathcal{V}^{\times} , \tag{4.100}$$

which generate a commutator upon acting on ρ, see (4.99). We assume that the initial-density operator ρ_0 for the combined electron–phonon system can be written as

$$\rho_0 = \rho_{\text{electron}} \otimes \rho_{\text{phonon}} , \tag{4.101}$$

which is reasonable at 0 K, where there are no real phonons and only virtual phonons are involved in the dynamical processes. We can then choose

$$\rho_{\text{phonon}} = |0\rangle\langle 0| , \tag{4.102}$$

where $|0\rangle$ is the vacuum-state ket for phonons:

$$a_{\boldsymbol{q}}|0\rangle = 0 \quad \text{for any } \boldsymbol{q} . \tag{4.103}$$

The phonon vacuum average will be denoted by an upper bar or angular brackets:

$$\overline{\rho}(t) \equiv \langle 0|\rho(t)|0\rangle \equiv \langle \rho(t)\rangle_{\text{av}} . \tag{4.104}$$

Using a time-dependent perturbation theory and taking a phonon-average, we obtain (Problem 4.4.1) from (4.99)

$$\frac{\partial \overline{\rho}(t)}{\partial t} = -\frac{\lambda^2}{\hbar^2} \int_0^t \mathrm{d}\tau \langle \mathcal{V}^{\times} \exp(-\mathrm{i}\tau\hbar^{-1}\mathcal{H}_0^{\times})\mathcal{V}^{\times}\rho(t-\tau)\rangle_{\text{av}} . \tag{4.105}$$

In the *weak-coupling approximation*, we may calculate the phonon-exchange effect to the lowest (second) order in λ, so that we obtain (Problem 4.4.2)

$$\begin{aligned} &\lambda^2 \langle \mathcal{V}^{\times} \exp(-\mathrm{i}\tau\hbar^{-1}\mathcal{H}_0^{\times})\mathcal{V}^{\times}\rho(t-\tau)\rangle \\ &= \lambda^2 \langle \mathcal{V}^{\times} \exp(-\mathrm{i}\tau\hbar^{-1}\mathcal{H}_0^{\times})\mathcal{V}^{\times}\rangle_{\text{av}}\, \overline{\rho}(t-\tau) . \end{aligned} \tag{4.106}$$

3) The quantum Liouville operator (4.100) corresponds to the Hamiltonian $\mathcal{H} = \mathcal{H}_0 + \lambda\mathcal{V}$ in (4.95).

In the Markovian approximation we may replace $\overline{\rho}(t-\tau)$ by $\overline{\rho}(t)$ and take the upper limit t of the τ-integration to ∞. Using these two approximations, we obtain from (4.105)

$$\frac{\partial \overline{\rho}(t)}{\partial t} = \mathrm{i}\lambda^2\hbar^{-1} \lim_{a\to 0} \langle \mathcal{V}^\times (\mathcal{H}_0^\times - \mathrm{i}a)^{-1} \mathcal{V}^\times \rangle_{\mathrm{av}} \overline{\rho}(t) \,, \quad a > 0 \,. \tag{4.107}$$

Let us now take momentum-state matrix elements of (4.107). The lhs is

$$\frac{\partial}{\partial t} \langle \boldsymbol{k}_1 s_1, \boldsymbol{k}_2 s_2 | \rho(t) | \boldsymbol{k}_3 s_3, \boldsymbol{k}_4 s_4 \rangle \equiv \frac{\partial}{\partial t} \rho(1,2;3,4,t) \,, \tag{4.108}$$

where we dropped the upper bar indicating the phonon average. The rhs requires more sophisticated computations since the Liouville operators $(\mathcal{V}^\times, \mathcal{H}_0^\times)$ are involved. After lengthy but straightforward calculations, we obtain (Problem 4.4.3) from (4.107)

$$\begin{aligned} \frac{\partial}{\partial t} \rho(1,2;3,4,t) = \sum_{\boldsymbol{k}_5, s_5} \sum_{\boldsymbol{k}_6, s_6} -\mathrm{i}\hbar^{-1} [\langle 1,2 | \mathcal{V}_{\mathrm{eff}} | 5,6 \rangle \rho(5,6;3,4,t) \\ - \rho(1,2;5,6,t) \langle 5,6 | \mathcal{V}_{\mathrm{eff}} | 3,4 \rangle] \,, \end{aligned} \tag{4.109}$$

where $\langle 1,2|\mathcal{V}_{\mathrm{eff}}|3,4\rangle$ is given by

$$\begin{aligned} \langle 1,2 | \mathcal{V}_{\mathrm{eff}} | 3,4 \rangle \equiv |V_q|^2 \frac{\hbar\omega_q}{(\varepsilon_1 - \varepsilon_3)^2 - \hbar^2 \omega_q^2} \\ \times \delta_{\boldsymbol{k}_3 + \boldsymbol{k}_4, \boldsymbol{k}_1 + \boldsymbol{k}_2} \delta_{\boldsymbol{k}_1 - \boldsymbol{k}_3, \boldsymbol{q}} \delta_{s_1, s_3} \delta_{s_2, s_4} \,. \end{aligned} \tag{4.110}$$

The Kronecker delta $\delta_{\boldsymbol{k}_3+\boldsymbol{k}_4,\boldsymbol{k}_1+\boldsymbol{k}_2}$ in (4.110) means that the net momentum is conserved, since the phonon-exchange interaction is an internal interaction.

For comparison, consider the electron-gas system characterized by the Hamiltonian $\mathcal{H}_{\mathrm{C}}$ in (4.96). The two-electron density matrix ρ_{C} for this system changes, following

$$\begin{aligned} \frac{\partial}{\partial t} \rho_{\mathrm{C}}(1,2;3,4,t) = \sum_{\boldsymbol{k}_5, s_5} \sum_{\boldsymbol{k}_6, s_6} -\mathrm{i}\hbar^{-1} [\langle 1,2 | \mathcal{V}_{\mathrm{C}} | 5,6 \rangle \rho_{\mathrm{C}}(5,6;3,4,t) \\ - \rho_{\mathrm{C}}(1,2;5,6,t) \langle 5,6 | \mathcal{V}_{\mathrm{C}} | 3,4 \rangle] \,, \end{aligned} \tag{4.111}$$

which is of the same form as (4.109). The only differences are in the interaction matrix elements. Comparison between (4.97) and (4.110) yields (Problem 4.4.4)

$$\frac{4\pi e^2 k_0}{V} \frac{1}{q^2} \qquad \text{(Coulomb interaction)} \,, \tag{4.112}$$

$$|V_q|^2 \frac{\hbar\omega_q}{(\varepsilon_{\boldsymbol{k}_1+\boldsymbol{q}} - \varepsilon_{\boldsymbol{k}_1})^2 - \hbar^2\omega_q^2} \qquad \text{(phonon-exchange interaction)} \,. \tag{4.113}$$

In our derivation, the weak-coupling and the Markovian approximation were used. The Markovian approximation is justified in the steady-state condition in which the effect of the duration of interaction can be neglected. The electron mass is four orders of magnitude smaller than the lattice-ion mass, and hence the coupling between the electron and ionic motion must be small by the mass mismatch. Thus, expression (4.113) is highly accurate for the effective phonon-exchange interaction at 0 K. This expression has remarkable features. First, it depends on the phonon energy $\hbar\omega_q$. Second, it depends on the electron energy difference $\varepsilon_{k_1+q}-\varepsilon_{k_1}$ before and after the transition. Third, if

$$|\varepsilon_{k_1+q} - \varepsilon_{k_1}| < \hbar\omega_q \,, \tag{4.114}$$

the effective interaction is *attractive*. Fourth, the attraction is greatest when $\varepsilon_{k_1+q} - \varepsilon_{k_1} = 0$, that is, when the phonon momentum $\boldsymbol{q}$ is parallel to the constant energy (Fermi) surface. A bound electron pair, called a *Cooper pair*, may be formed due to the phonon-exchange attraction as demonstrated by Cooper [14], which will be discussed in the following chapter.

Problem 4.4.1. Derive (4.105).

Problem 4.4.2. Derive (4.106).

Problem 4.4.3. Derive (4.109) along with (4.110).

Problem 4.4.4. Derive the expressions (4.112) and (4.113).

References

1 Kibble, T.W.B. (1966) *Classical Mechanics*, McGraw-Hill, Maidenhead, England, pp. 166–171.
2 Walker, C.B. (1956) *Phys. Rev.*, **103**, 547.
3 Debye, P. (1912) *Ann. Phys.*, **39**, 789.
4 Hove, L. van (1953) *Phys. Rev.*, **89**, 1189.
5 Srivastava, G.P. (1990) *The Physics of Phonons*, Adam Hilger, Bristol, pp. 41–44.
6 Fujita, S., Pientka, J., and Suzuki, A. (2012) *Mod. Phys. Lett. B*, **26**, 1 250 091.
7 Bardeen, J., Cooper, L.N., and Schrieffer, J.R. (1957) *Phys. Rev.*, **108**, 1175.
8 Frölich, H. (1950) *Phys. Rev.*, **79**, 845.
9 Frölich, H. (1950) *Proc. R. Soc.* A, **215**, 291.
10 Feynman, R.P. (1972) *Statistical Mechanics*, Addison-Wesley, Reading, MA.
11 Feynman, R.P. (1961) *Quantum Electrodynamics*, Addison-Wesley, Reading, MA.
12 Fujita, S. and Godoy, S. (1996) *Quantum Statistical Theory of Superconductivity*, Plenum, New York, pp. 150–153.
13 Fujita, S. and Godoy, S. (2001) *Theory of High Temperature Superconductivity*, Kluwer, Dordrecht, pp. 54–58.
14 Cooper, L.N. (1956) *Phys. Rev.*, **104**, 1189.
15 S. Fujita and Ito, K. (2007) *Quantum Theory of Conducting Matter*, Springer, New York, pp. 106–107.
16 Kittel, C. (1986) *Introduction to Solid State Physics*, 6th edn, John Wiley & Sons, Inc., New York, p. 23, Table 3.
17 Einstein, A. (1907) *Ann. Phys.*, **22**, 180.
18 Harrison, W.A. (1980) *Solid State Theory*, Dover, New York, pp. 390–393.

5
Electrical Conductivity of Multiwalled Nanotubes

The electrical conductivity of carbon nanotubes varies, depending on the temperature, radius, and pitch of the sample. In the majority of cases, the resistance decreases with increasing temperature, indicating a thermally activated process. The standard band theory based on the Wigner–Seitz (WS) cell model predicts a gapless semiconductor, which does not account for the thermal activation. A new band model in which an "electron" ("hole") has an orthogonal unit cell size for graphene is proposed. The normal charge carriers in graphene transport are "electrons" and "holes." The "electron" ("hole") wavepackets extend over the unit cell and carry the charges $-e$ $(+e)$. Thermally activated "electrons" or "holes" are shown to generate the observed temperature behavior of the conductivity in multiwalled nanotubes.

5.1
Introduction

In 1991 Iijima [1] discovered carbon nanotubes in the soot created in an electric discharge between two carbon electrodes. A nanotube (nt) can be considered as a single sheet of graphite, called *graphene*, that is rolled up into a tube. A *single-wall nanotube* (swnt) has a radius of 5–10 Å while a *multiwalled nanotube* (mwnt) rolled like wallpaper has a size exceeding 10 nm (= 100 Å) in radius and microns in length. Nanotubes have remarkable mechanical properties that can be exploited to strengthen materials. And since they are composed entirely of carbon, nanotubes are light and also have a low specific heat. Ebbesen *et al.* [2] measured the electrical conductivity of individual nanotubes. In the majority of cases, the resistance R decreases with increasing temperature T while the resistance R for a normal metal like copper (Cu) increases with T. This temperature behavior in nanotubes indicates a *thermally activated process*. Schönenberger and Forró [3] reviewed many aspects of carbon nanotubes; this review and [2] contain many important references. The current band theory based on the Wigner–Seitz (WS) cell model [4] predicts a gapless semiconductor for graphene and cannot explain the observed T-behavior. A new theory is required. The WS model is suited for the study of the ground state energy of a crystal [5]. To treat electron motion for a honeycomb lattice, we must introduce a different unit cell [5]. We present a new theoretical model for Bloch electron dynamics [6].

Electrical Conduction in Graphene and Nanotubes, First Edition. S. Fujita and A. Suzuki.

5.2 Graphene

Following Ashcroft and Mermin [5], we adopt the semiclassical electron dynamics in solids. In the semiclassical (wavepacket) theory, it is necessary to introduce a k-vector:

$$k = k_x \hat{e}_x + k_y \hat{e}_y + k_z \hat{e}_z \,, \quad (\hat{e}_x, \hat{e}_y, \hat{e}_z\text{: Cartesian unit vectors}) \tag{5.1}$$

since the k-vector is involved in the semiclassical equation of motion:

$$\hbar \dot{\boldsymbol{k}} \equiv \hbar \frac{\mathrm{d}\boldsymbol{k}}{\mathrm{d}t} = q(\boldsymbol{E} + \boldsymbol{v} \times \boldsymbol{B}) \,, \tag{5.2}$$

where $\boldsymbol{E}$ and $\boldsymbol{B}$ are the electric and magnetic fields, respectively, and the vector

$$\boldsymbol{v} \equiv \frac{1}{\hbar} \frac{\partial \varepsilon}{\partial \boldsymbol{k}} \tag{5.3}$$

is the electron velocity where ε is the energy. The choice of the Cartesian axes and the unit cell for a simple cubic (sc) crystal is obvious. We choose the center of the cube as the origin and take the x-, y-, and z-axis along the cube sides. Two-dimensional crystals such as graphene can also be treated similarly, with only the z-component being dropped. We must choose an orthogonal unit cell for the honeycomb lattice, as shown below.

Graphene forms a 2D honeycomb lattice. The WS unit cell is a rhombus (shaded area) as shown in Figure 5.1a. The potential energy $\mathcal{V}(\boldsymbol{r})$ is *lattice periodic*:

$$\mathcal{V}(\boldsymbol{r} + \boldsymbol{R}_{mn}) = \mathcal{V}(\boldsymbol{r}) \,, \tag{5.4}$$

where

$$\boldsymbol{R}_{mn} \equiv m\boldsymbol{a}_1 + n\boldsymbol{a}_2 \,, \tag{5.5}$$

are *Bravais vectors* with the primitive vectors $(\boldsymbol{a}_1, \boldsymbol{a}_2)$ and integers (m, n).[1] In the field theoretical formulation the field point $\boldsymbol{r}$ is given by

$$\boldsymbol{r} = \boldsymbol{r}' + \boldsymbol{R}_{mn} \,, \tag{5.6}$$

where $\boldsymbol{r}'$ is the point defined within the standard unit cell. Equation (5.4) describes the *2D lattice periodicity* but does *not* establish the k-space, which is explained below.

We first consider an electron in a simple square (sq) lattice. The Schrödinger wave equation is

$$\mathrm{i}\hbar \frac{\partial}{\partial t} \psi(\boldsymbol{r}) = -\frac{\hbar^2}{2m^*} \nabla^2 \psi(\boldsymbol{r}) + \mathcal{V}_{\mathrm{sq}}(\boldsymbol{r}) \psi(\boldsymbol{r}) \,. \tag{5.7}$$

1) Referring to Figure 5.1a the primitive lattice vectors can be written as $\boldsymbol{a}_1 = (1/2, \sqrt{3}/2)a_0$, $\boldsymbol{a}_2 = (-1/2, \sqrt{3}/2)a_0$ in the x-y coordinates, where $a_0 = |\boldsymbol{a}_1| = |\boldsymbol{a}_2| = 2.46$ Å is the lattice constant of graphene [7].

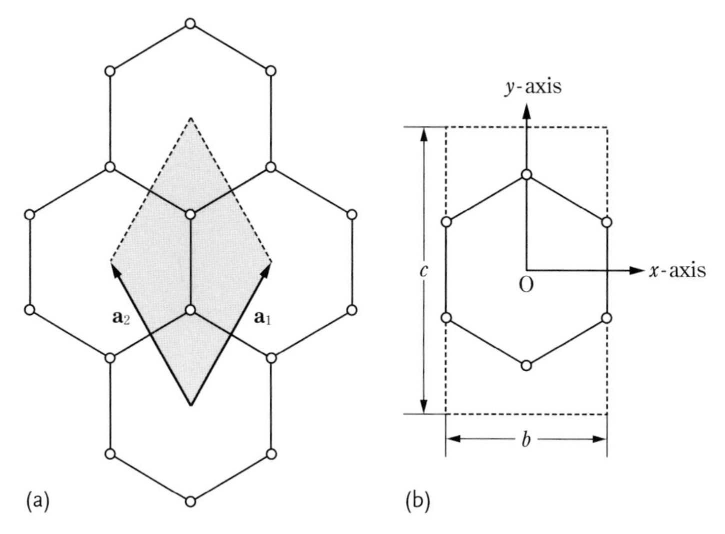

Figure 5.1 (a) The WS unit cell, a rhombus (shaded area) for graphene. (b) The orthogonal unit cell, a rectangle (dotted lines) with side lengths (b, c). Open circles (∘) indicate the C^+ ions and the lines between open circles indicate the chemical bonds.

The Bravais vector for the sq lattice $\boldsymbol{R}_{mn}^{(0)}$ is

$$\boldsymbol{R}_{mn}^{(0)} \equiv m\boldsymbol{a}_x + n\boldsymbol{a}_y = ma\hat{\boldsymbol{e}}_x + na\hat{\boldsymbol{e}}_y \quad (a = \text{lattice constant}) . \tag{5.8}$$

The system is lattice periodic:

$$\mathcal{V}_{\text{sq}}\left(\boldsymbol{r} + \boldsymbol{R}_{mn}^{(0)}\right) = \mathcal{V}_{\text{sq}}(\boldsymbol{r}) . \tag{5.9}$$

If we choose a set of Cartesian coordinates (x, y) along the sq lattice, then the Laplacian term in (5.7) is given by

$$\nabla^2 \psi(x, y) = \left(\frac{\partial^2}{\partial x^2} + \frac{\partial^2}{\partial y^2}\right) \psi(x, y) . \tag{5.10}$$

If we choose a periodic sq boundary with the side length Na (N: integer), then there are 2D Fourier transforms and (2D) k-vectors.

We now go back to the original graphene system. If we choose the x-axis along either $\boldsymbol{a}_1$ or $\boldsymbol{a}_2$, then the potential energy field $\mathcal{V}(\boldsymbol{r})$ is periodic in the x-direction, but it is aperiodic in the y-direction. For an infinite lattice the periodic boundary is the only acceptable boundary condition for the Fourier transformation. Then, there is *no* 2D $\boldsymbol{k}$-space spanned by 2D k-vectors. If we omit the kinetic energy term, then we can still use (5.4) and obtain the ground state energy (except the zero-point energy).

We now choose the *orthogonal unit cell* as shown in Figure 5.1b for graphene [6]. The unit cell (rectangle) has side lengths:

$$b = \sqrt{3}a_0 , \quad c = 3a_0 , \tag{5.11}$$

where a_0 is the nearest-neighbor distance between two C^+ ions. The unit cell has four (4) C^+ atoms. The system is lattice periodic in the x- and y-directions, and hence there are 2D k-spaces.

The importance of using the Cartesian unit cell is also underscored by the stability of the lattice structure. If a lattice is in a stable equilibrium, then the excitation energy of an electron (and a phonon) must be quadratic in each Cartesian component of the momentum. This property is assured if the lattice has an inversion symmetry. The honeycomb lattice clearly has a reflection (mirror) symmetry about the x*-and* y-axis in Figure 5.1b. The reflection symmetry can be discussed only in terms of the Cartesian frame of coordinates.

We shall further discuss lattice stability and reflection symmetry in detail in Section 5.3.

The "electron" ("hole") in the present text is defined as a *quasielectron* that has an energy higher (lower) than the Fermi energy ε_F and "electrons" ("holes") are excited on the positive (negative) side of the Fermi surface with the convention that the positive normal vector at the surface points in the energy-increasing direction.

The "electron" (wavepacket) may move up or down along the y-axis to the neighboring hexagon sites passing over one C^+. The positively charged C^+ acts as a welcoming (favorable) potential valley for the negatively charged "electron" while the same C^+ act as a hindering potential hill for the positively charged "hole." The "hole," however, can move horizontally along the x-axis without meeting the hindering potential hills. Thus, the easy channel direction for the "electrons" ("holes") are along the y-(x-)axes, see Figure 5.1b.

Let us consider the system (graphene) at 0 K. If we put an electron in the crystal, then the electron should occupy the center O of the Brillouin zone, where the lowest energy lies. Additional electrons occupy points neighboring the center O in consideration of Pauli's exclusion principle. The electron distribution is *lattice periodic* over the entire crystal in accordance with the Bloch theorem [8].

Carbon (C) is a quadrivalent metal. The first few low-lying energy bands are completely filled. The uppermost partially filled bands are important for discussion of transport properties. We consider such a band. The (2D) Fermi surface, which defines the boundary between the filled and unfilled k-space (area), is *not* a circle since the x-y symmetry is broken ($b \neq c$). The effective mass of the "electron" is lighter in the y-direction than perpendicular to it. Hence, the motion of the electron is intrinsically anisotropic. The negatively charged "electron" stays close to the positive C^+ ions while the "hole" is farther away from the C^+ ions. Hence, the gain in Coulomb interaction is greater for the "electron." That is, the "electron" is more easily activated. Thus, the "electrons" are the majority carriers at zero gate voltage.

We may represent the activation energy difference by [6]

$$\varepsilon_1 < \varepsilon_2 \,. \tag{5.12}$$

The thermally activated (or excited) electron densities are given by

$$n_j(T) = n_j e^{-\varepsilon_j/(k_B T)} \,, \tag{5.13}$$

where $j = 1$ and 2 represent the "electron" and "hole," respectively. The prefactor n_j is the density at the high-temperature limit.

5.3
Lattice Stability and Reflection Symmetry

In 1956 Lee and Yang published a historic paper [9] on parity nonconservation for neutrinos after examining the space inversion property of the massless Dirac relativistic wave equation. Inversion and reflection symmetry properties are also important in solid state physics. They play important roles for the stability of crystal lattices in which electrons and phonons move. On the basis of the reflection symmetry properties, we show that the monoclinic (mcl) crystal has a one-dimensional (1D) k-space, and the triclinic (tcl) crystal has no k-vectors for electrons. For phonons a tcl crystal has three disjoint sets of 1D nonorthogonal k-vectors. The phonons' motion is highly directional, and there are no spherical waves formed.

There are seven crystal systems, as seen in the book by Ashcroft and Mermin [5]. They are the cubic (cub), tetragonal (tet), orthorhombic (orc), monoclinic (mcl), rhombohedral (rhl), hexagonal (hex), and triclinic (tcl) systems. Arsenic (As) and Bismuth (Bi) form rhl crystals. A rhl crystal can be obtained by stretching (or contracting) the three body-diagonal distances from a sc crystal. But the body-diagonal directions remain orthogonal to each other after any stretching. Hence, if an orthogonal unit cell with the Cartesian axes along the body diagonals containing six corner atoms is chosen, then the system is periodic along the x-, y-, and z-axes passing the center. Thus, the system can be regarded as an orc crystal, and hence it has a 3D k-space. Diamond (C), silicon (Si), and germanium (Ge) form diamond (dia) crystals. A dia lattice can be decomposed into two face-centered cubic (fcc) sublattices, and can therefore be treated similarly to a cub crystal. A number of elements including graphite form hex crystals. hex crystals can be treated similarly to orc crystals by choosing orthogonal unit cells. See below for the case of graphite.

A crystal lattice must be stable. If the lattice is symmetric under the space inversion:

$$\boldsymbol{r} \to -\boldsymbol{r} , \tag{5.14}$$

then the electron energy ε is quadratic in (k_x, k_y, k_z) near the origin. The five systems, cub, tet, orc, hex, rhl, have inversion symmetry. The mcl has a mirror symmetry only with respect to the x-y plane. Therefore, the mcl has a 1D k-space. No reflection symmetry is found for a tcl crystal. Therefore, the tcl must be an insulator.

A mcl crystal has a c-axis. It is reflection-symmetric with respect to the x-y plane perpendicular to the c-(z-)axis. It has only 1D k-vectors along the c-axis. A tcl crystal has no reflection symmetry and it has therefore no k-vectors. Hence, it is an intrinsic insulator.

In summary, cub, tet, orc, rhl, and hex crystals have 3D k-space spanned by 3D k-vectors. mcl crystals have 1D k-space. tcl crystals have no k-vectors.

This is a significant finding. The same results can also be obtained by using the Schrödinger equations as shown in Section 5.2

Let us now consider small oscillations for a system of atoms forming a sc lattice. Assume a longitudinal wave traveling along a cubic axis (say, the x-axis). Imagine hypothetical planes perpendicular to the x-axis containing atoms forming a square lattice. This plane has a mass per unit square of side length a (the lattice constant), equal to the atomic mass m. The plane is subjected to a restoring force per cm^2 equal to the Young modulus Y. The dynamics of a set of the parallel planes are similar to that of coupled harmonic oscillators.

Assume next a transverse wave traveling along the x-axis. The hypothetical planes containing many atoms are subjected to a restoring stress equal to the rigidity (shear) modulus S. The dynamics is also similar to the coupled harmonic oscillators in 1D.

Low-frequency phonons are those to which Debye's continuum solid model [10] applied. The wave equations are

$$\frac{\partial^2 u_i}{\partial t^2} = c_i^2 \nabla^2 u_i \,, \tag{5.15}$$

where $i = \mathrm{l}$ (longitudinal) or t (transverse). The longitudinal wave phase velocity c_i is

$$c_\mathrm{l} \equiv \sqrt{\frac{Y}{\rho}} \,, \tag{5.16}$$

where ρ is the mass density. The transverse wave phase velocity c_t is

$$c_\mathrm{t} \equiv \sqrt{\frac{S}{\rho}} \,. \tag{5.17}$$

The waves are superposable. Hence, the movement of phonons is not restricted only to the crystal's cubic directions. In short, there is a 3D k-vector, $\boldsymbol{k}$. The wave propagation is isotropic.

Consider now an orc crystal. We may choose a Cartesian coordinate (x, y, z) passing through the center of the unit cell. The small oscillations are similar to the case of a sc lattice. The dynamics of the parallel plates are the same but the restoring forces are different in x-, y-, z-directions. The plane waves have different phase velocities, depending on the directions. They are superposable since these waves are still solutions of the wave equations (5.15).

Phonons are quanta corresponding to the running plane wave modes of lattice vibrations. Phonons are bosons, and the energies are distributed, following the Planck distribution function:

$$f(\varepsilon) = \left[\exp\left(\frac{\varepsilon}{k_\mathrm{B} T}\right) + 1\right]^{-1} \,. \tag{5.18}$$

There is no activation energy as in the case of the "electrons." This arises from the boson nature of phonons. The temperature T alone determines the average number and the average energy.

Phonons and conduction electrons are generated based on the same lattice- and k-spaces. This is important when describing the electron–phonon interaction.

The "electrons" and "holes" by postulate have the same orthogonal unit cell size. Phonon size is much greater than the electron size. The low-energy phonons have small k and great wavelengths. The average energy of a fermionic electron is greater than a bosonic phonon by two or more orders of magnitude. This establishes the usual physical picture that a point-like electron runs and is occasionally scattered by a cloud-like phonon in the crystal.

We saw earlier that a mcl crystal has 1D k-vectors pointing along the c-axis for the electrons. There are similar 1D k-vectors for phonons. Besides, there are two other sets of 1D k-vectors. Plane waves running in the z-direction can be visualized by imagining the parallel plates, each containing a great number of atoms executing small *longitudinal* and *transverse* oscillations. Thus, plane waves proceeding upward or downward exist.

Consider an oblique net of points (atoms) viewed from the top as shown in Figure 5.2. Planes defined by the vector $\boldsymbol{a}$ and the x-axis are parallel and each plane contains a great number of atoms. Planes defined by the vector $\boldsymbol{b}$ and the c-axis are also parallel, and each contains a number of atoms also executing small oscillations. These three sets of 1D phonons stabilize the lattice.

We next consider a tcl crystal, which has no k-vectors for electrons. There are, however, three sets of 1D k-vectors for phonons as shown below. Take a primitive tcl unit cell. The opposing faces are parallel to each other. There are restoring forces characterized by the Young modulus Y and shear modulus S. Then, there are 1D k-vectors perpendicular to the faces. The set of 1D phonons can stabilize the lattice. These phonons in tcl are highly directional. There are no spherical waves formed. We used the lattice property that the facing planes are parallel. This parallel plane configuration is common to all seven crystal systems. A typical hex system, graphene, clearly has three sets of parallel material planes containing many atoms.

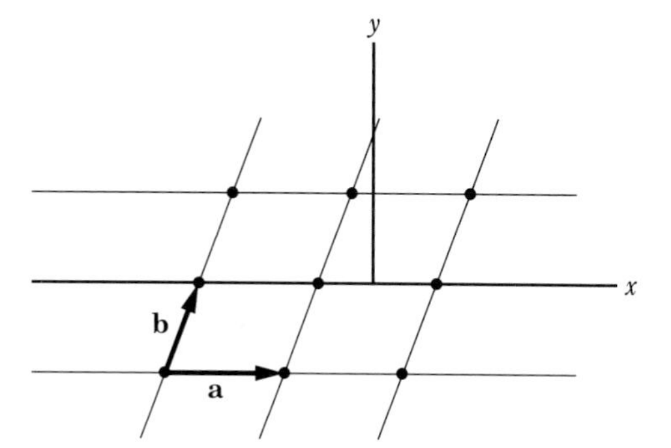

Figure 5.2 An oblique net with base vectors $(\boldsymbol{a}, \boldsymbol{b})$.

The rhl system similarly has parallel planes. The parallel material plane configuration is the basic condition for phonon generation and lattice stability.

In summary, we established based on the reflection symmetry properties of crystals that:

- The cub, tet, orc, rhl, and hex crystal systems have 3D k-spaces for electrons. The mcl system has a 1D k-space. The tcl crystal system has no k-vectors.
- The mcl and tcl crystal systems have 1D phonons, which are highly directional. No spherical phonon distributions are generated.
- The parallel material plane configuration is the basic condition for phonon generation and lattice stability.
- For rhl and hex crystals orthogonal unit cells must be chosen for electron dynamics.
- "Electrons" and "holes" have the same unit cell size, and they move with different effective masses. "Electrons" and "holes" in semiconductors are excited with different activation energies. Phonons can be excited with no activation energies.
- Both phonons and electrons are generated based on the same orthogonal unit cells. This fact is important when dealing with the electron–phonon interaction.
- Both electrons and phonons move as wavepackets. The electron size is the orthogonal unit cell size. The average phonon size is greater by two or more orders of magnitude at room temperature.

5.4
Single-Wall Nanotubes

A tube made of a graphene sheet (a single graphite layer) rolled up into a circular cylinder is called a *single-wall nanotube*. The charge may be transported by the channeling "electrons" and "holes" in the graphene wall. But the "holes" present in the inner surface of the swnt also contribute to charge transport. The carbon ions in the wall are positively charged. Hence, the positively charged "holes" move inside the tube wall. In contrast, the negatively charged "electrons" are attracted by the carbon wall and hence cannot travel in a straight line inside the tube wall. Because of this extra channel in the inner surface of the carbon wall, "holes" are the majority carriers in nanotubes although "electrons" are the dominant carriers in graphene. Moriyama *et al.* [11] observed electrical transport in swnts in the temperature range 2.6–200 K, and determined from a field effect (gate voltage) study that the carriers are "holes."

The conductivity was found to depend on the *helicity* or the *pitch* of the swnt. The *helical line* is defined as the line passing through the centers of the nearest neighbors of the C^+-hexagons. The *helical angle* φ is the angle between the helical line and the tube axis.[2)] The degree of *helicity* h may be defined as

$$h \equiv \cos \varphi \,. \tag{5.19}$$

For a macroscopically large graphene the conductivity does not show any directional dependence as we saw in Section 5.2. The conductivity σ in a swnt was experimentally found to depend on the helicity h [2]. This is a kind of finite size effect. The circumference, that is, 2π times the tube radius is finite while the tube length is macroscopic.

5.5 Multiwalled Nanotubes

A graphene sheet can be rolled like wallpaper to produce a multiwalled nanotube (see Figure 1.2). The radius of a mwnt tube can reach 100 Å. The "holes" run in shells between the walls. The shell area is greatest at the greatest radius. The "hole" current flows should therefore be greatest in the outermost shells, which is what is observed. The "hole" current in the shells should also be an activated process and the "hole" mass m_3^* is different from the graphene "hole" mass m_2^* in the carbon wall.

We are now ready to discuss the electrical conductivity of nanotubes. There are four currents carried by [6]

(a) "Electrons" moving in graphene walls with mass m_1^* and density $n_1 \exp(-\varepsilon_1/(k_B T))$, running through the channels $\langle 110 \rangle \equiv [110], [011]$, and $[101]$.
(b) "Holes" moving inside graphene walls with mass m_2^* and density $n_2 \exp(-\varepsilon_2/(k_B T))$, running through the channels $\langle 100 \rangle \equiv [100], [010]$, and $[001]$.
(c) "Holes" moving in shells between the walls with the mass m_3^* and the densities $n_3 \exp(-\varepsilon_3/(k_B T))$ running in the tube-axis direction. The activation (or excitation) energy ε_3 and the effective mass m_3^* may vary with the radius of the shell and the pitch.
(d) Cooper pairs (pairons), which are formed by the phonon-exchange attraction.

In actuality, one of the currents may be dominant, and be observed.

In the normal Ohmic conduction the resistance R is proportional to the sample (tube) length. Then, the conductivity σ can be defined, and this σ is given by the *Drude formula*:

$$\sigma = \frac{q^2}{m^*} n \tau \equiv \frac{q^2}{m^*} n \frac{1}{\Gamma} \,, \tag{5.20}$$

2) The term "chirality" is used for the "helicity" in some literature. The chiral angle φ is then defined as the angle between the tube axis and the helical (chiral) line.

where q is the carrier charge ($\pm e$), m^* the effective mass, n the carrier density, and τ the relaxation (collision) time. The relaxation rate $\Gamma \equiv \tau^{-1}$ is the inverse of the relaxation time. If the impurities and phonons are scatterers, then the rate Γ is given by the sum of the impurity scattering rate Γ_{imp} and the phonon scattering rate $\Gamma_{\text{ph}}(T)$ (*Matthiessen's rule* [5]):

$$\Gamma = \Gamma_{\text{imp}} + \Gamma_{\text{ph}}(T) . \tag{5.21}$$

The impurity scattering rate Γ_{imp} is temperature-independent, and the phonon scattering rate $\Gamma_{\text{ph}}(T)$ is temperature (T)-dependent. The phonon rate $\Gamma_{\text{ph}}(T)$ is linear in T above 4 K:

$$\Gamma_{\text{ph}}(T) = aT , \quad a = \text{constant} . \tag{5.22}$$

The temperature dependence of the conductivity should arise from the carrier density $n(T)$ and the phonon scattering rate $\Gamma_{\text{ph}}(T)$. Writing the T-dependence explicitly, we obtain from (5.14)–(5.22)

$$\sigma = \sum_j \frac{e^2}{m_j^*} n_j \mathrm{e}^{-\varepsilon_j/(k_B T)} \frac{1}{\Gamma_{\text{imp}} + aT} . \tag{5.23}$$

Ebbesen *et al.* [2] measured the electrical conductivity of individual nanotubes of different radii and pitches. The data are reproduced in Figure 5.3. Sample data were detailed in Table 1 in [2]. Resistance R is seen to decrease with increasing temperature T for NT8 and NT2. They may be interpreted in terms of the three normal currents (a)–(c) mentioned above. NT4 and NT5 were analyzed using a semilogarithmic plot of R versus T^{-1} as shown in [2, Figure 3], which is reproduced in Figure 5.4.

The slopes yield activation energies ε of 0.1 and 0.3 eV for NT4 and NT5, respectively. The currents in NT5 should be due to the "electrons" running through the channels $\langle 110 \rangle$ in the graphene wall or the "holes" running in the channels $\langle 100 \rangle$. The Hall coefficient or the Seebeck coefficient measurements will answer this question. The lack of linearity in NT4 suggests that the relevant currents are due to the "holes" moving in shells along the tube axis. The activation energy ε_3 depends on the tube radius and circumference. Since the currents along the tube axis are from the shells between walls of different circumferences, there is a distribution of ε_3, which destroys strict linearity.

NT7a in Figure 5.3 shows a resistance increase starting above around 220 K. A possible cause for this is a "neck" Fermi surface where the electron effective mass becomes abnormally high, making the conductivity very small. If this is the case, the resistance should rise and fall as the temperature is increased further. This needs to be checked experimentally.

NT6 in Figure 5.3 shows a very small T-independent resistance at room temperature. A room-temperature superconductivity, if it exists, is a quite remarkable phenomenon. The superconducting state has a critical temperature T_c above which the normal currents flow. The data indicates that the critical temperature T_c is higher than 350 K. It is desirable to perform experiments to see if the resistance is recovered at higher temperatures.

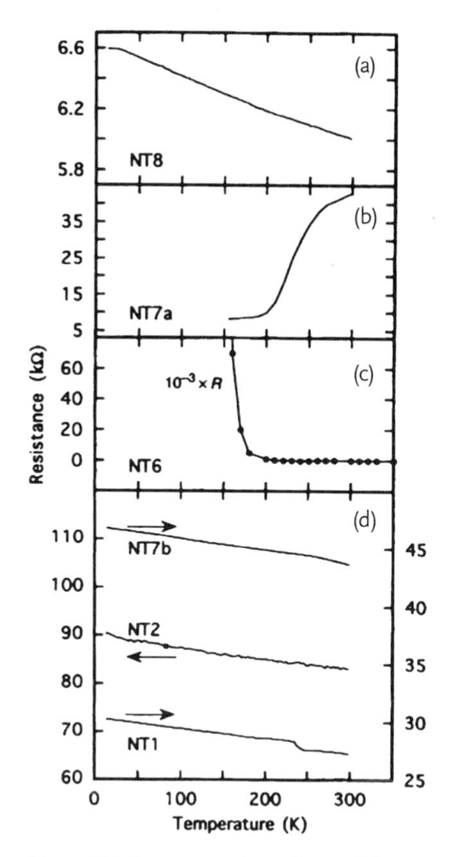

Figure 5.3 Temperature dependence of the resistance of different NTs after Ebbesen *et al.* [2]. (a) Nanotube NT8, (b) NT7a, (c) NT6, (d) NT1, NT2, NT7b.

5.6
Summary and Discussion

The conductivity of carbon nanotubes is varied, depending on the temperature T, the tube radius, and pitch. In the majority of cases, the conductivity σ decreases with increasing T, often called the *semiconductor-like T-behavior*. The current band theory based on the WS cell model predicts a gapless semiconductor, and cannot explain the observed T-behavior. A new band model suitable for the Bloch electron dynamics is presented, in which the "electron" ("hole") has not only the charge $-(+)e$ but also the rectangular size.

Applying our model to graphene and nanotubes, we have obtained the following results:

- In graphene "electrons" (1) are easier to excite than "holes" (2) with the activation energies $\varepsilon_1 < \varepsilon_2$. The majority carriers are "electrons" without external fields.

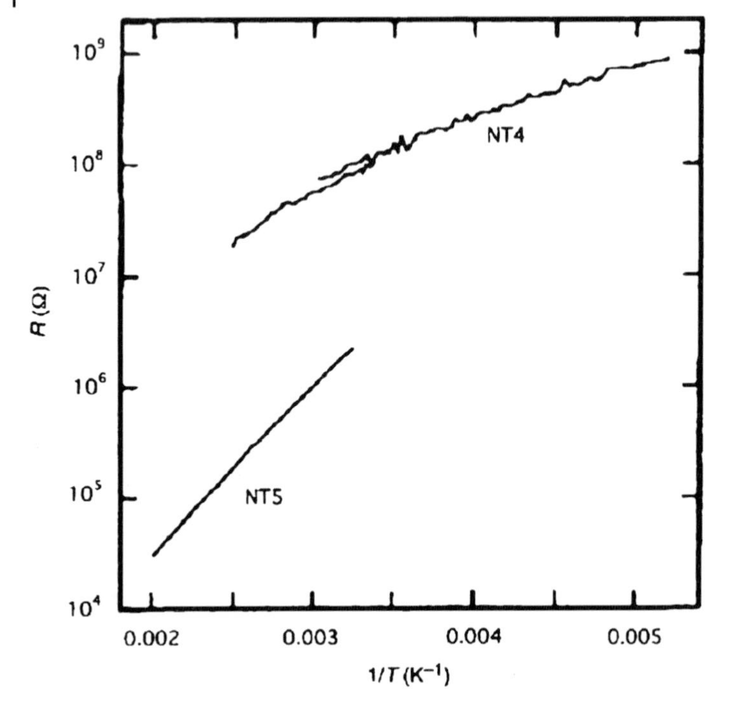

Figure 5.4 Semilogarithmic plot of the resistance R of NT4 and NT5 versus the inverse of the temperature after Ebbesen *et al.* [2].

- The "electron" motion is intrinsically anisotropic so that the effective mass m_1^* is much smaller in $\langle 110 \rangle$ than perpendicular to it.
- If the "electron" density is varied by the application of a gate voltage, then a "neck" Fermi surface develops.
- The "electron" ("hole") moves easily in the direction $\langle 110 \rangle$ ($\langle 100 \rangle$) with the effective mass m_1^* (m_2^*). The channeling "electrons" and "holes," however, generate isotropic currents.
- In swnts "electrons" and "holes" can run in the carbon wall in the same manner as in graphene.
- The tube length is macroscopic. But the circumference is finite, and it is measured in nanometers. This generates a helicity (pitch) dependence for the conductivity.
- The "holes" on the inner side of the carbon wall can contribute to the conduction. Because of this extra channel the majority carriers in swnts are "holes."
- In mwnt "holes" run in shells between carbon walls. The "hole" currents depend on the size of the shells. The currents are greatest for the outermost shells.
- Only "holes" can be excited in the shells with the activation energies ε_3. The ε_3 depend on the circumference. A distribution of ε_3 destroys the strict Arrhenius law.

The conductivity in mwnts varies widely. The behavior in most cases can be interpreted in terms of "electrons" and "holes" and using the Drude formula (5.20). Some nanotube samples show a nonlinear current–voltage behavior; a superconductivity is suspected, which will be treated in Chapter 8.

References

1 Iijima, S. (1991) *Nature*, **354**, 56.
2 Ebbesen, T.W., Lezec, H.J., Hiura, H., Bennett, J.W., Ghaemi, H.F., and Thio, T. (1996) *Nature*, **382**, 54.
3 Schönenberger, C. and Forró, L. *Multiwall carbon nanotubes*, http://physicsworld.com/cws/article/print/606, (as of April 5, 2013).
4 Wigner, E. and Seitz, F. (1933) *Phys. Rev.*, **43**, 804.
5 Ashcroft, N.W. and Mermin, N.D. (1981) *Solid State Physics*, Saunders College, Philadelphia, Chaps. 4, 5, 7, 8 and Chap. 16, p. 323.
6 Fujita, S. and Suzuki, A. (2010) *J. Appl. Phys*, **107**, 013711.
7 Saito, R., Dresselhaus, G., and Dresselhaus, M.S. (1998) *Physical Properties of Carbon Nanotubes*, Imperial College Press, London, pp. 35–58.
8 Bloch, F. (1928) *Z. Phys.*, **555**, 52.
9 Lee, T.D. and Yang, C.N. (1956) *Phy. Rev.*, **104**, 254.
10 Debye, P. (1912) *Ann. Phys.*, **39**, 719.
11 Moriyama, S., Toratani, K., Tsuya, D., Suzuki, M., Aoyagi, Y., and Ishibashi, K. (2004) *Physica E*, **24**, 46.

6
Semiconducting SWNTs

The conduction of a single-wall carbon nanotube depends on the pitch. If there are an integral number of carbon hexagons per pitch, then the system is periodic along the tube axis and allows "holes" (not "electrons") to move inside the tube. This case accounts for the semiconducting behavior with an activation energy of the order of 3 meV. There is a distribution of the activation energy since the pitch and the circumference can vary. In other cases, SWNT show metallic behaviors. "Electrons" and "holes" can move in the graphene wall (two dimensions). The conduction in the wall is the same as in graphene if the finiteness of the circumference is disregarded. Cooper pairs formed by phonon exchange attraction moving in the wall are shown to generate a temperature-independent conduction at low temperature (3–20 K).

6.1
Introduction

Iijima [1] found after his electron diffraction analysis that the carbon nanotubes range from 4 to 30 nm in diameter, and consisted of helical multiwalled tubes. A single-wall nanotube (SWNT) is about 1 nm in diameter and microns (μm) in length. Ebbesen *et al.* [2] measured the electrical conductivity σ of carbon nanotubes and found that σ varies depending on the temperature T, the tube radius r, and the pitch p. Experiments show that SWNTs can be either *semiconducting* or *metallic*, depending on how they are rolled up from the graphene sheets [3]. In this chapter we present a microscopic theory of the electrical conductivity of semiconducting SWNTs, starting with a graphene honeycomb lattice, developing Bloch electron dynamics based on a rectangular cell model [3], and using kinetic theory.

A SWNT can be formed by rolling a graphene sheet into a circular cylinder. The graphene which forms a honeycomb lattice is intrinsically anisotropic as we shall explain in more detail in Section 6.2. Moriyama *et al.* [4] fabricated 12 SWNT devices from one chip, and observed that two of the SWNT samples were semiconducting and the other ten metallic, the difference in the room-temperature resistance being of two to three orders of magnitude. The semiconducting SWNT samples show an activated-state temperature behavior. That is, the resistance decreases

Electrical Conduction in Graphene and Nanotubes, First Edition. S. Fujita and A. Suzuki.

with increasing temperature. Why do these two sets of samples show very different behavior? The answer to this question arises as follows. The line passing the centers of the nearest-neighbor carbon hexagons forms a helical line around the nanotube with a pitch p and a radius r. In Figure 6.1a, a section of the circular tube with pitch p is drawn. Its unrolled plane is shown in Figure 6.1b. The circumference $2\pi r$ likely contains an integral number m of carbon hexagons (units). The pitch p, however, may or may not contain an integral number n of units. The pitch is not controlled in the fabrication process. In the first alternative, the nanotube is periodic with the minimum period p along the tube axis. Then, there is a one-dimensional (1D) k-vector along the tube. A "hole" which has a positive charge $+e$ and a size of a unit ring of height p and radius r can move inside the positively charged carbon wall. An "electron" having a negative charge $-e$ and a similar size is attracted by the carbon wall of positive C^+ ions, and hence it cannot travel in a straight line inside the wall. Thus, there should be an extra "hole" channel current in a SWNT. Moriyama *et al.* [4] observed a "hole"-like current after examining the gate voltage effect. The system should have the lowest energy if the unit ring contains an integer set (m, n) of carbon hexagons, which may be attained after annealing at high temperatures. This should occur if the tube length is comparable with the circumference. The experimental tube length is much greater (thousand times) than the circumference $2\pi r$, and the pitch angle can be varied continuously. As has already been stated in the fabrication process the pitch is not controlled. The set of irrational numbers is greater in cardinality than the set of rational numbers. Then, the first case in which the unit contains an integer set (m, n) of hexagons must be in the minority. This case then generates a semiconducting transport behavior. We shall show later that the transport requires an activation energy. Fujita and Suzuki [7] showed that the "electrons" and "holes" must be activated based on the *rectangular cell model* for graphene.

Saito, Dresselhaus, and Dresselhaus [3] state that a SWNT is characterized by two integer indices (m, n), for example, $m = n$ for an armchair nanotube whereas $m = 0$ for a zigzag nanotube. If $n - m$ is a multiple of 3, then the SWNT is metallic. They then argued that approximately one third of SWNTs are metallic, and the other two thirds are semiconducting. This model is in variance with the

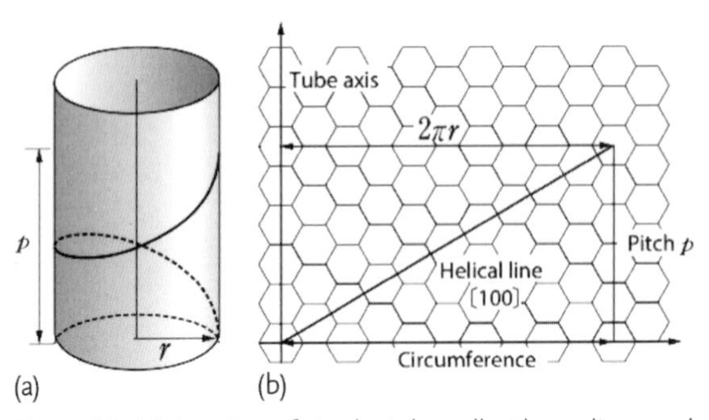

Figure 6.1 (a) A section of circular tube wall with a radius r and pitch p. (b) Its unrolled plane.

experimental observation by Moriyama *et al.* [4], where the majority of SWNTs are metallic. We must look for a different classification scheme.

If a SWNT contains an irrational number of carbon hexagons, which happens more often than not, then the system does not allow conduction along the tube axis. The system is still conductive since the conduction electrons ("electrons," "holes") can move inside the tube wall. This conduction is two-dimensional (2D), as opposed to 1D, as can be seen in the unrolled configuration, which is precisely the graphene honeycomb lattice. This means that the conduction in the carbon wall should be the same as the conduction in graphene if the effect of the finiteness of the tube radius is neglected.

We consider graphene in Section 6.2. The current band theory of the honeycomb crystal is based on the Wigner–Seitz (WS) cell model [3, 5]. The model applied to graphene predicts a gapless semiconductor, which is not observed. The WS model [5] is suitable for the study of the ground state energy of the crystal. To describe the Bloch electron dynamics [6] one must use a new theory based on a Cartesian unit cell which does not match the natural triangular crystal axes. Also phonons can be discussed naturally by using the Cartesian coordinate systems, and not with the triangular coordinate systems. The conduction electron moves as a wavepacket formed by Bloch waves as pointed out by Ashcroft and Mermin in their book [6]. This picture is fully incorporated into our theoretical model [7].

6.2 Single-Wall Nanotubes

Let us consider the long SWNT rolled with a graphene sheet. The charge may be transported by the channeling "electrons" and "holes" in the graphene wall. But the "holes" within the wall surface also contribute to the transport of charge. Because of this extra channel inside the carbon nanotube, "holes" are the majority carriers in nanotubes although "electrons" are the dominant carriers in graphene. Moriyama *et al.* [4] observed electrical transport in a semiconducting SWNT in the temperature range 2.6–200 K, and found from the field (gate voltage) effect study that the carriers are "hole"-like. Their data are reproduced in Figure 6.2.

The conductivity depends on the pitch of the SWNT. The helical line is defined as the line passing through the centers of the nearest neighbors of the C^+-hexagons. The helical angle φ is the angle between the helical line and the tube axis. The degree of helicity h may be defined as

$$h \equiv \cos \varphi \,. \tag{6.1}$$

The conductivity σ in (semiconducting) SWNTs depends on this helicity h. This is a kind of *finite size effect*. The circumference is finite while the tube length is macroscopic.

In a four-valence electron system such as graphene all electrons are bound to ions and there is no conduction at 0 K. If a "hole" having the charge $+e$ and the size of a unit cell is excited, then this "hole" can move along the tube axis with

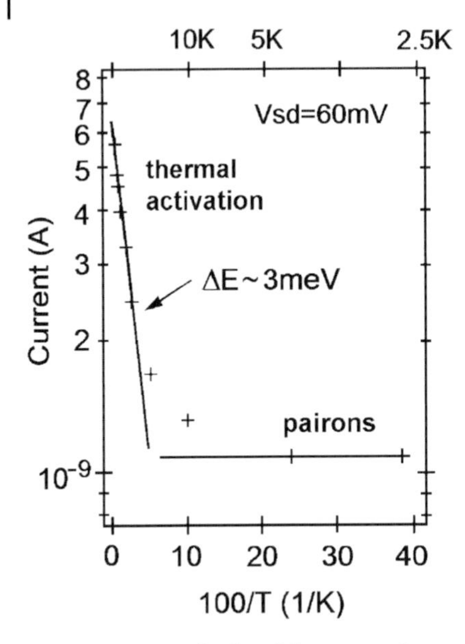

Figure 6.2 Log-scale plot of the currents in a semiconducting SWNT as a function of inverse temperature after Moriyama *et al.* [4].

the activation energy ε_3 and the effective mass m_3. Both ε_3 and m_3 depend on the radius and the pitch.

We are now ready to discuss the conductivity of a SWNT. There are four currents carried by

(a) "Electrons" moving in the graphene wall with mass m_1 and density

$$n_1 \exp\left(\frac{-\varepsilon_1}{k_B T}\right),$$

running through the channels $\langle 110 \rangle$ ($\equiv$ [110], [011], and [101]).

(b) "Holes" moving inside the graphene wall with the mass m_2 and the density

$$n_2 \exp\left(\frac{-\varepsilon_2}{k_B T}\right),$$

running through the channels $\langle 100 \rangle$ ($\equiv$ [100], [010], and [001]).

(c) "Holes" moving with the mass m_3 and the density

$$n_3 \exp\left(\frac{-\varepsilon_3}{k_B T}\right),$$

running in the tube-axis direction. The activation (or excitation) energy ε_3 and the effective mass m_3 vary with the radius and the pitch.

For a macroscopically large graphene the conductivity does not show any directional dependence. The same easy channels in which the "electron" runs with a small mass, may be assumed for other hexagonal directions, [011] and [101].

The currents run in the three channels $\langle 110 \rangle \equiv [110]$, [011], and [101]. The total current (magnitude) along the field direction μ is proportional to [10]

$$\sum_{\text{channels } \kappa} \cos^2(\mu, \kappa) = \cos^2 \vartheta + \cos^2\left(\vartheta + \frac{2\pi}{3}\right) + \cos^2\left(\vartheta - \frac{2\pi}{3}\right) = \frac{3}{2} . \tag{6.2}$$

Hence, the total current does not depend on the angle ϑ between the field direction μ and the channel current direction κ (Problem 6.2.1).

(d) Cooper pairs (pairons) formed by the phonon-exchange attraction, which move freely in the graphene wall.

In actuality, one of the currents may be dominant, and be observed. In the normal Ohmic conduction because of the conduction electrons the resistance is proportional to the sample (tube) length. Then, the conductivity σ is given by the *Drude formula*:

$$\sigma = \frac{nq^2}{m^*}\tau \equiv \frac{nq^2}{m^*}\frac{1}{\Gamma} , \tag{6.3}$$

where q is the carrier charge ($\pm e$), m^* the effective mass, n the carrier density, and τ the relaxation (collision) time. The relaxation rate $\Gamma \equiv \tau^{-1}$ is the inverse of the relaxation time. If the impurities and phonons are scatterers, then the rate Γ is the sum of the impurity scattering rate Γ_{imp} and the phonon scattering rate $\Gamma_{\text{ph}}(T)$:

$$\Gamma = \Gamma_{\text{imp}} + \Gamma_{\text{ph}}(T) . \tag{6.4}$$

The impurity scattering rate Γ_{imp} is temperature-independent and the phonon scattering rate $\Gamma_{\text{ph}}(T)$ is temperature (T)-dependent. The phonon scattering rate $\Gamma_{\text{ph}}(T)$ is linear in T above around 2 K:

$$\Gamma_{\text{ph}}(T) = aT , \quad a = \text{constant}, \quad (\text{above } 2\,\text{K}) . \tag{6.5}$$

The temperature dependence should arise from the carrier density $n(T)$ and the phonon scattering rate $\Gamma_{\text{ph}}(T)$. Writing the T-dependence explicitly, we obtain from Equations (6.3) and (6.5)

$$\sigma = \sum_j \frac{e^2}{m_j^*} n_j(T) \frac{1}{\Gamma_{\text{imp}} + aT} , \tag{6.6}$$

where the carrier density n is replaced by the thermally activated (or excited) electron densities $n_j(T)$:

$$n_j(T) = n_j \exp\left(-\frac{\varepsilon_j}{k_B T}\right) , \quad j = 1 \text{ "electron" and } j = 2 \text{ "hole"} . \tag{6.7}$$

The prefactor n_j is the density at high temperature.

Moriyama *et al.* [4] used the Arrhenius plot for the data above 20 K and obtained the activation energy

$$\varepsilon_3 \sim 3\,\text{meV} . \tag{6.8}$$

By studying the field (gate voltage) effect, the carriers were found to be hole-like. Thus, the major currents observed can be interpreted in terms of the "holes" moving within the tube wall.

This "hole" axial transport depends on the unit ring containing $m \times n$ hexagons. Since the pitch and the circumference have distributions, the activation energy ε_3 should also have a distribution. Hence, the obtained value in (6.8) must be regarded as the averaged value.

Liu *et al.* [11] systematically measured the resistance $\rho(T)$ of SWNTs under hydrostatic pressures, and fitted their data by using a 2D *variable range hopping* (vrh) theoretical formula [12]:

$$\rho(T) = \rho_0 \exp\left(\frac{T_0}{T}\right)^{1/3} . \tag{6.9}$$

Here ρ_0 is a (resistance) parameter and

$$T_0 = 525\,\text{K} \tag{6.10}$$

is a (temperature) fit parameter. Mott's vrh theory [12] is applicable when highly random disorders exist in the system. An individual SWNT (annealed) is unlikely to have such randomness. We take a different view here. The scattering is due to normally assumed impurities and phonons. But carriers ("holes") have a distribution in the unit cell size. Hence, the distribution of the activation energy introduces the flattening of the Arrhenius slope by the factor 1/3.

We now go back to the data shown in Figure 6.2. Below 20 K the currents observed are very small and they appear to approach a constant in the low-temperature limit (large T^{-1} limit). These currents, we believe, are due to the Cooper pairs.

The Cooper pairs (pairons) move in 2D with the *linear* dispersion relation [13]:

$$\varepsilon = c^{(j)} p , \tag{6.11}$$

$$c^{(j)} = \frac{2}{\pi} v_F^{(j)} , \tag{6.12}$$

where $v_F^{(j)}$ is the Fermi velocity of the "electron" ($j = 1$) and of the "hole" ($j = 2$).

Consider first "electron" pairs. The velocity $\boldsymbol{v}$ is given by (omitting superscript)

$$\boldsymbol{v} = \frac{\partial \varepsilon}{\partial \boldsymbol{p}} \quad \text{or} \quad v_x = \frac{\partial \varepsilon}{\partial p}\frac{\partial p}{\partial p_x} \equiv c\frac{p_x}{p} , \tag{6.13}$$

$$p \equiv \sqrt{p_x^2 + p_y^2} . \tag{6.14}$$

The equation of motion along the E-field (x-)direction is

$$\frac{\partial p_x}{\partial t} = q' E \,, \tag{6.15}$$

where q' is the charge $\pm 2e$ of a pairon. The solution of (6.15) is given by

$$p_x = q' E t + p_x^0 \,, \tag{6.16}$$

where p_x^0 is the initial momentum component. The current density j_p is calculated from q' (charge) $\times$ n_p (number density) $\times$ $\overline{v}$ (average velocity). The average velocity $\overline{v}$ is calculated by using (6.13) and (6.16) with the assumption that the pair is accelerated only for the collision time τ. We obtain

$$\begin{aligned} j_\mathrm{p} &\equiv q' n_\mathrm{p} \overline{v} \\ &= q' n_\mathrm{p} c \frac{1}{p} (q' E \tau) = q'^2 \frac{c}{p} n_\mathrm{p} E \tau \,. \end{aligned} \tag{6.17}$$

For stationary currents, the pairon density n_p is given by the Bose distribution function $f(\varepsilon_\mathrm{p})$

$$n_\mathrm{p} = f(\varepsilon_\mathrm{p}) \equiv \frac{1}{\exp(\varepsilon_\mathrm{p}/(k_\mathrm{B} T) - \alpha) - 1} \,, \tag{6.18}$$

where e^α is the fugacity. Integrating the current j_p over all 2D p-space, and using Ohm's law $j = \sigma E$ we obtain for the conductivity σ (Problem 6.2.2):

$$\sigma = \frac{q'^2 c}{(2\pi\hbar)^2} \int \mathrm{d}^2 p \, p^{-1} f(\varepsilon_\mathrm{p}) \tau \,. \tag{6.19}$$

In the temperature ranging between 2 and 20 K we may assume the Boltzmann distribution function for $f(\varepsilon_\mathrm{p})$:

$$f(\varepsilon_\mathrm{p}) \simeq \exp\left(\alpha - \frac{\varepsilon_\mathrm{p}}{k_\mathrm{B} T}\right) \,. \tag{6.20}$$

We assume that the relaxation time arises from the phonon scattering so that $\tau = (a T)^{-1}$, see (6.3)–(6.5). After performing the p-integration and setting $q' = -2e$ for the pairon charge, we obtain (Problem 6.2.3)

$$\sigma = \frac{2}{\pi} \frac{e^2 k_\mathrm{B}}{a \hbar^2} \mathrm{e}^\alpha \,. \tag{6.21}$$

We note that this σ is temperature-independent. If there are "electron" and "hole" pairs, they contribute additively to the conductivity. These pairons should undergo a Bose–Einstein condensation at a temperature lower than 2.2 K. We predict a superconducting state at lower temperatures.

Problem 6.2.1. Show that $\cos^2 \vartheta + \cos^2(\vartheta + 2\pi/3) + \cos^2(\vartheta - 2\pi/3) = 3/2$.

Problem 6.2.2. Derive (6.19).

Problem 6.2.3. Derive Equation (6.21).

6.3
Summary and Discussion

A SWNT is likely to have an integral number of carbon hexagons around the circumference. If each pitch contains an integral number of hexagons, then the system is periodic along the tube axis, and "holes" (not "electrons") can move along the tube axis. The system is semiconducting with an activation energy ε_3. This energy ε_3 has a distribution since both pitch and circumference have distributions. The pitch angle is not controlled in the fabrication process. There are numerous other cases where the pitch contains an irrational numbers of hexagons. In these cases the system shows a metallic behavior experimentally [14].

In the process of arriving at our main conclusion we have uncovered the following results:

- "Electrons" and "holes" can move in 2D in the carbon wall in the same manner as in graphene.
- The above implies that the conduction in the wall shows no pitch dependence for a long SWNT.
- The Cooper pairs are formed in the wall. They should undergo BEC at low temperatures, exhibiting a superconducting state.

A metallic SWNT will be treated in Chapter 8.

References

1 Iijima, S. (1991) *Nature*, **354**, 56.

2 Ebbesen, T.W., Lezec, H.J., Hiura, H., Bennett, J.W., Ghaemi, H.F., and Thio, T. (1996) *Nature*, **382**, 54.

3 Saito, R., Dresselhaus, G., and Dresselhaus, M.S. (1998) *Physical Properties of Carbon Nanotubes*, Imperial College Press, London, pp. 35–58.

4 Moriyama, S., Toratami, K., Tsuya, D., Suzuki., M., Aoyagi, Y., and Ishibashi, K. (2004) *Physica E*, **24**, 46.

5 Wigner, E. and Seitz, F. (1993) *Phys. Rev.*, **43**, 804.

6 Ashcroft, N.W. and Mermin, N.D. (1976) *Solid State Physics*, Saunders College, Philadelphia, Chaps. 4, 5, 7 and 8, and p. 217, pp. 91–93, pp. 228–229.

7 Fujita, S. and Suzuki, A. (2010) *J. Appl. Phys.*, **107**, 013 711.

8 Bloch, F. (1928) *Z. Phys.*, **555**, 52.

9 Fujita, S. and Ito, K. (2007) *Quantum Theory of Conducting Matter*, Springer, New York, pp. 106–107, pp. 85–90.

10 Fujita, S., Garcia, A., O'Leyar, D., Watanabe, S., and Burnett, T. (1989) *J. Phys. Chem. Solids*, **50**, 27.

11 Liu, B., Sundqvist, B., Andersson,O., Wågberg, T., Nyeanchi, E.B., Zhu, X.-M., and Zou, G. (2001) *Solid State Commun.*, **118**, 31.

12 Mott, N.F. (1987) *Conduction in Non-Crystalline Materials*, Oxford University Press, Oxford.

13 Fujita, S., Ito, K., and Godoy, S. (2009) *Quantum Theory of Conducting Matter; Superconductivity*, Springer, New York, pp. 77–79.

14 Tans, S.J., Devoret, M.H., Dai, H. Thess, A., Smalley, R.E., Geerligs, L.J., and Dekker, C. (1997) *Nature*, London, **386**, 474.

7
Superconductivity

We describe the basic properties of superconductors, occurrence of superconductors, theoretical background, and quantum statistical theory of superconductivity in this chapter.

7.1
Basic Properties of a Superconductor

Superconductivity is characterized by the following six basic properties: zero resistance, Meissner effect, magnetic flux quantization, Josephson effects, gaps in elementary excitation energy spectra, and sharp phase change. We shall briefly describe these properties in this section.

7.1.1
Zero Resistance

The phenomenon of superconductivity was discovered in 1911 by Kamerlingh Onnes [1], who measured extremely small electric resistance in mercury below a certain critical temperature T_c ($\approx$ 4.2 K). His data are reproduced in Figure 7.1. This *zero resistance* property can be confirmed by the never-decaying supercurrent ring experiment described in Section 7.1.3.

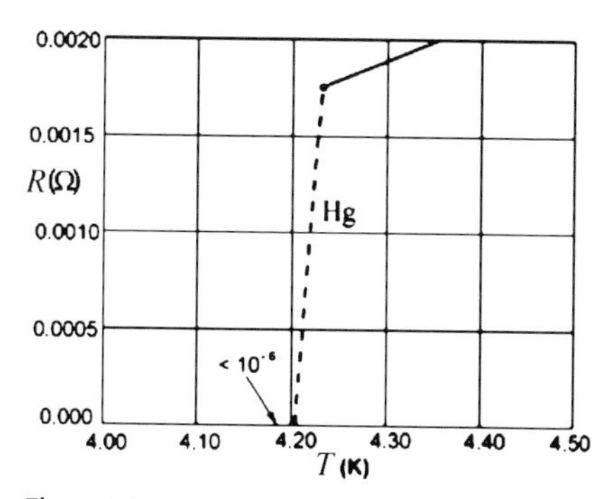

Figure 7.1 Resistance (in ohm) versus temperature (in kelvin), after Kamerlingh Onnes [1].

Electrical Conduction in Graphene and Nanotubes, First Edition. S. Fujita and A. Suzuki.

7.1.2
Meissner Effect

Substances that become superconducting at finite temperatures will be called *superconductors*. If a superconductor below T_c is placed under a weak magnetic field, it repels the magnetic field $\boldsymbol{B}$ completely from its interior as shown in Figure 7.2a. This is called the *Meissner effect*, which was discovered by Meissner and Ochsenfeld [2] in 1933.

The Meissner effect can be demonstrated dramatically by a floating magnet as shown in Figure 7.3. A small bar magnet above T_c simply rests on a superconductor dish. If the temperature is lowered below T_c, then the magnet will float as indicated. The gravitational force exerted on the magnet is balanced by the magnetic pressure (part of the electromagnetic stress tensor) due to the nonhomogeneous magnetic field (B-field) surrounding the magnet, which is represented by the magnetic flux lines.

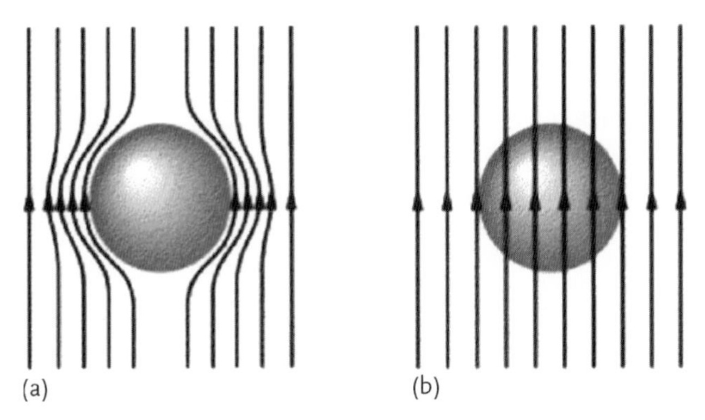

Figure 7.2 The Meissner effect. (a) A superconductor repels a weak magnetic field from its body below the transition temperature ($T < T_c$) while (b) at a temperature above the transition temperature ($T > T_c$) the magnetic field penetrates its body.

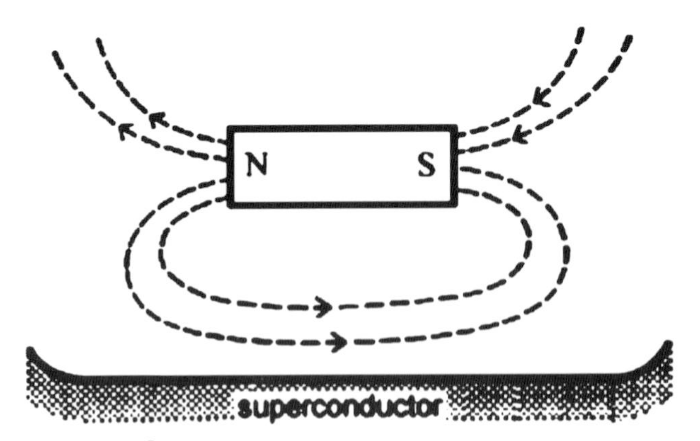

Figure 7.3 A floating magnet.

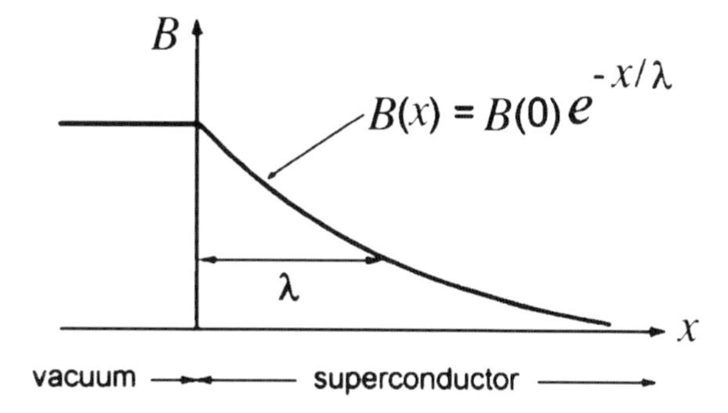

Figure 7.4 Penetration of the magnetic field into a superconductor slab. The penetration depth λ is of the order of 500 Å near 0 K.

Later more refined experiments reveal that a small magnetic field penetrates into a very thin surface layer of the superconductor. Consider the boundary of a semi-infinite slab. When an external field is applied parallel to the boundary, the B-field falls off exponentially:

$$B(x) = B(0)e^{-x/\lambda} , \tag{7.1}$$

as indicated in Figure 7.4. Here λ is called a *penetration depth*, which is of the order of 500 Å in most superconductors at lowest temperatures. Its small value on a macroscopic scale allows us to describe the superconductor as being perfectly diamagnetic. The penetration depth λ plays a very important role in the description of the magnetic properties.

7.1.3 Ring Supercurrent and Flux Quantization

Let us take a ring-shaped cylindrical superconductor. If a weak magnetic field $\boldsymbol{B}$ is applied along the ring axis and the temperature is lowered below T_c, then the field is expelled from the ring due to the Meissner effect. If the field is slowly reduced to zero, part of the magnetic flux lines can be trapped as shown in Figure 7.5. The magnetic moment generated is found to be maintained by a *never-decaying* supercurrent flowing around the ring [3].

More delicate experiments [4, 5] show that the *magnetic flux* enclosed by the ring is quantized as

$$\Phi = n\Phi_0 , \quad n = 0, 1, 2, \dots , \tag{7.2}$$

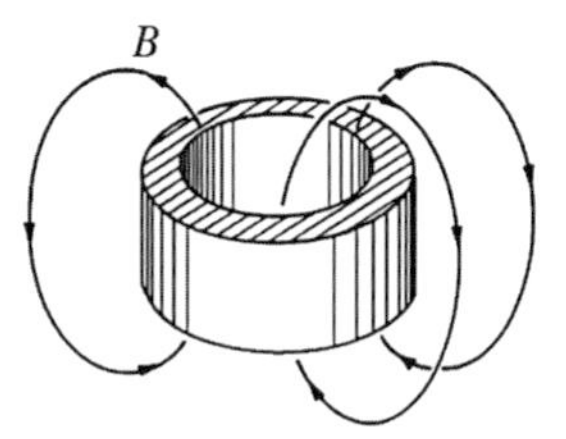

Figure 7.5 A set of magnetic flux lines are trapped in the ring.

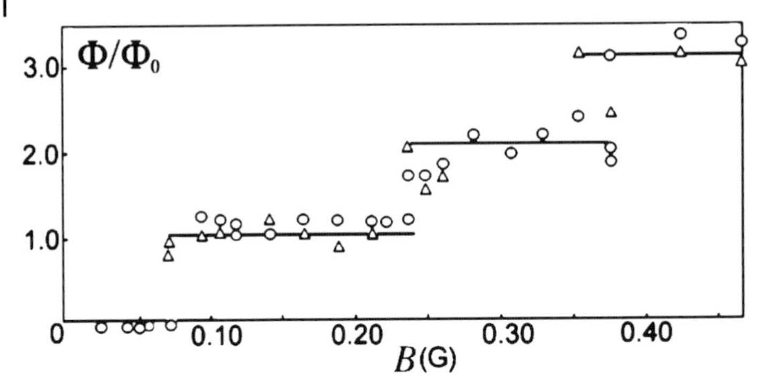

Figure 7.6 The magnetic flux quantization, after Deaver and Fairbank [4]. The two sets of data are shown as △ and ○.

$$\Phi_0 = \frac{h}{2e} = \frac{\pi\hbar}{e} = 2.07 \times 10^{-7}\,\mathrm{G\,cm}^2\,. \tag{7.3}$$

Φ_0 is called a (Cooper pair) *flux quantum*. The experimental data obtained by Deaver and Fairbank [4] are shown in Figure 7.6. The superconductor exhibits a quantum state represented by the quantum number n.

7.1.4
Josephson Effects

Let us take two superconductors (S_1, S_2) separated by an oxide layer of thickness on the order of 10 Å, called a *Josephson junction*. This system as part of a circuit including a battery is shown in Figure 7.7. Above T_c, the two superconductors, S_1 and S_2, and the junction I all show potential drops. If the temperature is lowered beyond T_c, the potential drops in S_1 and S_2 disappear because of zero resistance. The potential drop across junction I also disappears! In other words, the supercurrent runs through junction I with no energy loss. Josephson predicted [6], and later experiments [7] confirmed, this *Josephson tunneling*, also called a *DC Josephson effect*.

Let us take a closed loop superconductor containing two similar Josephson junctions and make a circuit as shown in Figure 7.8. Below T_c, the supercurrent I

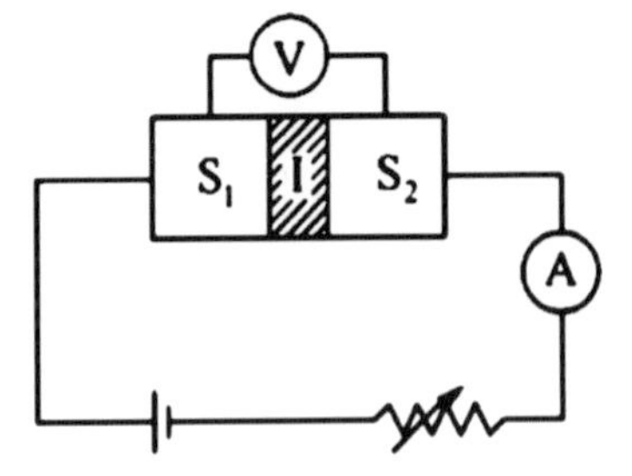

Figure 7.7 Two superconductors S_1 and S_2, and a Josephson junction I are connected to a battery.

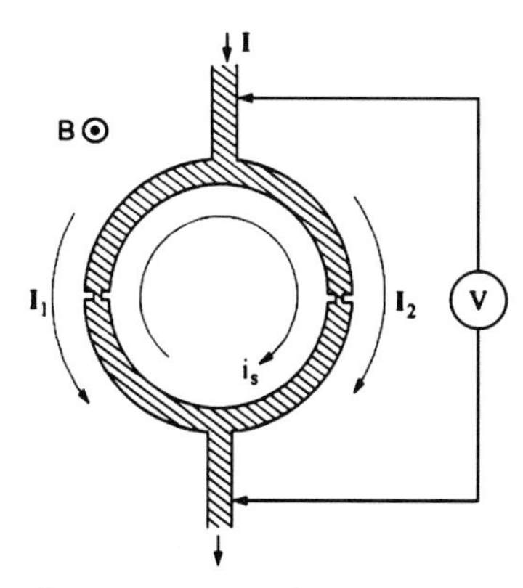

Figure 7.8 Superconducting quantum interference device (SQUID).

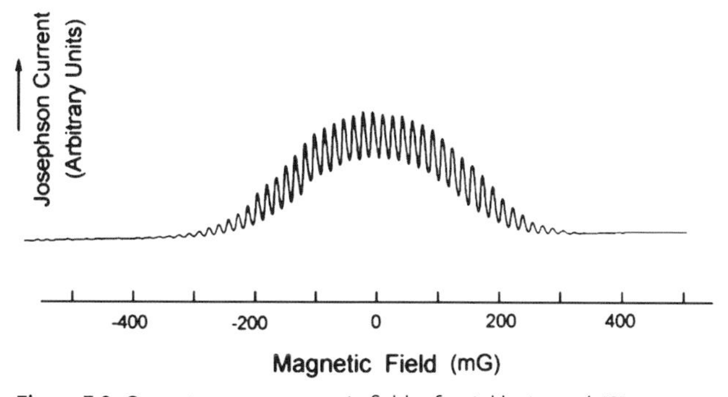

Figure 7.9 Current versus magnetic field, after Jaklevic *et al.* [8].

branches out into I_1 and I_2. We now apply a magnetic field $\boldsymbol{B}$ perpendicular to the loop. The magnetic flux can go through the junctions, and the field can be changed continuously. The total current I is found to have an oscillatory component:

$$I = I^{(0)} \cos\left(\frac{\pi \Phi}{\Phi_0}\right) , \quad (I^{(0)} = \text{constant}) , \tag{7.4}$$

where Φ is the magnetic flux enclosed by the loop, indicating that the two supercurrents I_1 and I_2, macroscopically separated ($\sim$ 1 mm), interfere just as two laser beams coming from the same source. This is called a *Josephson interference*. A sketch of the interference pattern [8] is shown in Figure 7.9.

The circuit in Figure 7.8 can be used to detect an extremely weak magnetic field. This device is called the *Superconducting Quantum Interference Device* (SQUID).

In the thermodynamic equilibrium, there can be no currents, super or normal. Thus, we must deal with a nonequilibrium condition when discussing the basic properties of superconductors such as zero resistance, flux quantization, and Josephson effects. All of these arise from the supercurrents that dominate the transport and magnetic phenomena. When a superconductor is used to form a circuit with a battery, and a steady state is established, all current passing the superconductor are supercurrents. Normal currents due to the moving electrons and other charged particles do not show up because no voltage difference can be developed in a homogeneous superconductor.

7.1.5
Energy Gap

If a continuous band of the excitation energy is separated by a finite gap ε_g from the discrete ground state energy level as shown in Figure 7.10, then this gap can be detected by photo-absorption [9, 10], quantum tunneling [11], heat capacity [12], and other experiments. This energy gap ε_g is found to be temperature-dependent. The energy gap $\varepsilon_g(T)$ as determined from the tunneling experiments [13] is shown in Figure 7.11. The energy gap is zero at T_c, and reaches a maximum value $\varepsilon_g(0)$ as the temperature approaches 0 K.

7.1.6
Sharp Phase Change

The superconducting transition is a sharp phase change. In Figure 7.12, the data of the electronic heat capacity C_{el} plotted as C_{el}/T against T, as reported by Loram

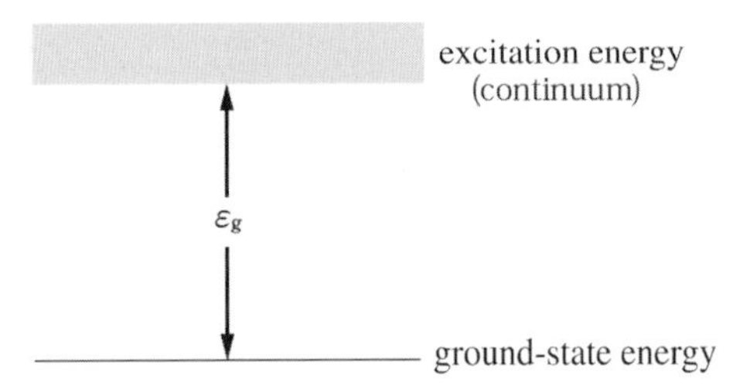

Figure 7.10 Excitation energy spectrum with a gap.

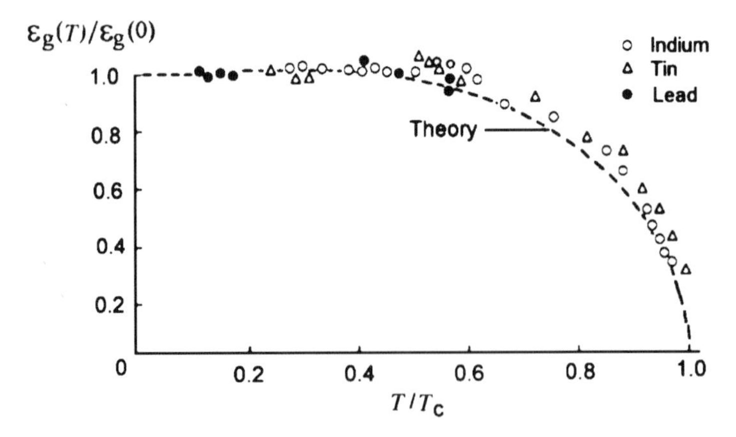

Figure 7.11 The energy gap $\varepsilon_g(T)$ versus temperature, as determined by tunneling experiments, after Giaver and Megerle [13].

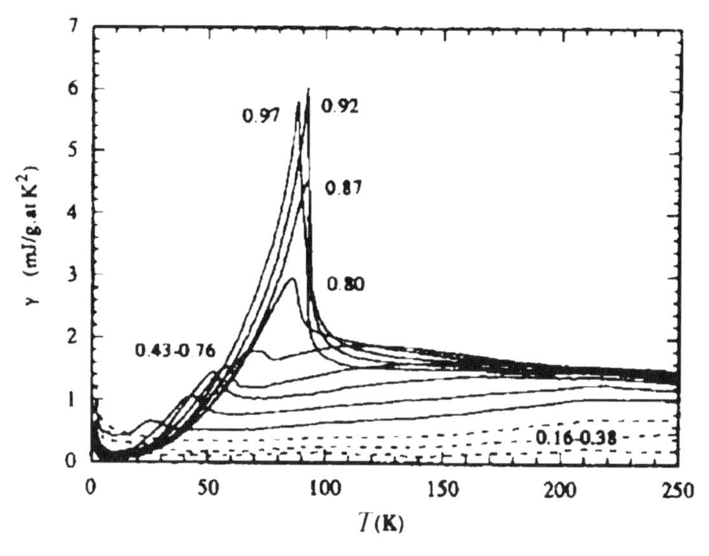

Figure 7.12 Electronic heat capacity C_{el}/T over temperature T, that is $\gamma(x, T) \equiv C_{el}/T$ is plotted against T for $YBa_2Cu_3O_{6+x}$ relative to $YBa_2Cu_3O_6$. Values of x are 0.16, 0.29, 0.38, 0.43, 0.48, 0.57, 0.67, 0.76, 0.80, 0.87, 0.92, and 0.97, after Loram *et. al.* [14].

et al. [14] for YBA_2CuO_{6+x} (2D superconductor) with the x-values, are shown. The data at $x = 0.92$ have the highest T_c. There are no latent heat and no discontinuity in C_{el} at T_c. Below T_c there is a complex long-range order, which may be treated by the Ginzburg–Landau theory [15]. For a 3D superconductor such as lead (Pb), there is a jump in the heat capacity and the phase change is of the second order (no latent heat).

7.2 Occurrence of a Superconductor

The occurrence of superconductors is discussed in this section.

7.2.1 Elemental Superconductors

More than 40 elements become superconducting at the lowest temperatures. Table 7.1 shows the critical temperature T_c and the critical magnetic fields B_c at 0 K. Most nonmagnetic metals tend to be superconductors, with notable exceptions being monovalent metals such as Li, Na, K, Cu, Ag, and Au. Some metals can become superconductors under applied pressures and/or in thin films, and these are indicated by asterisks in Table 7.1.

Table 7.1 Superconductivity parameters of the elements.

Li	Be											B	C	N	O	Ne
Na	Mg										$T_c =$ $B_c =$	Al 1.18 105	Si*	P	S	Ar
K	Ca	Sc	Ti 0.39 100	V 5.38 1420	Cr	Mn	Fe	Co	Ni	Cu	Zn 0.87 53	Ga 1.09 51	Ge*	As	Se*	Kr
Rb	Sr	Y*	Zr 0.54 47	Nb 9.20 1980	Mo 0.92 95	Tc 7.77 1410	Ru 0.51 70	Rh	Pd	Ag	Cd 3.40	In 3.40 293	Sn 3.72 309	Sb*	Te*	Xe
Cs*	Ba*	La 6.00 1100	Hf	Ta 4.48 830	W 0.01 1.07	Re 1.69 198	Os 0.65 65	Ir 0.14 19	Pt	Au	Hg 4.15 412	Tl 2.39 171	Pb 7.19 803	Bi*	Po	Rn
Fr	Ra	Ac														

Ce*	Pr	Nd	Pm	Sm	Eu	Gd	Tb	Dy	Ho	Er	Tm	Yb	Lu
Th 1.36 1.62	Pa 1.4	U 0.68	Np	Pu	Am	Cm	Bk	Cf	Es	Fm	Md	No	Lw

An asterisk (*) denotes superconductivity in thin films or under high pressure.
Transition temperature T_c in kelvin and critical magnetic field B_c at 0 K in gauss.

7.2.2
Compound Superconductors

Thousands of metallic compounds are found to be superconductors. A selection of compound superconductors with critical temperature T_c are shown in Table 7.2. Note: the critical temperature T_c tends to be higher in compounds than in elements. Nb_3Ge has the highest T_c ($\sim$ 23 K).

Compound superconductors exhibit type II magnetic behavior different from that of type I elemental superconductors. A very weak magnetic field is expelled from the body (the Meissner effect) just as by the type I superconductor. If the field

Table 7.2 Critical temperatures of selected compounds.

Compound	T_c (K)	Compound	T_c (K)
Nb_3Ge	23.0	MoN	12.0
$Nb_3(Al_{0.8}Ge_{0.2})$	20.9	V_3Ga	16.5
Nb_3Sn	18.9	V_3Si	17.1
Nb_3Al	17.5	UCo	1.70
Nb_3Au	11.5	Ti_2Co	3.44
NbN	16.0	La_3In	10.4

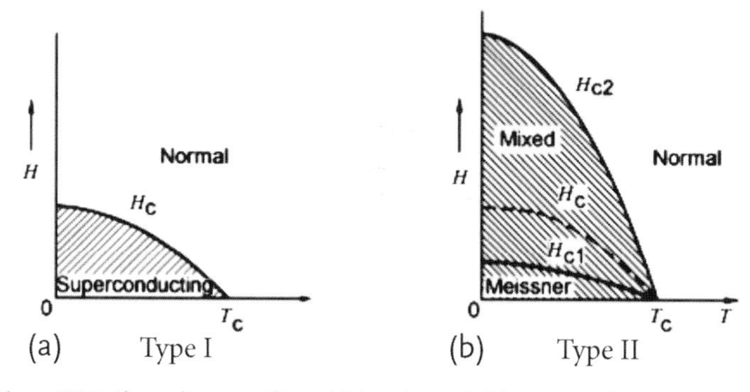

Figure 7.13 Phase diagram of type I (a) and type II (b) superconductors.

is raised beyond the lower critical field H_{c1}, the body allows a partial penetration of the field, still remaining in the superconducting state. A further field increase turns the body to a normal state upon passing the upper critical field H_{c2}. Between H_{c1} and H_{c2}, the superconductor is in a *mixed state* in which magnetic flux lines surrounded by supercurrents, called *vertices*, penetrate the body. The critical fields vs. temperature are shown in Figure 7.13. The upper critical field H_{c2} can be very high ($20\,\mathrm{T} = 2.0 \times 10^5\,\mathrm{G}$ for Nb_3Sn). Also the critical temperature T_c tends to be high for high-H_{c2} superconductors. These properties make compound superconductors useful for devices and magnets.

7.2.3
High-T_c Superconductors

In 1986 Bendnorz and Müller [16] reported the discovery of the first *cuprate superconductors*, also called *high-temperature superconductors* (HTSC). Since then, many investigations have been carried out on the high-T_c superconductors including YBACuO with $T_c \sim 94\,\mathrm{K}$ [17]. The boiling point of abundantly available and inexpensive liquid nitrogen (N) is 77 K. So the potential application of HTSC's, which are of type II, appears to be great. The superconducting state of these conductors is essentially the same as that of elemental superconductors.

7.3
Theoretical Survey

We review briefly the current theories of superconductivity.

7.3.1
The Cause of Superconductivity

At present superconductivity in solids is believed to be caused by a phonon-exchange attraction. When a phonon is exchanged between two electrons, a bound electron pair, called a *Cooper pair* (pairon), is formed [18].

The exchange of a boson (phonon) between the two fermions (electrons) can be pictured as the emission of a boson by a fermion and the subsequent absorption of the boson by another fermion. To describe the emission and absorption of the boson requires a new theory, called a *second-quantization formulation* in which creation and annihilation operators are introduced. The second-quantization formulation is summarized in Appendix A.1. The electron–phonon interaction and the phonon-exchange attraction are discussed in Chapter 4 (Sections 4.3 and 4.4).

7.3.2
The Bardeen–Cooper–Schrieffer Theory

In 1957 *Bardeen, Cooper,* and *Schrieffer* (BCS) published a classic paper [19] that is regarded as one of the most important theoretical works in the twentieth century. The Nobel Prize for physics in 1972 was shared by Bardeen, Cooper, and Schrieffer for this work. We shall briefly review the BCS theory.

In spite of the Coulomb repulsion among electrons there exists a sharp Fermi surface for the normal state of a conductor, as described by the Fermi liquid model of Landau [20, 21] (see Chapter 3, Section 3.4). The phonon-exchange attraction can bind pairs of electrons near the Fermi surface within a distance (energy) equal to Planck constant $\hbar$ times the Debye frequency ω_D. The bound electron pairs, each having antiparallel spins and charge (magnitude) $2e$, are called *Cooper pairs* (*Pairons*). "Cooper pair" and "pairon" both denote the same entity. When we emphasize the quasiparticle aspect rather than the two electron composition aspect, we use the term "pairon" more often in this text.

BCS started with a Hamiltonian $\mathcal{H}_{BCS}$ in the form:

$$\mathcal{H}_{BCS} = \sum_{\substack{k \\ \varepsilon_k > 0}} \sum_{s} \varepsilon_k c^\dagger_{ks} c_{ks} + \sum_{\substack{k \\ \varepsilon_k < 0}} \sum_{s} |\varepsilon_k| c_{ks} c^\dagger_{ks} + \frac{1}{2} \sum_{k_1} \sum_{s_1} \cdots \sum_{k_4} \sum_{s_4} \langle 1, 2|\mathcal{V}|3, 4\rangle c^\dagger_1 c^\dagger_2 c_4 c_3 , \tag{7.5}$$

where $\varepsilon_{k_1} \equiv \varepsilon_1$ is the kinetic energy of a *free* electron measured relative to the Fermi energy ε_F, and $c^\dagger_{k_1 s_1} (c_{k_1 s_1}) \equiv c^\dagger_1 (c_1)$ are creation (annihilation) operators satisfying the Fermi anticommutation rules:

$$\left\{c_{ks}, c^\dagger_{k's'}\right\} = c_{ks} c^\dagger_{k's'} + c^\dagger_{k's'} c_{ks} = \delta_{k,k'} \delta_{s,s'} ,$$
$$\{c_{ks}, c_{k's'}\} = \left\{c^\dagger_{ks}, c^\dagger_{k's'}\right\} = 0 . \tag{7.6}$$

The first (second) sum on the rhs of (7.5) represents the total kinetic energy of "electrons" with positive ε_k ("holes" with negative ε_k). The matrix element $\langle 1, 2|\mathcal{V}|3, 4\rangle$ denotes the net interaction arising from the virtual exchange of a phonon and the Coulomb repulsion between electrons. Specifically,

$$\langle 1, 2|\mathcal{V}|3, 4\rangle = \begin{cases} -\mathcal{V}_0 V^{-1} \delta_{k_1+k_2, k_3+k_4} \delta_{s_1,s_3} \delta_{s_2,s_4} & \text{if } |\varepsilon_{max}| < \hbar\omega_D \\ 0 & \text{otherwise} \end{cases} , \tag{7.7}$$

where $\mathcal{V}_0$ is a constant (energy) and V is a sample volume.

Starting with the Hamiltonian (7.5), BCS obtained an expression W for the ground state energy:

$$W = \hbar\omega_D \mathcal{D}(0) w_0 = N_0 w_0 \,, \tag{7.8}$$

where

$$w_0 = \frac{-2\hbar\omega_D}{\exp\left(\frac{2}{v_0 \mathcal{D}(0)}\right) - 1} \,, \quad (v_0 \equiv \mathcal{V}_0 V^{-1}) \,, \tag{7.9}$$

is the pairon ground state energy,

$$N_0 \equiv \hbar\omega_D \mathcal{D}(0) \tag{7.10}$$

the total number of pairons, and $\mathcal{D}(0)$ the density of states (DOS) per spin at the Fermi energy. In the variational calculation of the ground state energy, BCS found that the *unpaired electrons,* often called the *quasielectrons*[1], not joining the ground pairons that form the supercondensate, have the energy

$$E_k = \left(\varepsilon_k^2 + \Delta^2\right)^{1/2} \,, \quad \varepsilon_k = \frac{|\boldsymbol{k}|^2 - |\boldsymbol{k}_F|^2}{2m} \,, \tag{7.11}$$

where ε_k is the kinetic energy of a free "electron" relative to the Fermi energy.

The energy constant Δ, called the *quasielectron energy gap*, in (7.11) is greatest at 0 K and decreases to zero as the temperature is raised to the critical temperature T_c. BCS further showed that the energy gap at 0 K, $\Delta(T = 0) \equiv \Delta_0$ and the critical temperature T_c are related (in the weak coupling limit) by

$$2\Delta_0 = 3.53 k_B T_c \,. \tag{7.12}$$

These findings of (7.8)–(7.12) are among the most important results obtained in the BCS theory. A large body of theoretical and experimental work followed several years after the BCS theory. By 1964 the general consensus was that the BCS theory is an essentially correct theory of superconductivity.

BCS assumed the Hamiltonian (7.5) containing *"electron"* and *"hole"* kinetic energies. They also assumed a *spherical Fermi surface*. However, these two assumptions contradict each other. If a Fermi sphere whose inside (outside) is filled with electrons is assumed, then there are "electrons" ("holes") only (see Chapter 3, Section 3.5). Besides this logical inconsistency, if a free electron model having a spherical Fermi surface is assumed, then the question of why metals such as sodium (Na) and potassium (K) remain normal to the lowest temperatures cannot be answered. We must incorporate the band structures of electrons more explicitly. We shall discuss a generalization of the BCS Hamiltonian in Sections 7.4 and 7.7.

1) In the text, quasielectrons (quasiholes) are distinctly denoted by "electrons" ("holes") possessing charge (magnitude) e that circulate counterclockwise (clockwise) viewed from the tip of the applied magnetic field vector $\boldsymbol{B}$.

7.3.3
Quantum Statistical Theory

In a quantum statistical theory one starts with a reasonable Hamiltonian and derives everything from this, following step-by-step calculations. Only Heisenberg's equation of motion (quantum mechanics), Pauli's exclusion principle (quantum statistics), and Boltzmann's statistical principle (grand canonical ensemble theory) are assumed.

The major superconducting properties were enumerated in Section 7.1. The purpose of a microscopic theory is to explain all of these from first principles, starting with a reasonable Hamiltonian. Besides, one must answer basic questions such as:

- What causes superconductivity? The answer is the *phonon-exchange attraction.* We have already discussed this interaction in Chapter 4. It generates Cooper pairs [18], called *pairons* for short, under certain conditions.
- Why do impurities that must exist in any superconductor not hinder the supercurrent? Why is the supercurrent stable against an applied voltage? Why does increasing the magnetic field destroy the superconducting state?
- Why does the supercurrent dominate the normal current in the steady state?
- What is the supercondensate whose motion generates the supercurrent? How does magnetic-flux quantization arise? Josephson interference indicates that two supercurrents can interfere macroscopically just as two lasers from the same source. Where does this property come from?
- Below the critical temperature T_c, there is a profound change in the behavior of the electrons as shown by Bardeen, Cooper, and Schrieffer in their classical work [19]. What is the cause of the energy gap? Why does the energy gap depend on the temperature? Can the gap $\Delta(T)$ be observed directly?
- Phonons can be exchanged between any electrons at all times and at all temperatures. The phonon-exchange attraction can bind a pair of quasielectrons to form moving (or excited) pairons. What is the energy of excited pairons? How do the moving pairons affect the low-temperature behavior of the superconductor?
- All superconductors behave alike below T_c. Why does the law of corresponding states work here? Why is the supercurrent temperature- and material-independent?
- What is the nature of the superconducting transition? Does the transition depend on dimensionality?
- About half of all elemental metals are superconductors. Why does Na remain normal down to 0 K? What is the criterion for superconductivity? What is the connection between superconductivity and band structures?
- Compound, organic, and high-T superconductors in general show type II magnetic behaviors. Why do they behave differently compared with type I elemental superconductors?
- All superconductors exhibit six basic properties: (1) zero resistance, (2) Meissner effects, (3) flux quantization, (4) Josephson effects, (5) gaps in the elementary

excitation energy spectra, and (6) sharp phase change. Can a quantum statistical theory explain all types of superconductors in a unified manner?

- Below 2.2 K, liquid He^4 exhibits a superfluid phase in which the superfluid can flow without a viscous resistance, the flow property remarkably similar to the supercurrent. Why and how does this similarity arise?

We shall discuss a generalization of the BCS Hamiltonian incorporating the band structures of electrons more explicitly in the following sections.

7.4
Quantum Statistical Theory of Superconductivity

Fujita and his group developed a quantum statistical theory of superconductivity in a series of papers [22–26]. We present this theory in the following sections.

We construct a generalized BCS Hamiltonian which contains the kinetic energies of "electrons" and "holes," and the pairing Hamiltonian arising from phonon-exchange attraction and Coulomb repulsion. We follow the original BCS theory to construct a many-pairon ground state and find a ground state energy.

7.4.1
The Generalized BCS Hamiltonian

BCS assumed a Hamiltonian containing "electron" and "hole" kinetic energies and a pairing interaction, see (7.5). They also assumed a spherical Fermi surface. But if we assume a free electron model, we cannot explain why only some, and not all, metals are superconductors. We must incorporate the *band structures of electrons* explicitly. In this section we set up and discuss a generalized BCS Hamiltonian developed by Fujita and his group [22–26].

We assume that

- In spite of the Coulomb interaction, there exists a sharp Fermi surface at 0 K for the normal state of a conductor (the Fermi liquid model, see Chapter 3).
- The phonon exchange attraction can bind Cooper pairs near the Fermi surface within a distance (energy) equal to the Planck's constance $\hbar$ times the Debye frequency ω_{D}.
- "Electrons" and "holes" have *different effective masses* (magnitude). Thus $\varepsilon_{k}^{(1)} \neq \varepsilon_{k}^{(2)}$.
- The pairing interaction strengths V_{ij} among *and* between "electron" (1) and "hole" (2) pairons are different so that

$$\langle 1,2;i|\mathcal{V}|3,4;j\rangle = \begin{cases} -\mathcal{V}_{ij}\,V^{-1}\delta_{k_1+k_2,k_3+k_4}\,\delta_{s_1,s_3}\,\delta_{s_2,s_4} & \text{if } |\varepsilon_{\max}| < \hbar\omega_{\mathrm{D}} \\ 0 & \text{otherwise}\,. \end{cases} \tag{7.13}$$

We note that "electrons" and "holes" are different *quasiparticles*, which will be denoted distinctly.

In the ground state there are no currents for any system. To describe a ring supercurrent that can run indefinitely at 0 K, we must introduce *moving pairons*, that is, pairons with finite center of mass (CM) momenta. Creation (annihilation) operators for "electron" (1) and "hole" (2) pairons are respectively defined by[2)]

$$B_{12}^{(1)\dagger} \equiv B_{k_1\uparrow k_2\downarrow}^{(1)\dagger} \equiv c_1^{(1)\dagger} c_2^{(1)\dagger}\,, \quad B_{34}^{(2)\dagger} = c_4^{(2)\dagger} c_3^{(2)\dagger}\,. \tag{7.14}$$

The *pairon operators* are denoted by B's, which should not be confused with the magnetic field B $(= |\boldsymbol{B}|)$. Odd-numbered "electrons" carry up-spin $\uparrow$ and even-numbered "electrons" carry down-spin $\downarrow$. By using the Fermi anticommutation rules for c's and $c^\dagger$'s, the commutators among B's and $B^\dagger$'s are given by (Problem 7.4.1)

$$\left[B_{12}^{(j)}, B_{34}^{(j)}\right] \equiv B_{12}^{(j)} B_{34}^{(j)} - B_{34}^{(j)} B_{12}^{(j)} = 0\,, \quad \left[B_{12}^{(j)}\right]^2 = B_{12}^{(j)} B_{12}^{(j)} = 0\,, \tag{7.15}$$

$$\left[B_{12}^{(j)}, B_{34}^{(j)\dagger}\right] = \begin{cases} 1 - n_1^{(j)} - n_2^{(j)} & \text{if } \boldsymbol{k}_1 = \boldsymbol{k}_3 \text{ and } \boldsymbol{k}_2 = \boldsymbol{k}_4 \\ c_2^{(j)} c_4^{(j)\dagger} & \text{if } \boldsymbol{k}_1 = \boldsymbol{k}_3 \text{ and } \boldsymbol{k}_2 \neq \boldsymbol{k}_4 \\ c_1^{(j)} c_3^{(j)\dagger} & \text{if } \boldsymbol{k}_1 \neq \boldsymbol{k}_3 \text{ and } \boldsymbol{k}_2 = \boldsymbol{k}_4 \\ 0 & \text{otherwise}\,. \end{cases} \tag{7.16}$$

Pairon operators of different types j always commute:

$$\left[B^{(j)}, B^{(i)}\right] = 0 \quad \text{if} \quad j \neq i\,. \tag{7.17}$$

Here

$$n_1^{(j)} \equiv n_{k_1,s}^{(j)} \equiv c_{k_1\uparrow}^{(j)\dagger} c_{k_1\uparrow}^{(j)}\,, \quad n_2^{(j)} \equiv n_{k_2,s}^{(j)} \equiv c_{k_2\downarrow}^{(j)\dagger} c_{k_2\downarrow}^{(j)} \tag{7.18}$$

represent the *number operators* for "electrons" ($j = 1$) and "holes" ($j = 2$).

Let us now introduce the relative and net (or CM) momenta $(\boldsymbol{k}, \boldsymbol{q})$ such that

$$\left.\begin{array}{l} \boldsymbol{k} \equiv \frac{1}{2}(\boldsymbol{k}_1 - \boldsymbol{k}_2) \\ \boldsymbol{q} \equiv \boldsymbol{k}_1 + \boldsymbol{k}_2 \end{array}\right\} \Longleftrightarrow \left\{\begin{array}{l} \boldsymbol{k}_1 = \boldsymbol{k} + \frac{\boldsymbol{q}}{2} \\ \boldsymbol{k}_2 = -\boldsymbol{k} + \frac{\boldsymbol{q}}{2} \end{array}\right.\,. \tag{7.19}$$

The pairon annihilation operators are then alternatively represented by

$$B_{kq}^{\prime(1)} \equiv B_{k_1\uparrow k_2\downarrow}^{(1)} \equiv c_{-k+q/2\downarrow}^{(1)} c_{k+q/2\uparrow}^{(1)}\,, \quad B_{kq}^{\prime(2)} = c_{k+q/2\uparrow}^{(2)} c_{-k+q/2\downarrow}^{(2)}\,. \tag{7.20}$$

2) The second-quantized operators for a pair of "electrons" (1) and a pair of "holes" (2), called respectively the −pairon and the +pairon operators, are denoted by B's as defined in (7.14), where $c_1^{(j)\dagger} \equiv c_{1s}^{(j)\dagger} \equiv c_{k_1 s}^{(j)\dagger}$ ($c_1^{(j)} = c_{k_1 s}^{(j)}$) denotes the creation (annihilation) operator for "electron" ("hole"), $j = 1(2)$, satisfying the Fermi anticommutation rules:

$$\left\{c_{ks}, c_{k's'}^\dagger\right\} \equiv c_{ks} c_{k's'}^\dagger + c_{k's'}^\dagger c_{ks} = \delta_{k,k'}\delta_{s,s'}\,, \quad \{c_{ks}, c_{k's'}\} = 0\,,$$

where s represents up (down)-spin $\uparrow$ ($\downarrow$).

The prime on B's will be dropped hereafter. In the $\boldsymbol{k}$-$\boldsymbol{q}$ representation the commutation relations (7.15) and (7.16) are re-expressed as (Problem 7.4.2)

$$\left[B_{kq}^{(j)}, B_{k'q'}^{(i)}\right] = 0\,, \quad \left[B_{kq}^{(j)}\right]^2 = 0\,, \tag{7.21}$$

$$\left[B_{kq}^{(j)}, B_{k'q'}^{(i)\dagger}\right] = \begin{cases} (1 - n_{k+q/2\uparrow} - n_{-k+q/2\downarrow})\delta_{ji} & \text{if } \boldsymbol{k} = \boldsymbol{k}' \text{ and } \boldsymbol{q} = \boldsymbol{q}' \\ c_{-k+q/2\downarrow}^{(j)} c_{-k'+q'/2\downarrow}^{(j)\dagger} \delta_{ji} & \text{if } \boldsymbol{k} + \frac{\boldsymbol{q}}{2} = \boldsymbol{k}' + \frac{\boldsymbol{q}'}{2} \\ & \text{and } -\boldsymbol{k} + \frac{\boldsymbol{q}}{2} \neq -\boldsymbol{k}' + \frac{\boldsymbol{q}'}{2} \\ c_{k+q/2\uparrow}^{(j)} c_{k'+q'/2\uparrow}^{(j)\dagger} \delta_{ji} & \text{if } \boldsymbol{k} + \frac{\boldsymbol{q}}{2} \neq \boldsymbol{k}' + \frac{\boldsymbol{q}'}{2} \\ & \text{and } -\boldsymbol{k} + \frac{\boldsymbol{q}}{2} = -\boldsymbol{k}' + \frac{\boldsymbol{q}'}{2} \\ 0 & \text{otherwise}\,. \end{cases} \tag{7.22}$$

Using the new notation for pairons, we now write the *generalized BCS Hamiltonian* $\mathcal{H}$ (Problem 7.4.3):

$$\mathcal{H} = \sum_{k,s} \varepsilon_k^{(1)} n_{k,s}^{(1)} + \sum_{k,s} \varepsilon_k^{(2)} n_{k,s}^{(2)} - \sum_k{}' \sum_{k'}{}' \sum_q{}' \left(v_{11} B_{kq}^{(1)\dagger} B_{k'q}^{(1)} \right.$$
$$\left. + v_{12} B_{kq}^{(1)\dagger} B_{k'q}^{(2)\dagger} + v_{21} B_{kq}^{(2)} B_{k'q}^{(1)} + v_{22} B_{kq}^{(2)} B_{k'q}^{(2)\dagger} \right)\,, \tag{7.23}$$

where v_{ij} ($\equiv V^{-1}\mathcal{V}_{ij}$) is the pairing interaction strengths[3] per volume among and between "electron" (1) and "hole" (2) pairons. $n_{k,s}^{(1)}$ and $n_{k,s}^{(2)}$ are the number operators for "electrons" and "holes" respectively, defined by (7.18). It should be noted that the $\boldsymbol{q}$-summation is taken over all momenta including the zero momentum. The Hamiltonian (7.23), called *the generalized BCS Hamiltonian,* expresses the complete Hamiltonian for the system, which can describe *moving pairons* as well as stationary pairons. Here, the primes on the summation symbols indicate the restriction arising from the phonon-exchange attraction.

Problem 7.4.1. Verify (7.15) and (7.16).

Problem 7.4.2. Verify (7.21) and (7.22).

Problem 7.4.3. Derive (7.23).

3) $\mathcal{V}_{ij}$ denotes the interaction strengths arising from the phonon-exchange attraction (*that is,* the virtual exchange of a phonon and the Coulomb repulsion between "electrons"). The pairing interaction strength v_{ij} ($= \mathcal{V}_{ij}/V$) in the generalized BCS Hamiltonian (7.23) is reduced to v_0 ($= V_0/V$) in the BCS Hamiltonian (7.5) when the spherical Fermi surface is assumed. Note that $\mathcal{V}_{ij}$ should appear when the theory incorporates the band structures of electrons explicitly. The theory can be applied to (3, 2)D systems in a similar way.

7.5 The Cooper Pair Problem

In 1956 Cooper demonstrated [13] that, however weak the attraction may be, two electrons just above the *Fermi sea* can be bound. The binding energy is greater if the two electrons have opposite momenta $(\boldsymbol{p}, -\boldsymbol{p})$ and antiparallel spins $(\uparrow, \downarrow)$. The lowest bound energy w_0 is found to be

$$w_0 = \frac{-2\hbar\omega_D}{\exp(2/(v_0\mathcal{D}(0))) - 1}, \tag{7.24}$$

where ω_D is the Debye frequency, v_0 a positive constant characterizing the attraction, and $\mathcal{D}(0)$ the electron density of states per spin at the Fermi energy. If electrons having energy nearly opposite momenta $(\boldsymbol{p}, -\boldsymbol{p} + \boldsymbol{q})$ are paired, the binding energy is less than $|w_0|$. For small q, which represents the net momentum (magnitude) of a pairon, the *energy-momentum relation*, also called the *dispersion relation*, is given in the form:

$$w_q = w_0 + cq < 0\,, \tag{7.25}$$

where $c/v_F = 1/2\ (2/\pi)$ for 3D (2D) and $v_F \equiv (2\varepsilon_F/m^*)^{1/2}$ is the Fermi speed. Equations (7.24) and (7.25) play very important roles in the theory of superconductivity. We shall derive these equations in this and the next sections.

Two electrons near the Fermi surface can gain attraction by exchanging a phonon. This attraction can generate a bound electron pair. We shall look for the ground state energy of the Cooper pair (pairon). We anticipate that the energy is lowest for the pairon with zero net momentum. Moving pairons will be considered in the following section.

We consider a 2D system. This will simplify the concept and calculations. The 3D case can be treated similarly. Let us take two electrons just above the Fermi surface (circle), one electron having momentum $\boldsymbol{k}$ and up-spin $\uparrow$ and the other having momentum $-\boldsymbol{k}$ and down-spin $\downarrow$, see Figure 7.14. We measure the energy relative to the Fermi energy ε_F:

$$\varepsilon_{\boldsymbol{k}} \equiv \varepsilon(|\boldsymbol{k}|) = \frac{|\boldsymbol{k}|^2 - |\boldsymbol{k}_F|^2}{2m}\,. \tag{7.26}$$

The sum of the kinetic energies of the two electrons is $2\varepsilon_{\boldsymbol{k}}$. By exchanging a phonon, the pair's momenta change from $(\boldsymbol{k}, -\boldsymbol{k})$ to $(\boldsymbol{k}', -\boldsymbol{k}')$. This process lowers the energy of the pair.

The energy-eigenvalue equation for the 2D case is obtained from [13]:

$$w_0 A(\boldsymbol{k}) = 2\varepsilon_{\boldsymbol{k}} A(\boldsymbol{k}) - \frac{v_0}{(2\pi\hbar)^2} \int' \mathrm{d}^2k' A(\boldsymbol{k}')\,, \tag{7.27}$$

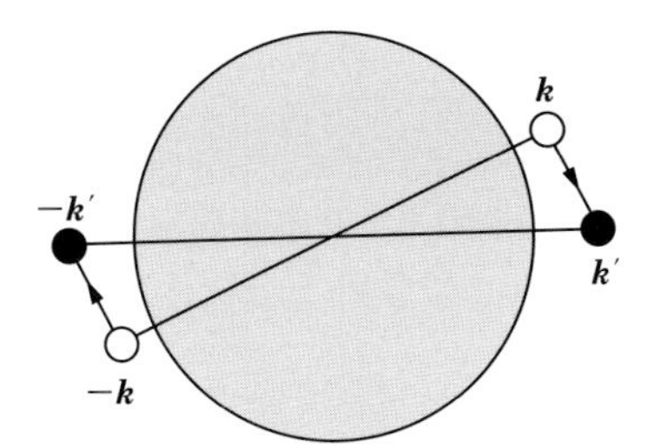

Figure 7.14 A stationary Cooper pair having zero net momentum.

where w_0 is the pairon ground state energy, and $A(\boldsymbol{k})$ the wavefunction; the prime on the integral sign means the restriction:

$$0 < \varepsilon_{\boldsymbol{k}} < \hbar\omega_{\mathrm{D}} . \tag{7.28}$$

Equation (7.27) can be solved simply as follows. Consider the integral:

$$C \equiv \frac{v_0}{(2\pi\hbar)^2} \int' \mathrm{d}^2 k' A(\boldsymbol{k}') , \tag{7.29}$$

which is a constant. Assume that the energy w_0 is negative:

$$w_0 < 0 . \tag{7.30}$$

Then, $2\varepsilon_{\boldsymbol{k}} - w_0 = 2\varepsilon_{\boldsymbol{k}} + |w_0| > 0$. After rearranging the terms in (7.27) and dividing the result by $2\varepsilon_{\boldsymbol{k}} + |w_0|$, we obtain

$$A(\boldsymbol{k}) = \frac{1}{2\varepsilon_{\boldsymbol{k}} + |w_0|} C . \tag{7.31}$$

Substituting this expression into (7.29) and dropping the common factor C, we obtain

$$1 = \frac{v_0}{(2\pi\hbar)^2} \int' \mathrm{d}^2 k \frac{1}{2\varepsilon_{\boldsymbol{k}} + |w_0|} . \tag{7.32}$$

By introducing the density of states at the Fermi energy, $\mathcal{D}(0)$, we can evaluate the k-integral as follows:

$$1 = v_0 \mathcal{D}(0) \int_0^{\hbar\omega_{\mathrm{D}}} \mathrm{d}\varepsilon \frac{1}{2\varepsilon + |w_0|} = \frac{1}{2} v_0 \mathcal{D}(0) \ln\left(\frac{2\hbar\omega_{\mathrm{D}}}{|w_0|} + 1\right) .$$

Solving this equation, we obtain (Problem 7.5.1)

$$w_0 = \frac{-2\hbar\omega_{\mathrm{D}}}{\exp(2/(v_0 \mathcal{D}(0))) - 1} . \tag{7.33}$$

We thus find a *negative energy* for the stationary pairon. The v_0-dependence of the energy w_0 is noteworthy. Since $\exp(2/x)$ cannot be expanded in powers of $x = v_0\mathcal{D}(0)$, the energy w_0 cannot be obtained by a perturbation (v_0-expansion) method. We note that formula (7.33) also holds for the 3D case with a 3D density of states.[4]

Problem 7.5.1. Verify (7.33).

7.6 Moving Pairons

The phonon-exchange attraction is in action for any pair of electrons near the Fermi surface. In general the bound pair has a *net* momentum, and hence, it moves. Such a pair is called a *moving pairon*. The energy w_q of a moving pairon for the 2D case can be obtained from a generalization of (7.27):

$$w_q a(\boldsymbol{k}, \boldsymbol{q}) = \left[\varepsilon\left(\left|\boldsymbol{k} + \frac{\boldsymbol{q}}{2}\right|\right) + \varepsilon\left(\left|-\boldsymbol{k} + \frac{\boldsymbol{q}}{2}\right|\right)\right] a(\boldsymbol{k}, \boldsymbol{q}) - \frac{v_0}{(2\pi\hbar)^2}\int' \mathrm{d}^2k'\, a(\boldsymbol{k}', \boldsymbol{q}) \,, \tag{7.34}$$

which is *Cooper's equation*, Eq. (1) of his 1956 Physical Review [18]. The prime on the k'-integral means the resrtiction on the integration domain arising from the phonon-exchange attraction, see below. We note that the net momentum $\boldsymbol{q}$ is a constant of motion, which arises from the fact that the phonon exchange is an internal process, and hence cannot change the net momentum. The pair wavefunctions $a(\boldsymbol{k}, \boldsymbol{q})$ are coupled with respect to the other variable $\boldsymbol{k}$, meaning that the exact (or energy-eigenstate) pairon wavefunctions are superpositions of the pair wavefunctions $a(\boldsymbol{k}, \boldsymbol{q})$.

The Cooper eigenvalue equation (7.34) can be derived, starting with a many-body Hamiltonian. A derivation of Cooper's equation (7.34) is given in Appendix A.3. We note that (7.34) is reduced to (7.27) in the small q-limit. The latter equation was solved in the previous section.

Let us solve (7.34) for w_q by using the same technique to obtain the pairon ground state energy (*that is*, the energy of a stationary pairon). We assume that the energy w_q is negative: $w_q < 0$. Then, $\varepsilon(|\boldsymbol{k} + \boldsymbol{q}/2|) + \varepsilon(|-\boldsymbol{k} + \boldsymbol{q}/2|) - w_q > 0$. Rearranging the terms in (7.34) and dividing by $\varepsilon(|\boldsymbol{k} + \boldsymbol{q}/2|) + \varepsilon(|-\boldsymbol{k} + \boldsymbol{q}/2|) - w_q$,

4) The density of states in the momentum $\boldsymbol{k}$-space for D dimensions, $\mathcal{D}^{(\mathrm{D})}(k)$, is generally defined through

$$\mathcal{D}^{(\mathrm{D})}(k)\mathrm{d}k = g\frac{L^{\mathrm{D}}}{(2\pi\hbar)^{\mathrm{D}}}\mathrm{d}^{\mathrm{D}}k \,,$$

where L and g $(= 2s + 1)$ denote the normalization length and the spin degeneracy factor, respectively. This formula is valid in the bulk limit for any shape of a box L^{D}. For electrons, $s = 1/2$ and g is given by $g = 2$.

we obtain

$$a(\boldsymbol{k}, \boldsymbol{q}) = \left[\varepsilon\left(\left|\boldsymbol{k} + \frac{\boldsymbol{q}}{2}\right|\right) + \varepsilon\left(\left|-\boldsymbol{k} + \frac{\boldsymbol{q}}{2}\right|\right) - w_q\right]^{-1} C(\boldsymbol{q}) , \tag{7.35}$$

where

$$C(\boldsymbol{q}) \equiv \frac{v_0}{(2\pi\hbar)^2} \int' \mathrm{d}^2 k' a(\boldsymbol{k}', \boldsymbol{q}) , \tag{7.36}$$

which is k-independent.

Introducing (7.35) in (7.34), and dropping the common factor $C(\boldsymbol{q})$, we obtain

$$1 = \frac{v_0}{(2\pi\hbar)^2} \int' \mathrm{d}^2 k \left[\varepsilon\left(\left|\boldsymbol{k} + \frac{\boldsymbol{q}}{2}\right|\right) + \varepsilon\left(\left|-\boldsymbol{k} + \frac{\boldsymbol{q}}{2}\right|\right) + |w_q|\right]^{-1} . \tag{7.37}$$

We now assume a 2D free electron model. The Fermi surface is a circle of the radius (momentum):

$$k_\mathrm{F} \equiv |\boldsymbol{k}_\mathrm{F}| = (2m_1\varepsilon_\mathrm{F})^{1/2} , \tag{7.38}$$

where m_1 represents the effective mass of an "electron." The energy $\varepsilon(|\boldsymbol{k}|)$ is given by

$$\varepsilon(|\boldsymbol{k}|) \equiv \varepsilon_k = \frac{|\boldsymbol{k}|^2 - |\boldsymbol{k}_\mathrm{F}|^2}{2m_1} . \tag{7.39}$$

The prime on the k-integral means the restriction:

$$0 < \varepsilon\left(\left|\boldsymbol{k} + \frac{\boldsymbol{q}}{2}\right|\right) , \quad \varepsilon\left(\left|-\boldsymbol{k} + \frac{\boldsymbol{q}}{2}\right|\right) < \hbar\omega_\mathrm{D} . \tag{7.40}$$

We may choose the z-axis along $\boldsymbol{q}$ as shown in Figure 7.15. We assume a small q and keep terms up to the first order in q. The k-integral can then be expressed by (Problem 7.6.1)

$$\begin{aligned} \frac{(2\pi\hbar)^2}{v_0} &= 4 \int_0^{\pi/2} \mathrm{d}\theta \int_{k_\mathrm{F}+\frac{1}{2}q\cos\theta}^{k_\mathrm{F}+k_\mathrm{D}-\frac{1}{2}q\cos\theta} \frac{k\mathrm{d}k}{|w_q| + \left(k^2 - k_\mathrm{F}^2\right)/m_1} \\ &= 2m_1 \int_0^{\pi/2} \mathrm{d}\theta \ln\left|\frac{|w_q| + 2\hbar\omega_\mathrm{D} - v_\mathrm{F} q\cos\theta}{|w_q| + v_\mathrm{F} q\cos\theta}\right| , \end{aligned} \tag{7.41}$$

$$k_\mathrm{D} \equiv m_1\hbar\omega_\mathrm{D} k_\mathrm{F}^{-1} , \tag{7.42}$$

where we retained the linear term in $k_\mathrm{D}/k_\mathrm{F}$ only since $k_\mathrm{D} \ll k_\mathrm{F}$. After performing the θ-integration, we obtain (Problem 7.6.2)

$$w_q = w_0 + \frac{2}{\pi} v_\mathrm{F} q \quad (2\mathrm{D}) , \tag{7.43}$$

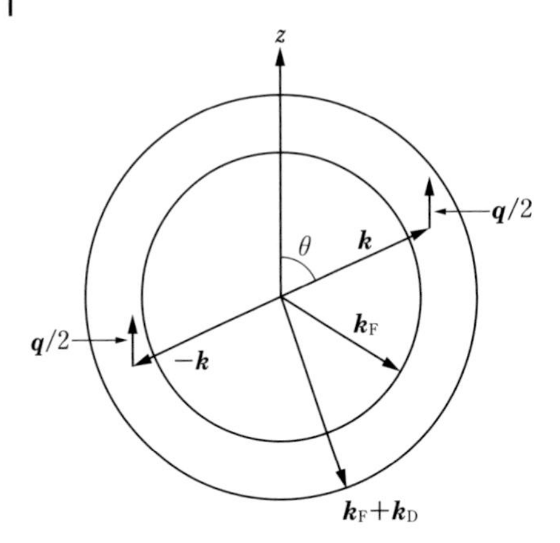

Figure 7.15 The range of the interaction variables (k, θ) is restricted to the circular shell of thickness k_D.

where w_0 is given by (7.33). A similar result for a 3D case,

$$w_q = w_0 + \frac{1}{2} v_F q \quad \text{(3D)} \tag{7.44}$$

was first obtained by Cooper (but unpublished). It is recorded in Schrieffer's book [27], (2.15). It should be noted that the lowest bound energies w_0 in (7.44) for 2D and in (7.44) for 3D are of the same form:

$$w_0 = \frac{-2\hbar\omega_D}{\exp(2/(v_0\mathcal{D}(0))) - 1} .$$

It depends, however, on the specific form of the density of states $\mathcal{D}(0)$ at the Fermi momenta for 2D and 3D, respectively. As expected, we can see that the zero-momentum pairon has the lowest energy. The excitation energy is continuous with *no* energy gap. The energy w_q increases *linearly* with momentum q for small q, rather than quadratically. This arises since the pairon density of states is strongly reduced with increasing momentum q, and this behavior dominates the q^2-increase of the kinetic energy. Pairons move like massless particles with a common speed $2/\pi v_F$ for two dimensions and $v_F/2$ for three dimensions. The linear dispersion relation[5] plays a vital role in the Bose–Einstein condensation (BEC) of pairons (see Section 7.9).

Problem 7.6.1. Verify (7.41). Hint: Use the diagram in Figure 7.15.

Problem 7.6.2. Derive an energy-momentum relation (7.43) for 2D.

5) The linear dispersion relation of the form $w_q = w_0 + cq$, is given with the assumption of a Fermi sphere, a Fermi circle. and a Fermi bar for 3, 2, and 1D, respectively. Here the speed c is given in terms of the Fermi speed v_F by $c/v_F = 1/2, 2/\pi$ and 1 for 3, 2 and 1D, respectively.

7.7
The BCS Ground State

7.7.1
The Reduced Generalized BCS Hamiltonian

At 0 K there are only ±ground state pairons, that is, pairons having the lowest energies. The ground state $|\Psi\rangle$ for the system may then be constructed based on the *reduced Hamiltonian* $\mathcal{H}_{\text{red}}$, which can be represented in terms of pairon operators, b's, only[6] (Problem 7.7.1):

$$\mathcal{H}_{\text{red}} = \sum_k 2\varepsilon_k^{(1)} b_k^{(1)\dagger} b_k^{(1)} + \sum_k 2\varepsilon_k^{(2)} b_k^{(2)\dagger} b_k^{(2)} - \sum_k{}' \sum_{k'}{}' \left(v_{11} b_k^{(1)\dagger} b_{k'}^{(1)} \right.$$
$$\left. + v_{12} b_k^{(1)\dagger} b_{k'}^{(2)\dagger} + v_{21} b_k^{(2)} b_{k'}^{(1)} + v_{22} b_k^{(2)} b_{k'}^{(2)\dagger} \right) , \tag{7.45}$$

where $v_{ij} \equiv V^{-1}\mathcal{V}_{ij}$ and $b^{(j)}$ are pair annihilation operators. The primes ($'$) on the summation symbols indicate the restriction[7] arising from the phonon-exchange attraction. The reduced Hamiltonian $\mathcal{H}_{\text{red}}$ is bilinear in pairon operators $(b, b^\dagger)$, and can be diagonalized exactly. Note that corresponding to (7.21) and (7.22), the commutation relations for pair operators, b's, are given by (Problem 7.7.2)

$$\left[b_k^{(j)}, b_{k'}^{(i)} \right] \equiv b_k^{(j)} b_{k'}^{(i)} - b_{k'}^{(i)} b_k^{(j)} = 0 , \quad \left[b_k^{(j)} \right]^2 \equiv b_k^{(j)} b_k^{(j)} = 0 ,$$
$$\left[b_k^{(j)}, b_{k'}^{(i)\dagger} \right] = \left(1 - n_{k\uparrow}^{(j)} - n_{-k\downarrow}^{(j)} \right) \delta_{k,k'} \delta_{j,i} , \quad j, i = 1, 2 . \tag{7.46}$$

For the sake of argument, let us drop the interaction Hamiltonian altogether in (7.45). We then have the first two sums representing the kinetic energies of "electrons" and "holes." We note that these energies $(\varepsilon_k^{(1)}, \varepsilon_k^{(2)})$ are positive by definition. Then the lowest energy of this system, called the *Bloch system*, is zero, and the corresponding ground state is characterized by zero-momentum "electrons" and "holes." This state will be called the physical *vacuum state*. In the theoretical developments in this section, we look for the ground state of the generalized BCS system whose energy is negative.

We now examine the physical meaning of the interaction strengths v_{ij} $(\equiv V^{-1}\mathcal{V}_{ij})$. Noting that the exchange of a phonon can pair-create to pair-

6) The reduced Hamiltonian $\mathcal{H}_{\text{red}}$ is reduced from the generalized BCS Hamiltonian (7.23), where the pairon operators, B's, are replaced in the following way:

$$B_{kq}^{\prime(1)} \to B_{k0}^{\prime(1)} = c_{-k\downarrow}^{(1)} c_{k\uparrow}^{(1)} \equiv b_k^{(1)} ,$$
$$B_{kq}^{\prime(2)} \to B_{k0}^{\prime(2)} = c_{k\uparrow}^{(2)} c_{-k\downarrow}^{(2)} \equiv b_k^{(2)} ,$$

since at 0 K, there are no phonons excited, meaning that the phonon momenta $q = 0$. Then, the pair annihilation (creation) operators for "electrons" (1) and "holes" (2) are respectively expressed in terms of c's ($c^\dagger$'s).

7) The restrictions for the summation symbols in (7.45) are given by $0 < \varepsilon_k^{(1)} \equiv \varepsilon_k < \hbar\omega_{\text{D}}$ for "electrons" and $0 < \varepsilon_k^{(2)} \equiv |\varepsilon_k| < \hbar\omega_{\text{D}}$ for "holes", respectively.

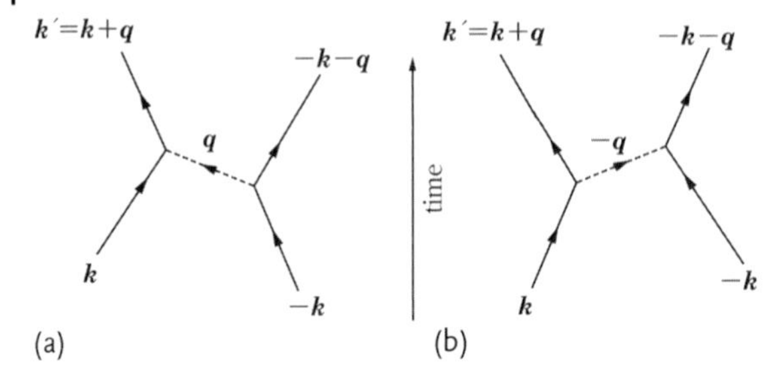

Figure 7.16 (a,b) Two Feynman diagrams representing a phonon exchange between two electrons.

annihilate "electron" ("hole") pairons, let us examine parts of the interaction terms in (7.45):

$$-v_{11} b_{\boldsymbol{k}}^{(1)\dagger} b_{\boldsymbol{k}'}^{(1)} , \quad -v_{22} b_{\boldsymbol{k}}^{(2)} b_{\boldsymbol{k}'}^{(2)\dagger} . \tag{7.47}$$

The first term generates a transition of the electron pair from $(\boldsymbol{k}\uparrow, -\boldsymbol{k}\downarrow)$ to $(\boldsymbol{k}'\uparrow, -\boldsymbol{k}'\downarrow)$. This transition is presented by the $\boldsymbol{k}$-space diagram in Figure 7.14. Such a transition may be generated by the emission of a virtual phonon with momentum $\boldsymbol{q} = \boldsymbol{k}' - \boldsymbol{k}(-\boldsymbol{q})$ by the down(up)-spin "electron" and subsequent absorption by the up(down)-spin "electron" as shown in Figure 7.16a,b. These two processes are distinct, but yield the same net transition. As we saw earlier the phonon exchange generates an attractive change of states between two "electrons" whose energies are nearly the same. The Coulomb interaction generates a repulsive correlation. The effect of this interaction is included in the strength v_{11}. Similarly, the exchange of a phonon induces a change of states between two "holes," and it is represented by the second term in (7.47). The exchange of a phonon can also pair-create or pair-annihilate "electron" ("hole") pairons, called $-(+)$ pairons, and the effects of these processes are represented by

$$-v_{12} b_{\boldsymbol{k}}^{(1)\dagger} b_{\boldsymbol{k}'}^{(2)\dagger} , \quad -v_{21} b_{\boldsymbol{k}}^{(1)} b_{\boldsymbol{k}'}^{(2)} . \tag{7.48}$$

These two processes are indicated by $\boldsymbol{k}$-space diagrams in Figure 7.17. The same processes can be represented by Feynman diagrams in Figure 7.18, where the time flows upwards by convention. Accordingly, "electrons" ("holes") proceed in the positive (negative) time directions. A phonon is electrically neutral; hence the total charge before and after the phonon exchange must be the same. The interaction Hamiltonians in (7.47) and (7.48) all conserve charge.

For type I elemental superconductors, the interaction strengths are all equal to each other;

$$v_{11} = v_{22} = v_{12} = v_{21} \equiv v_0 \quad \text{(type I)} . \tag{7.49}$$

In high-T_c superconductors, the interaction strengths v_{ij} are not equal because the Coulomb repulsion is not negligible and inequalities

$$v_{11} = v_{22} < v_{212} = v_{21} \quad (\text{high-}T_c) \tag{7.50}$$

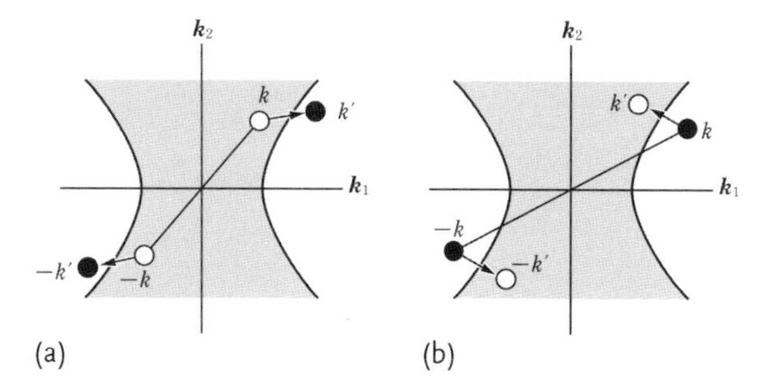

Figure 7.17 k-space diagrams representing (a) pair creation of ground pairons and (b) pair annihilation.

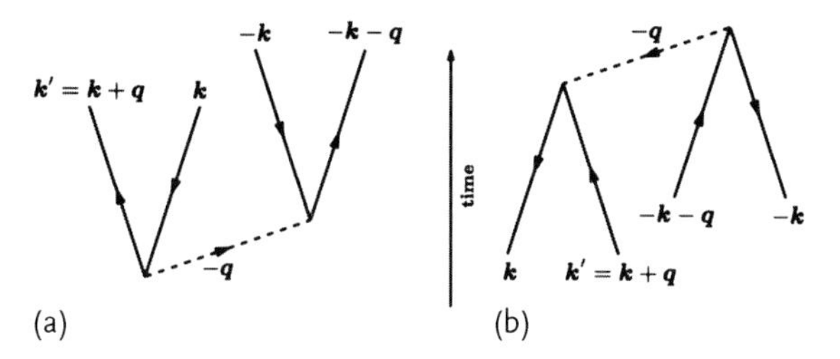

Figure 7.18 Feynmann diagrams representing (a) pair-creation of ±ground pairons from the physical vacuum, and (b) pair annihilation.

hold. For further discussion of the generalized BCS Hamiltonian, see original papers [22–26].

7.7.2 The Ground State

We now look for the ground state of the generalized BCS system. At 0 K there are only ± ground pairons. The ground state $|\Psi\rangle$ for the system may then be constructed based on the reduced Hamiltonian $\mathcal{H}_{\text{red}}$ in (7.45). Following BCS [12], we assume that the normalized ground state ket $|\Psi\rangle$ can be written as

$$|\Psi\rangle = \prod_{k}{}' \frac{1 + g_k^{(1)} b_k^{(1)\dagger}}{\left(1 + \left|g_k^{(1)}\right|^2\right)^{1/2}} \prod_{k'}{}' \frac{1 + g_{k'}^{(2)} b_{k'}^{(2)\dagger}}{\left(1 + \left|g_{k'}^{(2)}\right|^2\right)^{1/2}} |0\rangle . \tag{7.51}$$

Here the ket $|0\rangle$ by definition satisfies

$$c_{ks}^{(j)}|0\rangle \equiv c_{ks}^{(j)}|\phi_1\rangle|\phi_2\rangle = 0 . \tag{7.52}$$

It represents the *physical vacuum state* for "electrons" (1) and "holes" (2), that is, the ket $|0\rangle \equiv |\phi_1\rangle|\phi_2\rangle$ represents the *ground state of the Bloch system* with no "electrons" and no "holes" present. In (7.51) the product variables $\boldsymbol{k}$ (and $\boldsymbol{k}'$) extend over the

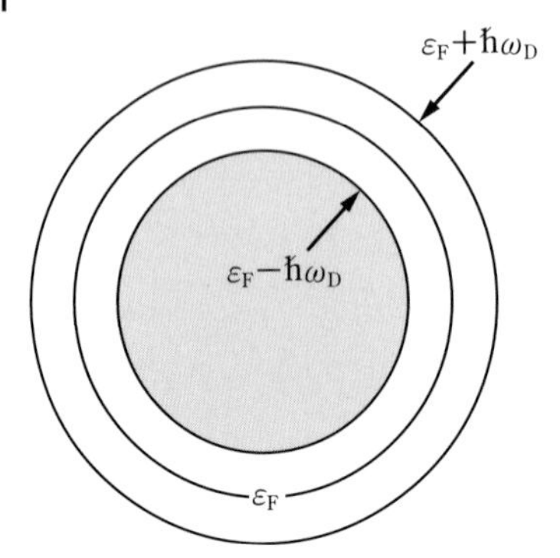

Figure 7.19 The k-space shell in 2D where pairons of both charge types are generated and intercorrelated.

region of the momenta whose associate energies are bounded: $0 < \varepsilon_k^{(1)}, \varepsilon_k^{(2)} < \hbar\omega_D$, and this limitation is indicated by the primes on the product symbols. The k-space shell in which pairons are generated and intercorrelated are shown in Figure 7.19.

Since $[b_k^{(j)\dagger}]^2 = 0$ [see (7.46)], only two terms appear for each $\boldsymbol{k}$ (or $\boldsymbol{k}'$) in the ground state ket (7.51). The quantity $|g_k^{(j)\dagger}|^2$ represents the probability that the pair states $(\boldsymbol{k}\uparrow, -\boldsymbol{k}\downarrow)$ are occupied. By expanding the product, we can see that the BCS ground state $|\Psi\rangle$ contains the zero-pairon state $|0\rangle$, one-pairon states $b^{(j)\dagger}|0\rangle$, two-pairon states $b^{(i)\dagger}b^{(j)\dagger}|0\rangle, \ldots$ The ket $|\Psi\rangle$ is normalized such that

$$\langle\Psi|\Psi\rangle = 1 \,. \tag{7.53}$$

In the case where there is only one state $\boldsymbol{k}$ in the product, we obtain

$$\langle\Psi|\Psi\rangle = \langle 0|\frac{1 + g_k^{(1)} b_k^{(1)}}{\left(1 + \left|g_k^{(1)}\right|^2\right)^{1/2}} \cdot \frac{1 + g_k^{(1)} b_k^{(1)\dagger}}{\left(1 + \left|g_k^{(1)}\right|^2\right)^{1/2}}|0\rangle = 1 \,.$$

The general case can be worked out similarly (Problem 7.7.3).

Since the ground state wave function has no nodes, we may choose $g_k^{(j)}$ to be nonnegative with no loss of rigor: $g_k^{(j)} \geq 0$. We now determine $\{g_k^{(j)}\}$ such that the ground state energy

$$W \equiv \langle\Psi|\mathcal{H}_{\text{red}}|\Psi\rangle \tag{7.54}$$

has a *minimum* value. This may be formulated by the extremum condition:

$$\delta W = \delta\langle\Psi|\mathcal{H}_{\text{red}}|\Psi\rangle = 0 \,. \tag{7.55}$$

The extremum problem meant by (7.55) with respect to the variation in g's can more effectively be solved by working with variations in the real probability amplitudes u's and v's defined by

$$u_k^{(j)} \equiv \left(1 + g_k^{(j)2}\right)^{-1/2} , \quad v_k^{(j)} \equiv g_k^{(j)} \left(1 + g_k^{(j)2}\right)^{-1/2} , \quad u_k^{(j)2} + v_k^{(j)2} = 1 . \tag{7.56}$$

The normalized ket $|\Psi\rangle$ can then be expressed by

$$|\Psi\rangle = \prod_k{}' \left(u_k^{(1)} + v_k^{(1)} b_k^{(1)\dagger}\right) \prod_{k'}{}' \left(u_{k'}^{(2)} + v_{k'}^{(2)} b_{k'}^{(2)\dagger}\right) |0\rangle . \tag{7.57}$$

The energy W can be written from (7.54) as (Problem 7.7.4)

$$W = \sum_k{}' 2\varepsilon_k^{(1)} v_k^{(1)2} + \sum_{k'}{}' 2\varepsilon_{k'}^{(2)} v_{k'}^{(2)2} - \sum_k{}' \sum_{k'}{}' \sum_i \sum_j v_{ij} u_k^{(i)} v_k^{(i)} u_{k'}^{(j)} v_{k'}^{(j)} . \tag{7.58}$$

Taking the variations in v's and u's, and noting that $u_k^{(j)} \delta u_k^{(j)} + v_k^{(j)} \delta v_k^{(j)} = 0$, we obtain from (7.55) and (7.58) (Problem 7.7.5)

$$2\varepsilon_k^{(j)} u_k^{(j)} v_k^{(j)} - \left(u_k^{(j)2} - v_k^{(j)2}\right) \sum_{k'}{}' \left(v_{j1} u_{k'}^{(1)} v_{k'}^{(1)} + v_{j2} u_{k'}^{(2)} v_{k'}^{(2)}\right) = 0 . \tag{7.59}$$

To simply treat these equations subject to the equations in (7.56), we introduce a set of energy parameters:

$$\Delta_k^{(j)} , \quad E_k^{(j)} \equiv \left(\varepsilon_k^{(j)2} + \Delta_k^{(j)2}\right)^{1/2} \tag{7.60}$$

such that (Problem 7.7.6)

$$u_k^{(j)2} - v_k^{(j)2} = \frac{\varepsilon_k^{(j)}}{E_k^{(j)}} , \quad u_k^{(j)} v_k^{(j)} = \frac{\Delta_k^{(j)}}{2E_k^{(j)}} . \tag{7.61}$$

Then, Equation (7.59) can be re-expressed as

$$\Delta_k^{(j)} = \sum_{k'}{}' \sum_{i=1}^{2} v_{ij} \frac{\Delta_{k'}^{(i)}}{2E_{k'}^{(i)}} . \tag{7.62}$$

Since the rhs of (7.62) does not depend on $\boldsymbol{k}$, the *energy gaps*

$$\Delta_k^{(j)} \equiv \Delta_j \tag{7.63}$$

are independent of $\boldsymbol{k}$. Hence, we can simplify (7.62) to

$$\Delta_j = \sum_{k'}{}' \sum_i v_{ij} \frac{\Delta_i}{2E_{k'}^{(i)}} . \tag{7.64}$$

These are called *generalized energy gap equations*. As we shall see later, $E_k^{(i)}$ is the energy of an *unpaired electron* (or *"electron"*). These "electrons" have energy gaps $\Delta^{(j)}$ relative to the Fermi energy as shown in Figure 7.20. Notice that there are in general two types of energy gaps: "electron"($j = 1$) and "hole" ($j = 2$) energy gaps, (Δ_1, Δ_2).

Using (7.58) along with (7.61)–(7.64), we calculate the energy W and obtain (Problem 7.7.7)

$$W = \sum_k{}' \sum_j 2\varepsilon_k^{(j)} v_k^{(j)2} - \sum_k{}' \sum_{k'}{}' \sum_i \sum_j v_{ij} u_k^{(i)} v_k^{(i)} u_{k'}^{(j)} v_{k'}^{(j)}$$
$$= \sum_k{}' \sum_j \left[\varepsilon_k^{(j)} \left(1 - \frac{\varepsilon_k^{(j)}}{E_k^{(j)}} \right) - \frac{\Delta_j^2}{2E_k^{(j)}} \right] . \qquad (7.65)$$

In the bulk limit the sums over $\boldsymbol{k}$ are converted into energy integrals, yielding

$$W = \sum_{j=1}^{2} \mathcal{D}_j(0) \int_0^{\hbar\omega_D} d\varepsilon \left[\varepsilon - \frac{\varepsilon^2}{\left(\varepsilon^2 + \Delta_j^2\right)^{1/2}} - \frac{\Delta_j^2}{2\left(\varepsilon^2 + \Delta_j^2\right)^{1/2}} \right] . \qquad (7.66)$$

The ground state $|\Psi\rangle$, from (7.51), is a superposition of many-pairon states. Each component state can be obtained from the physical vacuum state $|0\rangle$ by pair-creation and/or pair-annihilation of $\pm$pairons and pair states change through a succession of phonon exchanges. Since the phonon-exchange processes, as represented by (7.48), can pair-create (or pair-annihilate) $\pm$pairons simultaneously from the physical vacuum, the supercondensate is composed of *equal numbers* of $\pm$pairons. We can see from Figure 7.18 that the maximum numbers of $+(-)$pairons are given by $1/2\hbar\omega_D \mathcal{D}_1(0)$ $(1/2\hbar\omega_D \mathcal{D}_2(0))$. We must then have

$$\mathcal{D}_1(0) = \mathcal{D}_2(0) \equiv \mathcal{D}(0) , \qquad (7.67)$$

which will be justified by the assumption that the supercondensate is generated only on part of the Fermi surface (see Section 7.8.8).

Using (7.67), we obtain from (7.66) (Problem 7.7.8)

$$W = \frac{1}{2} N_0 (w_1 + w_2) , \quad w_i \equiv \hbar\omega_D \left\{ 1 - \left[1 + \left(\frac{\Delta_i}{\hbar\omega_D} \right)^2 \right]^{1/2} \right\} < 0 . \qquad (7.68)$$

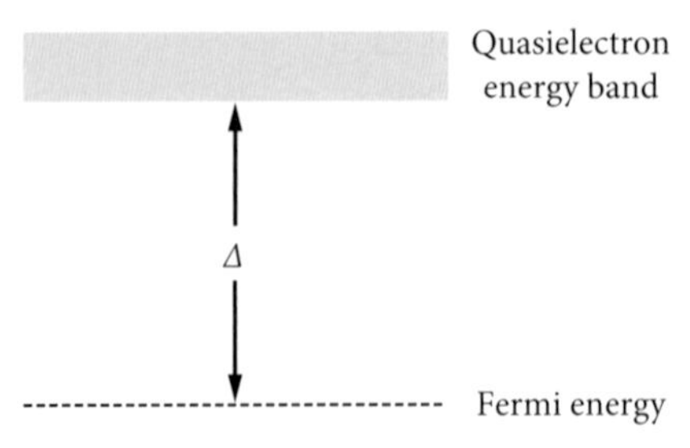

Figure 7.20 Quasielectrons have an energy gap Δ relative to the Fermi energy.

We thus find that the ground state energy of the generalized BCS system is negative, that is, the energy is *lower* than that of the Bloch system. Further note that the binding energy $|w_i|$ per pairon may in general be different for different charge types.

Let us now find Δ_j from the gap equations (7.64). In the bulk limit, these equations are simplified to

$$\begin{aligned}\Delta_j &= \frac{1}{2}v_{j1}\mathcal{D}(0)\int_0^{\hbar\omega_D} d\varepsilon \frac{\Delta_1}{\left(\varepsilon^2+\Delta_1^2\right)^{1/2}} + \frac{1}{2}v_{j2}\mathcal{D}(0)\int_0^{\hbar\omega_D} d\varepsilon \frac{\Delta_2}{\left(\varepsilon^2+\Delta_2^2\right)^{1/2}} \\ &= \frac{1}{2}v_{j1}\mathcal{D}(0)\Delta_1 \sinh^{-1}\left(\frac{\hbar\omega_D}{\Delta_1}\right) + \frac{1}{2}v_{j2}\mathcal{D}(0)\Delta_2 \sinh^{-1}\left(\frac{\hbar\omega_D}{\Delta_2}\right) .\end{aligned} \tag{7.69}$$

For type I elemental superconductors, we assume that the interaction strengths v_{ij} are all equal to each other: $v_{11} = v_{12} = v_{21} = v_{22} \equiv v_0$. We see from (7.69) that "electron" and "hole" energy gaps coincide:

$$\Delta_1 = \Delta_2 \equiv \Delta . \tag{7.70}$$

The generalized gap equations (7.64) are then reduced to a single equation:

$$\Delta = \sum_{k'}{}' \sum_i v_0 \frac{\Delta}{2E_{k'}^{(i)}} , \tag{7.71}$$

which is called the *BCS energy gap equation*. After dropping the common factor Δ and taking the bulk limit, we obtain (Problem 7.7.9)

$$1 = v_0\mathcal{D}(0)\sinh^{-1}\left(\frac{\hbar\omega_D}{\Delta}\right) . \tag{7.72}$$

Solving this we obtain (Problem 7.7.10)

$$\Delta = \frac{\hbar\omega_D}{\sinh(1/(v_0\mathcal{D}(0)))} \quad (= N_0 w_0) . \tag{7.73}$$

We now substitute (7.73) into (7.68) and calculate the ground state energy. After straightforward calculations, we obtain (Problem 7.7.11)

$$W = \frac{-2\mathcal{D}(0)\hbar^2\omega_D^2}{\exp(2/(v_0\mathcal{D}(0))) - 1} \quad (= N_0 w_0) . \tag{7.74}$$

Equations (7.73) and (7.74) are the famous BCS formulas for the energy gap and the ground state energy, respectively. They correspond respectively to (2.40) and (2.42) of the original paper [19]. We stress that these results are *exact*, obtained from the reduced BCS Hamiltonian $\mathcal{H}_{red}$ in (7.45) without using weak coupling limit ($v_0 \to 0$).

Problem 7.7.1. Verify (7.45).

Problem 7.7.2. Verify the commutation relations (7.46) for pair operators.

Problem 7.7.3. Verify (7.53) for the general ket $|\Psi\rangle$. Hint: Assume that there are only two $\boldsymbol{k}$-states in the product. If successful, then treat the general case.

Problem 7.7.4. Derive (7.58).

Problem 7.7.5. Derive (7.59).

Problem 7.7.6. Check the consistency of (7.56) and (7.61). Hint: Use the identity: $(u^2 + v^2)^2 - (u^2 - v^2)^2 = 4u^2v^2$.

Problem 7.7.7. Verify (7.65).

Problem 7.7.8. Verify (7.68).

Problem 7.7.9. Derive (7.72).

Problem 7.7.10. Verify (7.73)

Problem 7.7.11. Derive (7.74).

7.8 Remarks

We have uncovered several significant features of the ground state of the generalized BCS system.

7.8.1 The Nature of the Reduced Hamiltonian

The reduced Hamiltonian $\mathcal{H}_{\text{red}}$ in (7.45) has a different character from the normal starting Hamiltonian for a metal, which is composed of interacting electrons and ions. BCS envisioned that there are only zero-momentum pairons at 0 K. Only the basic ingredients to build up zero-momentum pairons are incorporated in the BCS Hamiltonian, see (7.5). Bloch "electrons" and "holes" are introduced from the outset. These particles are the elementary excitations in the normal state above the critical temperature.

7.8.2
Binding Energy per Pairon

We may rewrite (7.74) for the ground state energy in the form:

$$W = N_0 w_0 , \quad N_0 = \hbar \omega_D \mathcal{D}(0) , \quad w_0 = \frac{-2\hbar\omega_D}{\exp(2/(v_0 \mathcal{D}(0))) - 1} , \tag{7.75}$$

which can be interpreted as follows: the greatest total number of pairons generated consistent with the BCS Hamiltonian is equal to $\hbar\omega_D \mathcal{D}(0) = N_0$. Each pairon contributes a binding energy $|w_0|$. This energy $|w_0|$ can be measured directly by quantum tunneling experiments. Our interpretation of the ground state energy is quite natural, but it is distinct from that of the BCS theory, where the energy gap Δ is regarded as a measure of the binding energy. Our calculations do not support this view, see Section 7.8.4.

By the Meissner effect a superconductor expels a weak magnetic field $\boldsymbol{B}$ from its interior. The magnetic energy stored is higher in proportion to B^2 and the excluded volume than that for the uniform B-flux configuration. The difference in the energy for a macroscopic superconductor is given by

$$\frac{V B^2}{2\mu_0} . \tag{7.76}$$

If this energy exceeds the difference of the energy between super and normal conductors, $W_S - W_N$, which is equal to $|W_0|$, the superconducting state should break down. The minimum magnetic field B_c that destroys the superconducting state is the *critical field* at 0 K, $B_c(0) \equiv B_0$. We therefore obtain (Problem 7.8.1)

$$|W_S - W_N| = |W_0| = N_0 |w_0| = \frac{1}{2} V B_0^2 \mu_0^{-1} , \tag{7.77}$$

which gives a rigorous relation between the binding energy $|w_0|$ and the critical field B_0.

7.8.3
The Energy Gap

In the process of obtaining the ground state energy W by the variational calculation, we derived the energy-gap equation (7.64), which contains the energy parameters (see (7.60))

$$E_{\boldsymbol{k}}^{(j)} \equiv \left(\varepsilon_{\boldsymbol{k}}^{(j)2} + \Delta_j^2 \right)^{1/2} . \tag{7.78}$$

The fact that $E_{\boldsymbol{k}}^{(j)}$ represents the energy of a *quasielectron*, can be seen as follows [14]: the quasiparticle energy is defined to be the total excitation energy of the system when an extra particle is added to the system. From (7.58) we see that by negating the pair state $(\boldsymbol{k}\uparrow, -\boldsymbol{k}\downarrow)$, the energy is increased by (Problem 7.8.2)

$$-2\varepsilon_k^{(1)} v_k^{(1)2} + 2\left[\sum_{k'}{}' \left(v_{11} u_{k'}^{(1)} v_{k'}^{(1)} + v_{12} u_{k'}^{(2)} v_{k'}^{(2)}\right)\right] u_k^{(1)} v_k^{(1)}$$
$$= -2\varepsilon_k^{(1)} v_k^{(1)2} + 2\Delta_1 u_k^{(1)} v_k^{(1)} \,, \tag{7.79}$$

where we used (7.61) and (7.62). To this energy we must add the energy of $\varepsilon_k^{(1)}$ of the added "electron." Thus, the total excitation energy $\Delta\varepsilon$ is (Problem 7.8.3)

$$\Delta\varepsilon = \varepsilon_k^{(1)}\left(1 - 2v_k^{(1)2}\right) + 2\Delta_1 u_k^{(1)} v_k^{(1)} = E_k^{(1)} \,. \tag{7.80}$$

Thus, the unpaired electron (or "electron") has the energy $E_k^{(1)}$ as shown in Figure 7.20. Note that the validity domain for the above statement is $0 < \varepsilon_k^{(1)} < \hbar\omega_{\mathrm{D}}$.

7.8.4
The Energy Gap Equation

The reduced Hamiltonian $\mathcal{H}_{\text{red}}$ was expressed in terms of zero-momentum pairon operators, b's, only as in (7.45). The ground state ket $|\Psi\rangle$ in (7.51) contains b's only. Yet in the energy-gap equations, which follow from the extremum condition for the ground state energy, the energies of the quasielectron, $E_k^{(j)}$, appear unexpectedly. Generally speaking the physics is lost in the variational calculation. We shall derive the gap equation from a different angle by using the *equation-of-motion method* (see Appendix A.2).

Let us rederive the gap equations (7.64) by using the equation-of-motion method.

The supercondensate is made up of *ground pairons*, which can be described in terms of the pairon operators, b's. We calculate $[\mathcal{H}_{\text{red}}, b_k^{(1)\dagger}]$ and $[\mathcal{H}_{\text{red}}, b_k^{(2)}]$ to obtain (Problem 7.8.4)

$$\left[\mathcal{H}_{\text{red}}, b_k^{(1)\dagger}\right] = E_1 b_k^{(1)\dagger}$$
$$= 2\varepsilon_k^{(1)} b_k^{(1)\dagger} - \left(v_{11}\sum_{k'}{}' b_{k'}^{(1)\dagger} + v_{12}\sum_{k'}{}' b_{k'}^{(2)\dagger}\right)\left(1 - n_{k\uparrow}^{(1)} - n_{-k\downarrow}^{(1)}\right) \,, \tag{7.81}$$

$$\left[\mathcal{H}_{\text{red}}, b_k^{(2)\dagger}\right] = -E_2 b_k^{(2)}$$
$$= -2\varepsilon_k^{(2)} b_k^{(2)} - \left(v_{21}\sum_{k'}{}' b_{k'}^{(1)\dagger} + v_{22}\sum_{k'}{}' b_{k'}^{(2)}\right)\left(1 - n_{k\uparrow}^{(2)} - n_{-k\downarrow}^{(2)}\right) \,, \tag{7.82}$$

where $\mathcal{H}_{\text{red}}$ is the reduced Hamiltonian (7.45). Equations (7.81) and (7.82) indicate that the dynamics of ground pairons depends on the presence of "electrons" (quasielectrons) describable in terms of $n^{(j)}$. We now multiply (7.81) from the right by $\phi_1\rho_0$, where ϕ_1 is the electron pairon energy-state annihilation operator, and take a grand ensemble trace. From the first term on the rhs, we obtain

$$2\varepsilon_k^{(1)}\mathrm{TR}\left\{b_k^{(1)\dagger}\phi_1\rho_0\right\} = 2\varepsilon_k^{(1)}\langle\Psi|b_k^{(1)\dagger}\phi_1|\Psi\rangle = 2\varepsilon_k^{(1)}u_k^{(1)}v_k^{(1)}F_1\,, \tag{7.83}$$

where

$$F_1 \equiv \langle\Psi|\phi_1|\Psi\rangle\,. \tag{7.84}$$

From the first and second sums, we obtain (Problem 7.8.5)

$$-\sum_{k'}{}' \left(v_{11}u_{k'}^{(1)}v_{k'}^{(1)} + v_{12}u_{k'}^{(2)}v_{k'}^{(2)}\right)\left(1 - v_k^{(1)2} - v_{-k}^{(1)2}\right)F_1\,. \tag{7.85}$$

Since we are looking for the ground state energy, the eigenvalues E_1 and E_2 are zero:

$$E_1 = E_2 = 0\,. \tag{7.86}$$

Collecting all contributions, we obtain

$$\left[2\varepsilon_k^{(1)}u_k^{(1)}v_k^{(1)} - \left(u_k^{(1)2} - v_k^{(1)2}\right)\sum_{k'}{}'\left(v_{11}u_{k'}^{(1)}v_{k'}^{(1)} + v_{12}u_{k'}^{(2)}v_{k'}^{(2)}\right)\right]F_1 = 0\,, \tag{7.87}$$

where we used (Problem 7.8.6)

$$u_{-k}^{(1)} = v_k^{(1)}\,. \tag{7.88}$$

Since $F_1 \equiv \langle\Psi|\phi_1|\Psi\rangle \neq 0$, we obtain from (7.87)

$$2\varepsilon_k^{(1)}u_k^{(1)}v_k^{(1)} - \left(u_k^{(1)2} - v_k^{(1)2}\right)\sum_{k'}{}'\left(v_{11}u_{k'}^{(1)}v_{k'}^{(1)} + v_{12}u_{k'}^{(2)}v_{k'}^{(2)}\right) = 0\,, \tag{7.89}$$

which is just one of the equations in (7.59), the equations equivalent to the energy gap equations (7.64).

The ground state of $|\Psi\rangle$ of the BCS system is a superposition of many-pairon states and therefore quantities like $\langle\Psi|b_k^{(j)\dagger}|\Psi\rangle$, $\langle\Psi|b_k^{(j)\dagger}b_{k'}^{(j)\dagger}|\Psi\rangle, \ldots$, that connect states of different particle numbers do not vanish. In this sense the state $|\Psi\rangle$ can be defined in a grand ensemble.

We emphasize that the supercondensate state is made up of equal numbers of $\pm$pairons, and this state is reachable from the physical vacuum by a succession of phonon exchanges. Since pair creation and pair annihilation of pairons actually lower the system energy [see (7.48)] and since pairons are bosons, all pairons available in the system are condensed into the zero-momentum state; the maximum number of pairons is $\hbar\omega_D\mathcal{D}(0)$. The number of condensed pairons at any one instant may fluctuate around the equilibrium value; such fluctuations are in fact more favorable.

7.8.5
Neutral Supercondensate

The supercondensate composed of equal numbers of $\pm$pairons is electrically *neutral.* This neutrality explains the stability of the superconducting state against a weak electric field because no Lorentz electric force can be exerted on the supercondensate. The stability is analogous to that of a stationary excited atomic state, say, the 2p-state of a neutral hydrogen atom.

A neutral supercondensate is supported by experiments. If a superconducting wire S is used as part of a circuit connected to a battery, as shown in Figure 7.21, then the wire S, having no resistance, generates no potential drop. If a low-frequency AC voltage is applied to it, its response becomes more complicated. But the behavior can be accounted for if we assume that it has a normal component with a finite resistance and a super part. This is the *two fluid model* [28, 29]. Super part or supercondensate, decreases with rising temperature and vanishes at T_c. The normal part may be composed of any charged elementary excitations including quasielectrons and excited pairons. At any rate, analysis of all experiments indicate that the *supercondensate is not accelerated by the electric force.* This must be so. Otherwise the supercondensate would gain energy without limit since the supercurrent is slowed down by neither impurities nor phonons, and a stationary state would never have been observed in the circuit.

7.8.6
Cooper Pairs (Pairons)

The concept of pairons is inherent in the BCS theory, which is most clearly seen in the reduced Hamiltonian $\mathcal{H}_{\text{red}}$, expressed in terms of pairon operators, b's, only. The direct evidence for the fact that a Cooper pair is a bound quasiparticle having charge (magnitude) $2e$ comes from flux quantization experiments, see Figures 7.5 and 7.6.

7.8.7
Formation of a Supercondensate and Occurrence of Superconductors

We discuss the formation of a supercondensate based on the band structures of electrons and phonons. Let us first take Pb, which forms a face-centered cubic (fcc)

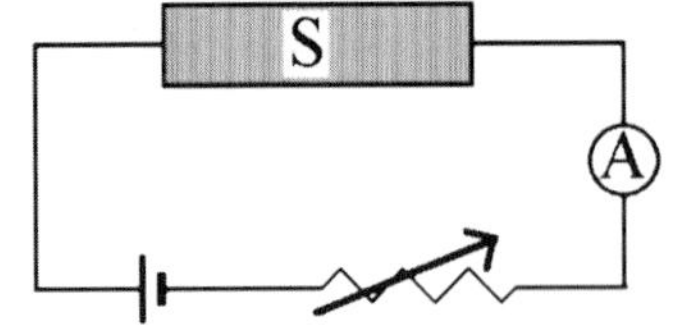

Figure 7.21 A circuit containing a superconductor (S), battery, and resistance.

lattice and which is a superconductor. This metal is known to have a neck-like hyperboloidal Fermi surface represented by

$$E = \frac{p_1^2}{2m_1} + \frac{p_2^2}{2m_2} + \frac{p_3^2}{2m_3}, \quad (m_1, m_2, m_3) = (1.18, 0.244, -8.71)m, \quad (7.90)$$

where m is the electron mass. See, for example, [30], Section 13.4.

We postulate that the supercondensate composed of $\pm$ground pairons is generated near the "necks." The electron transitions are subject to Pauli's exclusion principle, and hence creating pairons requires a high degree of symmetry in the Fermi surface. A typical way of generating pairons of both charge types by one phonon exchange near the neck is shown in Figure 7.18. Only parts of "electrons" and "holes" near the specific part of the Fermi surface are involved in the formation of the supercondensate. The numbers of $\pm$pairons, which are mutually equal by construction, may both then be represented by $\hbar\omega_D\mathcal{D}(0)/2$, which justifies (7.67). Next take aluminum (Al), which is also a known fcc superconductor. Its Fermi surface contains inverted double caps. Acoustic phonons with small momenta may generate a supercondensate near the inverted double caps. Supercondensation occurs independently of the lattice structure as long as "electrons" and "holes" are present in the system. Beryllium (Be) forms a hexagonal closed packed (hcp) crystal. Its Fermi surface in the second zone has necks. Thus, Be is a superconductor. Tungsten (W) is a body-centered cubic (bcc) metal, and its Fermi surface has necks. This metal also is a superconductor. In summary, type I elemental superconductors should have hyperboloidal Fermi surfaces favorable for the creation of $\pm$pairons mediated by small-momentum phonons. All of the elemental superconductors whose Fermi surface is known appear to satisfy this condition.

To test further let us consider a few more examples. A monovalent metal, such as Na, has a spherical Fermi surface within the first Brillouin zone. Such a metal cannot become superconducting at any temperature since it does not have "holes" to begin with; it cannot have $+$pairons and, therefore, cannot form a neutral supercondensate. A monovalent fcc metal like Cu has a set of necks at the Brillouin boundary. This neck is forced by the inversion symmetry of the lattice, see Figure 3.7, Chapter 3. The region of the hyperboloidal Fermi surface may be more severely restricted than those necks (unforced) in Pb. Thus, this metal may become superconducting at extremely low temperatures, which is not ruled out.

7.8.8
Blurred Fermi Surface

A normal metal has a sharp Fermi surface at 0 K. This fact manifests itself in the T-linear heat capacity generally observed at the lowest temperatures. The *T-linear law* is in fact the most important support for the Fermi liquid model. For a superconductor the Fermi surface is not sharp everywhere. To see this, let us solve (7.61) with respect to u_k^2 and v_k^2. We obtain

$$u_k^2 = \frac{1}{2}\left[1 + \frac{\varepsilon_k}{(\varepsilon_k^2 + \Delta^2)^{1/2}}\right], \quad v_k^2 = \frac{1}{2}\left[1 - \frac{\varepsilon_k}{(\varepsilon_k^2 + \Delta^2)^{1/2}}\right]. \tag{7.91}$$

Figure 7.22 shows a general behavior of v_k^2 near the Fermi energy. For the normal state $\Delta = 0$, there is a sharp boundary at $\varepsilon_k = 0$; but for a finite Δ, the quantity v_k^2 drops off to zero over a region of the order $(2 \sim 3)\Delta$. This v_k^2 represents the probability that the virtual electron pair at $(\boldsymbol{k}\uparrow, -\boldsymbol{k}\downarrow)$ participates in the formation of the supercondensate. It is *not* the probability that either electron of the pair occupies the state $\boldsymbol{k}$. Still, the diagram indicates the nature of the changed electron distribution in the ground state. The supercondensate is generated only near the necks and/or inverted double caps. Hence, these parts of the Fermi surface are blurred or fuzzy.

Problem 7.8.1. Derive (7.77).

Problem 7.8.2. Verify (7.79).

Problem 7.8.3. Verify (7.80).

Problem 7.8.4. Derive (7.81) and (7.82).

Problem 7.8.5. Verify (7.85).

Problem 7.8.6. Verify (7.88).

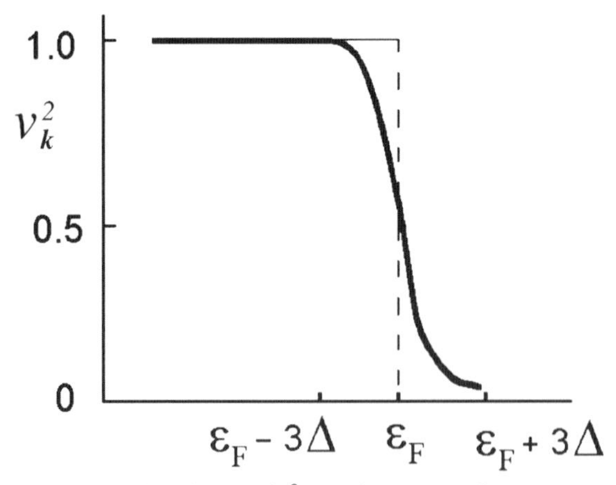

Figure 7.22 The behavior of v_k^2 near the Fermi surface.

7.9 Bose–Einstein Condensation in 2D

BCS [19] introduced *electron-pair operators*:

$$b_k^\dagger \equiv c_{k+}^\dagger c_{-k-}^\dagger \equiv c_k^\dagger c_{-k}^\dagger \,, \quad b_k \equiv c_{-k} c_k \,, \tag{7.92}$$

where $(c, c^\dagger)$ are electron operators (spin indices omitted) satisfying the Fermi anticommutation rules. They investigated the commutators among b and $b^\dagger$, which do not satisfy the usual Bose commutation rules. On the basis of these commutators and $b_k^2 = 0$, BCS did not consider the bosonic nature of the pairons. But the eigenvalues for $n_{12} \equiv c_{k_1}^\dagger c_{k_2}^\dagger c_{k_2} c_{k_1} \equiv c_1^\dagger c_2^\dagger c_2 c_1$ in the pair states $(\boldsymbol{k}_1, \boldsymbol{k}_2)$ are limited to 0 or 1 (fermion property), while the eigenvalues of the *total pair number* operator

$$n_0 \equiv \sum_k b_k^\dagger b_k \tag{7.93}$$

have no upper limit (bosonic property):

$$n_0' = 0, 1, 2, \ldots \tag{7.94}$$

The proof of (7.94) is given in Appendix A.4. Both fermionic and bosonic natures of the pairons must be used in the total description of superconductivity.

The most important signature of many bosons is the *Bose–Einstein Condensation*. Earlier we showed that the pairon moves with the linear dispersion relation:[8)]

$$w_\mathrm{p} = w_0 + cp \,, \tag{7.95}$$

where we designated the pairon net momentum by the more familiar p rather than q.

Let us consider a 2D system of free bosons having a linear dispersion relation: $\varepsilon = cp$, $c = 2/\pi v_\mathrm{F}$. The total number of bosons, N, and the Bose distribution function,

$$f_\mathrm{B}(\varepsilon; \beta, \mu) \equiv \frac{1}{\mathrm{e}^{\beta(\varepsilon-\mu)} - 1} \equiv f_\mathrm{B}(\varepsilon) \quad (\geq 0) \,, \tag{7.96}$$

are related by

$$N = \sum_p f_\mathrm{B}(\varepsilon_\mathrm{p}; \beta, \mu) = N_0 + \sum_{\substack{p \\ \varepsilon_\mathrm{p} > 0}}{}' f_\mathrm{B}(\varepsilon_\mathrm{p}) \,, \tag{7.97}$$

where

$$N_0 \equiv \left(\mathrm{e}^{-\beta\mu} - 1\right)^{-1} \tag{7.98}$$

8) See (7.43) and (7.44). The coefficient c in (7.95) is specifically given by $(2/\pi)v_\mathrm{F}$ for a 2D system and $(1/2)v_\mathrm{F}$ for a 3D system, respectively.

is the number of zero-momentum bosons. The prime on the summation in (7.97) indicates the omission of the zero-momentum state. We note that α is defined by $\alpha \equiv \beta\mu$, where μ is the chemical potential and $\beta \equiv 1/(k_B T)$. For notational convenience we write

$$\varepsilon = cp = \frac{2}{\pi} v_F p \quad (> 0) . \tag{7.99}$$

We divide (7.97) by the normalization area L^2, and take the *bulk limit*:

$$N \to \infty, \quad L \to \infty \quad \text{while} \quad N L^{-2} \equiv n . \tag{7.100}$$

We then obtain

$$n - n_0 \equiv \frac{1}{(2\pi\hbar)^2} \int d^2 p \, f_B(\varepsilon) , \tag{7.101}$$

where $n_0 \equiv N_0/L^2$ is the number density of zero-momentum bosons and n the total boson density. After performing the angular integration and changing integration variables, we obtain from (7.101) (Problem 7.9.1)

$$2\pi\hbar^2 c^2 \beta^2 (n - n_0) = \int_0^\infty dx \frac{x}{\lambda^{-1} e^x - 1} , \quad \left(x \equiv \beta\varepsilon , \quad \lambda \equiv e^{\beta\mu} \right) . \tag{7.102}$$

The fugacity λ is less than unity for all temperatures. After expanding the integrand in (7.102) in powers of $\lambda e^{-x} (< 1)$, and carrying out the x-integration, we obtain

$$n_x \equiv n - n_0 = \frac{k_B^2 T^2 \phi_2(\lambda)}{2\pi\hbar^2 c^2} , \tag{7.103}$$

$$\phi_m(\lambda) \equiv \sum_{k=1}^{\infty} \frac{\lambda^k}{k^m} , \quad (0 \le \lambda \le 1) . \tag{7.104}$$

We need $\phi_2(\lambda)$ here, but we introduced ϕ_m for later reference. Equation (7.103) gives a relation among λ, n, and T.

The function $\phi_2(\lambda)$ monotonically increases from zero to the maximum value $\phi_2(1) = 1.645$ as λ is raised from zero to one. In the low-temperature limit, $\lambda = 1$, $\phi_2(\lambda) = \phi_2(1) = 1.645$, and the density of excited bosons, n_x, varies like T^2 as seen in (7.103). This temperature behavior of n_x persists as long as the rhs of (7.103) is smaller than n; the *critical temperature* T_c occurs at $n = k_B^2 T_c^2 \phi_2(1)/2\pi\hbar^2 c^2$. Solving this, we obtain

$$k_B T_c = 1.954\hbar c n^{1/2} \quad \left(= 1.24\hbar v_F n^{1/2}\right) . \tag{7.105}$$

If the temperature is raised beyond T_c, the density of zero-momentum bosons, n_0, becomes vanishingly small, and the fugacity λ can be determined from

$$n = \frac{k_B T^2 \phi_2(\lambda)}{2\pi\hbar^2 c^2} , \quad T > T_c . \tag{7.106}$$

In summary, the fugacity λ is equal to unity in the condensed region: $T < T_c$, and it becomes smaller than unity for $T > T_c$, where its value is determined from (7.106).

The internal energy density u, that is, the thermal average of the system energy per unit area, is given by

$$u = \frac{1}{(2\pi\hbar)^2} \int d^2 p \, \varepsilon \, f_B(\varepsilon) \, . \tag{7.107}$$

This u can be calculated in a similar manner. We obtain (Problem 7.9.2)

$$u = \frac{\phi_3(\lambda)}{\pi\hbar^2 c^2 \beta^3} = 2nk_B \frac{T^3}{T_c^2} \frac{\phi_3(\lambda)}{\phi_2(1)} \, . \tag{7.108}$$

The molar heat capacity at constant density (volume), C_n, is given by

$$C \equiv C_n = R(nk_B)^{-1} \frac{\partial u(T, n)}{\partial T} \, , \tag{7.109}$$

where R is the gas constant. The partial derivative $\partial u/\partial t$ may be calculated through (Problem 7.9.3)

$$\begin{aligned} \frac{\partial u(T,n)}{\partial T} &= \frac{\partial u(T,\lambda)}{\partial T} + \frac{\partial u(T,\lambda)}{\partial \lambda} \frac{\partial \lambda(T,n)}{\partial T} \\ &= \frac{\partial u(T,\lambda)}{\partial T} - \frac{\partial u(T,\lambda)}{\partial \lambda} \frac{\partial n(T,\lambda)/\partial T}{\partial n(T,\lambda)/\partial \lambda} \, . \end{aligned} \tag{7.110}$$

All quantities (n, u, C) can now be expressed in terms of $\phi_m(\lambda)$. After straightforward calculations, the *molar heat capacity* C is given by (Problem 7.9.4)

$$\begin{aligned} C &= R(\pi\hbar^2 c^2 n k_B)^{-1} k_B^3 T^2 \left[3\phi_3(\lambda) - \frac{2\phi_2^2(\lambda)}{\phi_1(\lambda)} \right] \\ &= 6R \left(\frac{T}{T_c} \right)^2 \frac{\phi_3(\lambda)}{\phi_2(1)} - 4R \frac{\phi_2(\lambda)}{\phi_1(\lambda)} \, . \end{aligned} \tag{7.111}$$

In the condensed region $T < T_c$, the fugacity λ is unity. We observe that as $\lambda \to 1$,

$$\begin{aligned} &\phi_1 \to \sum_1^\infty k^{-1} = \infty \, , \qquad \phi_2(\lambda) \to \phi_2(1) = 1.645 \, , \\ &\phi_3(\lambda) \to \phi_3(1) = 1.202 \, , \quad \phi_4(\lambda) \to \phi_4(1) = 1.082 \, . \end{aligned} \tag{7.112}$$

Using these, we obtain from (7.108) and (7.111)

$$u = 2nk_B \frac{\phi_3(1)}{\phi_2(1)} \frac{T^3}{T_c^2} \, , \tag{7.113}$$

$$C = 6R \frac{\phi_3(1)}{\phi_2(1)} \left(\frac{T}{T_c} \right)^2 \, , \quad (T < T_c) \, . \tag{7.114}$$

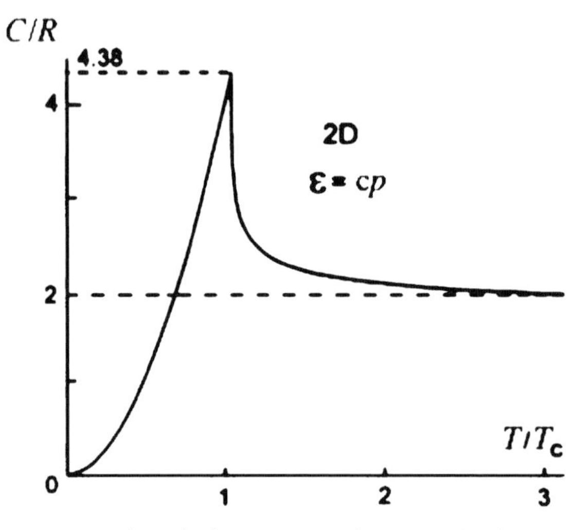

Figure 7.23 The molar heat capacity C for 2D massless bosons rises like $T * 2$, reaches 4.38 R at T_c, and then decreases to 2 R in the high-temperature limit.

Observe that the molar capacity C grows like T^2. Also note that the molar heat capacity C at T_c is given by

$$C(T_c) = C_{\max} = 6R\frac{\phi_3(1)}{\phi_2(1)} = 4.38\ R\ . \tag{7.115}$$

For $T > T_c$, the temperature dependence of λ, given by Equation (7.106), is quite complicated. We can numerically solve (7.111) for λ with a computer, and substitute the solution in (7.102) to obtain the temperature behavior of C. The result is shown in Figure 7.23. Equations (7.110) and (7.111) allow us not only to examine the analytical behavior of C near $T = T_c$ but also to obtain C without numerically computing the derivative $\partial u(T, n)/\partial T$.

In summary, the molar heat capacity C for a 2D *massless* boson rises like T^2 in the condensed region, reaches 4.38 R at $T = T_c$, and then decreases to the high-temperature limit value 2 R. The heat capacity changes continuously at $T = T_c$, but its temperature derivative $\partial C(T, n)/\partial T$ jumps at this point. The *order of phase transition* is defined to be that order of the derivative of the free energy F whose discontinuity appears for the first time. Since $C_V = T(\partial S/\partial T)_V = -T(\partial^2 F/\partial T^2)$, $\partial C_V/\partial T = -T(\partial^3 F/\partial T^3) - (\partial^2 F/\partial T^2)$, the BEC is a *third-order phase transition*. Note that the temperature behavior of the heat capacity C resembles that observed in YBCO shown in Figure 7.12. The condensation of massless bosons in 2D is noteworthy. This is not a violation of Hohenberg's theorem [31] that there can be no long range order in 2D, which is derived with the assumption of an f-sum rule representing the mass conservation law. In fact no BEC occurs in 2D for finite-mass bosons.

Problem 7.9.1. Verify (7.102).

Problem 7.9.2. Verify (7.108).

Problem 7.9.3. Prove (7.110).

Problem 7.9.4. Verify (7.111).

7.10 Discussion

The idea that the superconductivity is a manifestation of the BEC has long been suspected. The superconductivity in a metal and the superconductivity in liquid helium have many similarities. Both involve dissipationless flows, and they occur at very low temperatures. In particular, F. London treated superconductivity and superfluidity from the BEC point of view in his two-volume book [32].

The BEC temperature T_c in D dimensions can be found from

$$n_0 = \frac{N_0}{L^D} = \frac{1}{(2\pi\hbar)^D} \int \mathrm{d}^D p \frac{1}{\exp(\varepsilon/(k_B T_c)) - 1} , \qquad \varepsilon = cp . \tag{7.116}$$

Using (7.43) and (7.44) along with (7.107), we obtain (Problem 7.10.1)

$$T_c = \begin{cases} 1.01\hbar v_F k_B^{-1} n_0^{1/3} & \text{(3D)} \\ 1.24\hbar v_F k_B^{-1} n_0^{1/2} & \text{(2D)} . \end{cases} \tag{7.117}$$

The 2D BEC is noteworthy since the BEC of massive bosons ($\varepsilon = p^2/(2m)$) is known to occur in 3D only. The *interpairon distance* r_0 computed from (7.117) is

$$r_0 = \begin{cases} n_0^{-1/3} = 1.01\hbar v_F (k_B T_c)^{-1} & \text{(3D)} \\ n_0^{-1/2} = 1.24\hbar v_F (k_B T_c)^{-1} & \text{(2D)} . \end{cases} \tag{7.118}$$

The zero-temperature BCS pairon size [19] is given by

$$\xi_0 = \frac{\hbar v_F}{\pi \Delta} = 0.18 \frac{\hbar v_F}{k_B T_c} . \tag{7.119}$$

From the last two equations we obtain

$$\frac{r_0}{\xi_0} = \begin{cases} 5.6 & \text{(3D)} \\ 6.9 & \text{(2D)} , \end{cases} \tag{7.120}$$

indicating that the *condensed pairons do not overlap in space*. Hence, the free pairon model can be used to evaluate T_c.

The similarity in 2D and 3D BEC is most remarkable. In particular the critical temperature T_c depends on (v_F, r_0) nearly in the same manner. Now, the interpairon distance r_0 is different by the factor $10^2 \sim 10^3$ between 3D and 2D superconductors. The Fermi velocity v_F is different by the factor $10 \sim 10^2$. Hence, the *high* critical temperature in 2D superconductors is explained by the very short interpairon distance, partially compensated by a smaller Fermi velocity.

We stress that formulas (7.117) for the critical temperatures are distinct from the famous BCS formula (in the weak coupling limit):

$$k_B T_c = 1.13\hbar\omega_D \exp\left(\frac{1}{v_0 \mathcal{D}(0)}\right), \qquad (7.121)$$

where ω_D is the Debye frequency, v_0 the pairing strength, and $\mathcal{D}(0)$ the DOS per spin at the Fermi energy.

The pairon density n_0 and the Fermi velocity v_F appearing in formulas (7.117) can be determined experimentally from the data of the resistivity, the Hall coefficient, the Hall angle, the specific heat, and the superconducting temperature.

The *linear dispersion relation* can be probed by using Angle-Resolved Photoemission Spectroscopy (ARPES). Lanzara *et al.* [33] studied the dispersions in three different families of hole-doped copper oxides: $Bi_2Sr_2CaCu_2O_8$ (Bi2212), Pb-doped $Bi_2Sr_2Cu_6$ (Pb-Bi2201), and $La_{2-x}Sr_xCuO_4$ (LSCO). A summary of the data, reproduced from [33, Figure 1], is shown in Figure 7.24. The energy is measured down-

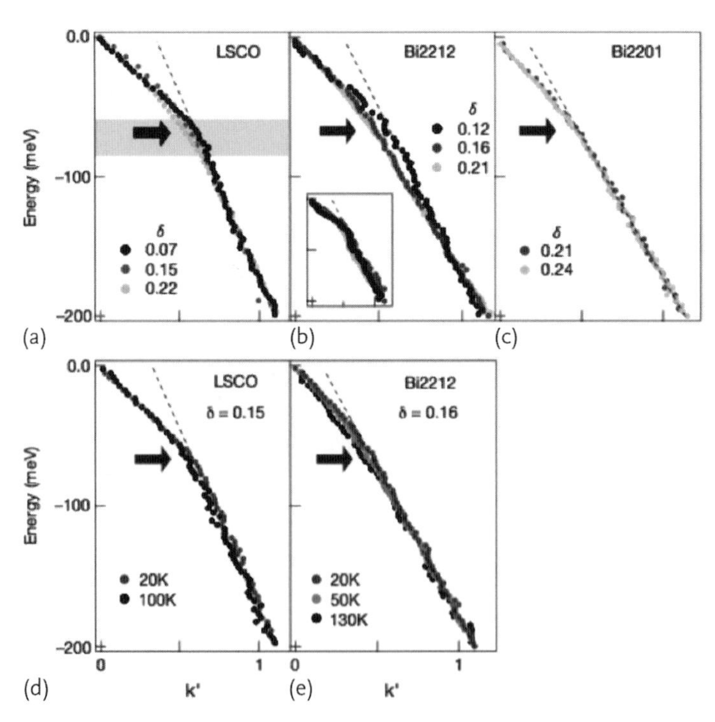

Figure 7.24 The quasiparticle dispersion relations derived from the momentum distribution curves along (0, 0)–(π, π) for (a) LSCO at 20 K, (b) Bi2212 at 20 K. Both materials are in superconducting states with the doping δ indicated. (c) The dispersion relation for Bi2201 at 30 K (normal state). The temperature dependence of the dispersion relation for LSCO at $\delta = 0.15$ (d) and Bi2212 at $\delta = 0.16$ (e). Note that the energy scale is measured downward. The arrows indicate slope changes in the curves. After Lanzara *et. al.* [33]

wards and the reduced momentum k is in the abscissa. See more detailed specification in the original reference. The data in Figure 7.24a,b are in the superconducting states while those in Figure 7.24c are in the normal state. Note that in all three cases the dispersion relation is *linear* for low k and quadratic for high k. The phonon energy has an upper limit of the order $\hbar\omega_D$ (Debye energy) and hence, the quasi-particle (pairon) mediated by the phonon exchange must have a finite energy. The change of the slopes, indicated by thick arrows, occurs around 50~80 meV, which are distinct from the superconducting energy gaps (10~50 meV). The energies 50~80 meV appear to correspond to the energy of the in-plane oxygen-stretching (breathing) longitudinal optical phonon.

Figure 7.24d,e indicate that the dispersion relations do not change above and below T_c for LSCO and Bi2212, respectively. The pairons have a linear dispersion relation with the same slope both below and above T_c. Thus, the ARPES fully supports our BEC picture of superconductivity. We stress that the pairons do not break up at T_c as thought in the original BCS theory.

Pairons can multiply occupy the same CM momentum state. They move freely as bosons via a linear dispersion relation: $\varepsilon = cp$. The system of free pairons (bosons) undergoes a BEC transition of the second (third) order in 3(2)D with the critical temperature

$$T_c = 1.01\hbar v_F k_B^{-1} n_0^{1/3} \left(1.24\hbar v_F k_B^{-1} n_0^{1/3}\right) ,$$

where n_0 is the pairon density. In general the critical temperature T_c for free bosons moving in D dimensions can be found from (7.116). The solutions for $D = 2$ and 3 are given in (7.118). For $D = 1$, (7.116) has no solution (Problem 7.10.2). In other words, there is *no* BEC in 1D. To see this, let us take a dispersion relation:

$$\varepsilon = ap^{\alpha} , \tag{7.122}$$

where a and α are constants. If we substitute this ε in (7.116), we can find a solution if $\alpha < 1$. The index α, however, must be greater than, or equal to, unity; otherwise the boson has an infinite speed at zero momentum.

Problem 7.10.1. Derive (7.117).

Problem 7.10.2. Prove that (7.116) has no solution (*that is*, no T_c) for $D = 1$.

References

1 Kammerlingh Onnes, H. (1991) *Acad. Wetenschappen*, Amsterdam, **14**, 113.
2 Meissner, W. and Ochsenfeld, R. (1933) *Naturwiss*, **21**, 787.
3 File, J. and Mills, R.G. (1933) *Phys. Rev. Lett.*, **10**, 93.
4 Deaver, B.S. and Fairbank, W.M. (1961) *Phys. Rev. Lett.*, **7**, 43.
5 Doll, R. and Nöbauer, M. (1961) *Phys. Rev. Lett.*, **7**, 51.
6 Josephson, B.D. (1962) *Phys. Lett.*, **1**, 231.

7 Anderson, P.W. and Lowell, J.M. (1963) *Phys. Rev. Lett.*, **10**, 486.
8 Jaklevie, R.C., Lambe, J., Marcereaw, J.E., and Silver, A.H. (1965) *Phy. Rev. A*, **140**, 1628.
9 Glover, R.E. III and Tinkham, M. (1967) *Phys. Rev.*, **108**, 243.
10 Biondi, M.A., and Garfunkelm, M. (1959) *Phys. Rev.*, **116**, 853.
11 Giaever, I. (1960) *Phys. Rev. Lett.*, **5**, 147, 464.
12 Phillips, N.E. (1959) *Phys. Rev.*, **114**, 676.
13 Giaever, I. and Megerle, K. (1961) *Phys. Rev.*, **122**, 1101.
14 Loram, I.W., Mirza, K.A., Cooper, J.R., Liang, W.Y., and Wade, J.M. (1994) *J. Superconduct.*, **7**, 347.
15 Ginzburg, V.L. and Landau, L.D. (1950) *J. Exp. Theor. Phys. (JETP) (USSR)*, **20**, 1064.
16 Bednorz, V.G. and Müller, K.A. (1986) *Z. Phys. B*, **64**, 189.
17 Wu, M.K., Ashburn, J.R., Torng, C.J., Hor, P.H., Meng, R.L., Gao, L., Huang, Z.J., Wang, Y.Q., and Chu, C.W. (1987) *Phys. Rev. Lett.*, **58**, 908.
18 Cooper, L.N. (1956) *Phys. Rev.*, **104**, 1189.
19 Bardeen, J., Cooper, L.N. and Schrieffer, J.R. (1957) *Phys. Rev.* **108**, 1175.
20 Landau, L.D. (1956) *J. Exp. Theor. Phys. (JETP)*, USSR, **30**, 1058.
21 Landau, L.D. (1957) *J. Exp. Theor. Phys. (JETP)*, USSR, **32**, 59.
22 Fujita, S. (1991) *J. Superconduct.*, **4**, 297.
23 Fujita, S. (1992) *J. Superconduct.*, **5**, 83.
24 Fujita, S. and Watanabe, S. (1992) *J. Superconduct.*, **5**, 219.
25 Fujita, S. and Watanabe, S. (1993) *J. Superconduct.*, **6**, 75.
26 Fujita, S. and Godoy, S. (1993) *J. Superconduct.*, **6**, 373.
27 Schrieffer, J.R. (1964) *Theory of Superconductuctivity*, Benjamin, New York.
28 Gortor, C.J. and Casimir, H.G.B. (1934) *Z. Phys.*, **35**, 963.
29 Gortor, C.J. and Casimir, H.G.B. (1934) *Z. Tech. Phys.*, **15**, 539.
30 Fujita, S. and Ito, K. (2007) *Quantum Theory of Conducting Matter*, Springer, New York.
31 Hohenberg, P.C. (1967) *Phys. Rev.*, **158**, 383.
32 London, F. (1964) *Superfluids, vol. I and II*, Dover, New York.
33 Lanzara, A., Bogdanob, P.V., Zhou, X.J., Keller, S.A., Feng, D.L., Lu, E.D., Yoshida, T., Eisaki, H., Fujimori, A., Kishio, K., Shimoyama, J.-I., Noda, T., Uchida, S., Hussain, Z., and Shen, Z.-X. (2001) *Nature*, **412**, 510.

8
Metallic (or Superconducting) SWNTs

A metallic (semiconducting) single-wall nanotube (SWNT) contains an irrational (integral) number of carbon hexagons in the pitch. The room-temperature conductivity is higher by two to three orders of magnitude in metallic nanotubes than in semiconducting nanotubes. Tans *et al.* [5] measured the electric currents in metallic single-wall carbon nanotubes under bias and gate voltages, and observed various non-Ohmic behaviors. The original authors interpreted their data in terms of a ballistic electron transport due to the Coulomb blockage on the electron carrier model. The mystery as to why a ballistic electron is not scattered by impurities and phonons is unexplained, however. An alternate interpretation is presented based on the Cooper pair (pairon) carrier model. Superconducting states are generated by the Bose–Einstein condensation (BEC) of the $\pm$pairons at momenta $2\pi\hbar n/L$, where L is the tube length and n a small integer. As the gate voltage changes the charge state of the tube, the superconducting states jump between different n. The normal current peak shapes appearing in the transition are temperature-dependent, which is shown to be caused by the electron–optical phonon interaction.

8.1
Introduction

Single-wall nanotubes (SWNTs) can be produced by rolling graphene sheets into circular cylinders of about 1 nm in diameter and microns (μm) in length [1, 2]. Electrical conduction in SWNTs depends on the pitch [3, 4] and can be classified into two groups: *semiconducting* or *metallic* [5, 6]. In our previous work [7] we have shown that this division into two groups arises as follows. A SWNT is likely to have an integral number of carbon hexagons around the circumference. If each pitch contains an integral number of hexagons, then the system is periodic along the tube axis, and "holes" (and not "electrons") can move along the tube. The system is semiconducting and its conduction is characterized by an activation energy ε_3. The energy ε_3 has a distribution since both pitch and circumference have distributions. The pitch angle is not controlled in the fabrication process. There are numerous other cases where the pitch contains an irrational number of hexagons. In these

Electrical Conduction in Graphene and Nanotubes, First Edition. S. Fujita and A. Suzuki.

cases the system shows a metallic behavior experimentally [5]. We primarily deal with these so-called metallic SWNTs in this chapter.

Tans *et al.* [5] measured the electric currents in metallic SWNTs under bias and gate voltages. Their data from [5, Figure 2], are reproduced in Figure 8.1. The currents versus the bias voltage are plotted at three gate voltages: A (88.2 mV), B (104.1 mV), and C (120.0 mV) (shown in Figure 8.1). Some significant features are:

a) A non-Ohmic behavior is observed for all, that is, the currents are not proportional to the bias voltage except for high voltages. The gate voltage charges the tube. The Coulomb (charging) energy of the charge Q of the system is represented by

$$E_{\text{Coul}} = \frac{Q^2}{2C} , \tag{8.1}$$

where C is the total capacitance of the tube.

b) The current near the origin appears to be constant for different gate voltages V_{gate}, traces A–C. This feature was confirmed by later experiments [8, 9]. The current does not change for small varying gate voltage in a metallic SWNT while the current (magnitude) decreases in a semiconducting SWNT.

c) The current at gate voltage $V_{\text{gate}} = 88.2\,\text{mV}$ (trace A) reverts to the normal resistive behavior after passing the critical bias voltages on both (positive and negative) sides. Similar behaviors are observed for $V_{\text{gate}} = 104.1\,\text{mV}$ (trace B) and $V_{\text{gate}} = 120.0\,\text{mV}$ (trace C).

d) The flat current is destroyed for higher bias voltages (magnitude). The critical bias voltage becomes smaller for higher gate voltages.

e) There is a restricted V_{gate}-range (view window in [5, Figure 3]) in which the horizontal stretch can be observed.

The original authors interpreted the flat currents near $V_{\text{bias}} = 0$ in Figure 8.1 in terms of ballistic electron transport based on the electron carrier model.

We propose a different interpretation of the data in Figure 8.1 based on the *pairon (Cooper pair) carrier model*. Pairons move as bosons, and hence they are generated with *no* activation energy factor (the main characteristic of metallic conduction). All features a)–e) listed above can be explained simply by the assumption that the tube is in the *superconducting state* as explained below.

The supercurrent runs without obeying Ohm's law. This explains feature a) above. The supercurrents can run without resistance due to phonon and impurity scattering and with no bias voltage. Bachtold *et al.* [10] observed by scanned probe microscopy that the currents run with no voltage change along the tube in metallic SWNTs. They also concluded that the currents run ballistically.

We shall show later that the system may be in a superconducting ground state, whose energy $-E_{\text{g}}$ is negative relative to the ground state energy of the Fermi liquid (electron) state. If the total energy E of the system is less than the condensation energy E_{g}:

$$E = K + E_{\text{Coul}} + E_{\phi} \leq E_{\text{g}} , \tag{8.2}$$

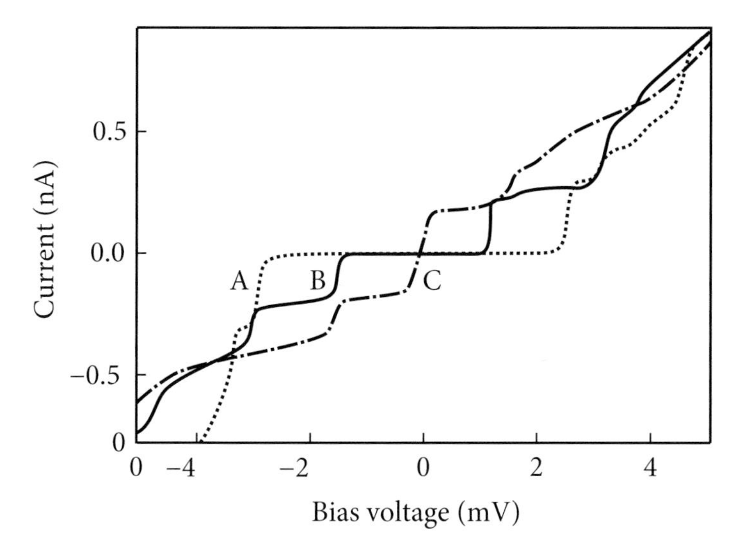

Figure 8.1 Current–voltage curves for nanotubes at gate voltages of 88.2 mV (trace A), 104.1 mV (trace B) and 120.0 mV (trace C) at $T = 5$ mK. (After Tans *et al.* in [5, Figure 2].)

where K is the kinetic energy of the conduction electrons and the pairons, and

$$E_\phi \equiv Q V_{\text{bias}} \tag{8.3}$$

is the Coulomb field energy, then the system is stable. The experiments shown in Figure 8.1 were done at 5 mK. We drop the kinetic energy K hereafter. The superconducting state is maintained and the currents run unchanged if the bias voltage V_{bias} is not too large so that the inequality (8.2) holds. This explains the horizontal stretch feature b).

If the bias voltage is high enough so that the inequality symbol in (8.2) is reversed, then normal currents revert and exhibit Ohmic behavior, which explains feature c).

Feature d) can be explained as follows. For higher V_{gate} there is more charge and hence the charges Q_{A}, Q_{B}, Q_{C} for the three cases (A, B, C in Figure 8.1) satisfy the inequalities:

$$Q_{\text{A}} < Q_{\text{B}} < Q_{\text{C}} \,. \tag{8.4}$$

The horizontal stretches are longer for smaller bias voltages. At the end of the stretch ($V_{\text{bias,max}}$) the system energy equals the condensation energy E_{g}. We then obtain from (8.2), after dropping the kinetic energy K,

$$\begin{aligned} E_{\phi,\max} &= Q V_{\text{bias,max}} \equiv Q V_{\max} \\ &= E_{\text{g}} - E_{\text{Coul}} = E_{\text{g}} - \frac{Q^2}{2C} \,. \end{aligned} \tag{8.5}$$

Using the inequalities (8.4), we then obtain

$$V_{\text{A,max}} > V_{\text{B,max}} > V_{\text{C,max}} \,, \tag{8.6}$$

which explains feature d) in the list above.

The horizontal stretch becomes shorter as the gate voltage V_{gate} is raised; it vanishes when V_{gate} is a little over 120.0 mV. The limit is given by

$$E_{\phi,max} = E_g - E_{Coul} = E_g - \frac{Q^2}{2C} = 0 . \tag{8.7}$$

If the charging energy E_{Coul} exceeds the condensation energy E_g, then there are no more supercurrents, which explains feature e).

Clearly, the important physical property in our pairon model is the condensation energy E_g.

In the currently prevailing theory it is argued that the electron (fermion) becomes ballistic at a certain quantum condition. But all fermions are known to be subject to scattering. It is difficult to justify why the ballistic electron is not scattered by impurities and phonons, which naturally exist in nanotubes. In contrast our theory is straightforward. We argue that the supercurrents, as is known, can run with no resistance (due to impurities and phonons).

To test the validity of the present theory we simply need to find the superconducting temperature by performing the conductivity measurements at higher temperatures.

Tans *et al.* [8, 11] noted that the measured currents are bistable at 0.1 and 0 nA. In our interpretation these currents are simply the supercurrents that arise from the Bose-condensed states at a finite momentum and zero momentum.

In Figure 8.2, current versus gate voltage at $V_{bias} = 30\ \mu V$ after Tans *et al.* [5] is shown. The delta function-like peaks are quasiperiodic. The flat currents near the origin are caused by the supercurrents due to the condensed pairons at a microscopic momentum

$$p = \frac{2\pi\hbar}{L} , \tag{8.8}$$

where L is the tube length. The peaks arise from the motion of noncondensed pairons. As the gate voltage is varied, the charge density changes, which makes the superconducting states jump among the different momenta

$$p_n = \frac{2\pi\hbar n}{L} , \tag{8.9}$$

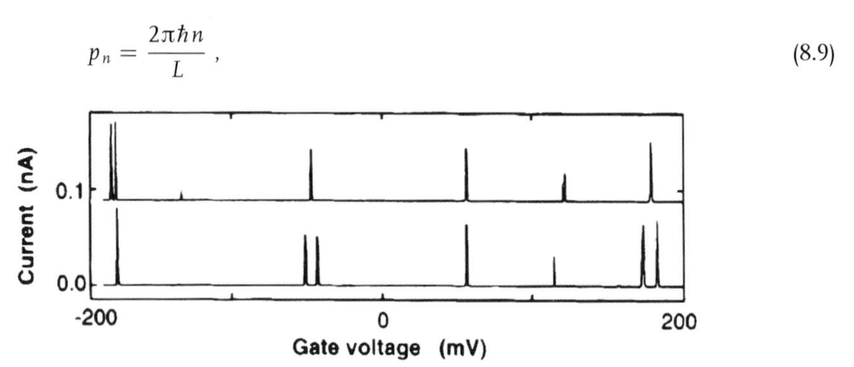

Figure 8.2 Current vs. gate voltage at $V_{bias} = 30\ \mu V$. (After Tans *et al.* in [5, Figure 2].)

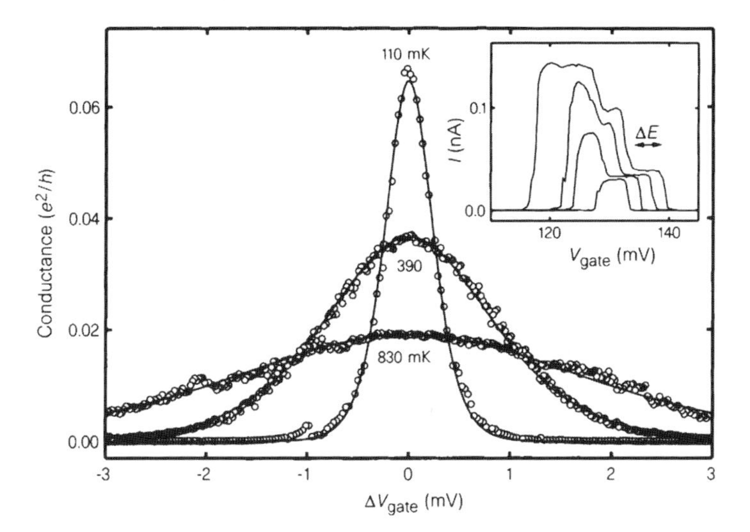

Figure 8.3 Conductance $G = I/V_{bias}$ versus ΔV_{gate} at low bias voltage $V_{bias} = 10\,\mu V$ and different temperatures. Solid lines are fits of $G \propto \cosh^{-2}(e\Delta V_{gate}/(\alpha 2k_B T))$, where ΔV_{gate} is the gate voltage difference. The factor α converts ΔV_{gate} into the corresponding electrostatic potential shift of the tube. (After Tans *et al.* in [5, Figure 3]).

where n is a small integer. In the transitions, the normal currents due to noncondensed pairons appear.

Tans *et al.* [5, 8] observed that the conductance G versus the gate voltage V_{gate} has peaks that are temperature-dependent as shown in Figure 8.3, reproduced from [5, Figure 3]. These authors obtained good fits with the assumption of the peak formula.

$$G \propto \cosh^{-2}\left(\frac{e\Delta V_{gate}}{\alpha 2k_B T}\right), \tag{8.10}$$

where factor α converts the gate voltage shift ΔV_{gate} into the corresponding electrostatic potential shift of the tube. See [5] for the definitions of ΔV_{gate} and α. We may re-interpret the data as follows.

The carriers in the normal currents (peaks) observed are noncondensed pairons, each having charges, $+2e$ or $-2e$, which are subject to phonon scattering.

The phonon scattering rate Γ is calculated by the standard formula:

$$\Gamma = n_{phonon} v_{pairon} A, \tag{8.11}$$

where v_{pairon} is the pairon speed and A the pairon-phonon scattering cross section, both of which are temperature-independent. We assume that optical phonons at the energy ε_0 with no dispersion are the predominant scatterers. Then, the phonon density n_{phonon} is given by the Planck distribution function:

$$\left[\exp\left(\frac{\varepsilon_0}{k_B T}\right) - 1\right]^{-1}. \tag{8.12}$$

At the experimental temperatures (110, 390, and 830 mK) the Planck distribution function can be approximated by the Boltzmann distribution function;

$$\left[\exp\left(\frac{\varepsilon_0}{k_B T}\right) - 1\right]^{-1} \simeq \exp\left(-\frac{\varepsilon_0}{k_B T}\right) . \tag{8.13}$$

In the experimental low-temperature range (110–830 mK) the peak-shape function (8.10) can be approximated as

$$\cosh^{-2}\left(\frac{e\Delta V_{gate}}{\alpha 2 k_B T}\right) \approx 4\exp\left(-\frac{e\Delta V_{gate}}{\alpha k_B T}\right) . \tag{8.14}$$

This has the same exponential form as formula (8.13), suggesting that the cause of the temperature-dependence is phonon scattering. This interpretation was also adopted by Saito, Dresselhaus, and Dresselhaus [6].

In the inset of Figure 8.3, the current versus the gate voltage is plotted at bias voltage $V_{bias} = 0.4, 0.8, 1.2$, and 1.6 mV. At $V_{bias} = 1.6$ mV, we see three steps (plateaus) separated by about 0.04 nA. We interpret these as three superconducting channels carried by the pairons condensed at three momenta. The lowest step is formed between $V_{gate} = 115$ and 140 mV and has the current value 0.04 nA. This step is also formed similarly at $V_{bias} = 0.4, 0.8$, and 1.2 mV. The second lowest step is formed between $V_{gate} = 115$ and 132 mV, and the third (highest) step is formed between $V_{gate} = 118$ and 128 mV. The supercurrents are additive. We predict that if the bias voltage is raised, then there should be more steps.

The center of mass (CM) of pairons move as bosons and their currents can be generated without activation. There are ±pairons, each having charge $\pm 2e$. The energy of the moving pairon is given by

$$\varepsilon_j = \frac{2}{\pi} v_F^{(j)} p , \tag{8.15}$$

where $v_F^{(j)}$, ($j = 1$ (electron), 2 (hole)), is the Fermi speed, which will be derived later in Section 8.4. Since the momentum (magnitude) p_n, see (8.9), is extremely small, the pairons in superconducting channels have negligible energies.

If the SWNT is unrolled, then we have a graphene sheet, which can be superconducting at a finite temperature. We first study the conductivity of graphene in Section 8.2, starting with the honeycomb lattice structure and introducing "electrons" and "holes" based on the Cartesian unit cell as distinct from the Wigner–Seitz (WS) unit cell (rhombus). Phonons are generated based on the same Cartesian unit cell. In Section 8.3 we construct a Hamiltonian suitable for the formation of the Cooper pairs. A linear energy-momentum (or dispersion) relation is derived in Section 8.4. The pairons moving with a linear dispersion relation in two dimensions undergo a Bose–Einstein condensation (BEC), which is shown in Section 8.5. Superconductivity in metallic SWNTs is discussed in Section 8.6. A summary is given in the last section.

8.2 Graphene

Before dealing with a metallic SWNT, let us consider graphene, in which carbon ions (C^+) occupy a two-dimensional (2D) honeycomb crystal lattice. The normal carriers in the transport of electrical charge are "electrons" and "holes." The "electron" ("hole") is a *quasielectron* that has an energy higher (lower) than the Fermi energy ε_F and those electrons (holes) are excited on the positive (negative) side of the Fermi surface with the convention that the positive normal vector at the surface points in the energy-increasing direction. The applied gate voltage can control the carrier charge type, $j = 1, 2$, electron (1) or hole (2), *and* the number density n_j. The "electron" ("hole") wavepacket is assumed to have a charge $-e$ $(+e)$ and has the size of a unit hexagon formed by positively charged carbon ions C^+. The positively charged "hole" tends to stay away from the C^+ ions. The center of mass of the "hole" (wave packet) is at the center of the hexagon and hence its charge is concentrated at the center of the hexagon. It moves easily with a small effective mass along the directions $\langle 100c\text{-axis}\rangle \equiv \langle 100\rangle$, where we used the conventional Miller indices for the hexagonal lattice with the omission of the c-axis index. The electron has a negative charge ($-e$) and a unit hexagon size. The negatively charged electron tends to stay close to the C^+ ions and the charge distribution is more concentrated near the C^+ hexagon. The CM of the "electrons" (wavepacket) is also at the center of the hexagon. It moves easily with a small effective mass along the directions $\langle 110\rangle$, see Figure 8.4 and below.

For a description of the electron motion in terms of the mass tensor, it is convenient to introduce Cartesian coordinates, which do not necessarily match the crystal's natural (triangular) axes. The "electron" (wavepacket) may move up or down in $\langle 110\rangle$ to the neighboring hexagon sites passing over one C^+. The positively charged C^+ acts as a welcoming (favorable) potential valley for the negatively charged "electron", while the same C^+ acts as a hindering potential hill for the positively charged "hole". The "hole", however, can move on a series of vacant sites

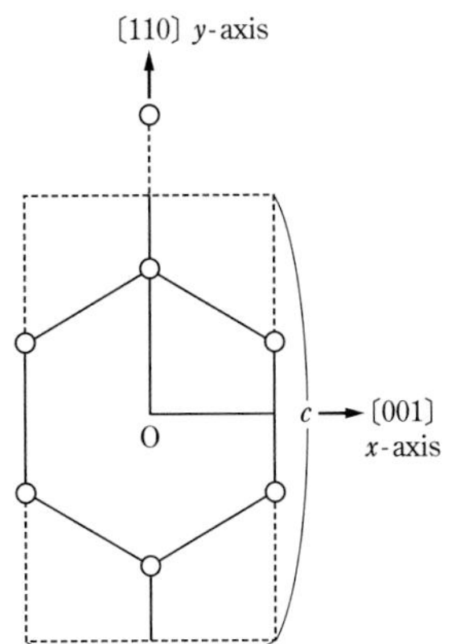

Figure 8.4 A rectangular (dotted line) unit cell of graphene. An open circle (○) indicates the positive ions, C^+.

in [001], each surrounded by six C^+, thus never meeting the hindering potential hills. Thus, the easy channel directions for the electrons and holes are $\langle 110 \rangle$ and $\langle 100 \rangle$, respectively.

We may choose a unit cell with side-length pair (b, c) as shown in Figure 8.4. Then the Brillouin zone in the k-space is unique: a rectangle with side lengths $(2\pi/b, 2\pi/c)$. We note that the lattice has inversion (mirror) symmetry with respect to the x- and y-axis.

Let us consider the system (graphene) at 0 K. If we put an electron in the crystal, then the electron should occupy the center O of the Brillouin zone, where the lowest energy lies. Additional electrons occupy the points neighboring O in consideration of Pauli's exclusion principle. The electron distribution is *lattice periodic* over the entire crystal in accordance with the Bloch theorem.

Graphene is a quadrivalent metal. The first few low-lying bands are completely filled. The uppermost partially filled bands are important for discussion of the transport properties. We consider such a band. The Fermi surface, which defines the boundary between the filled and unfilled k-spaces (area) is not a circle since the x-y symmetry is broken. The electron effective mass is lighter in the direction [110] than perpendicular to it. Hence, the electron motion is intrinsically anisotropic. If the electron number is raised by the gate voltage, then the Fermi surface should grow more quickly in this direction with increasing number of electrons. Because of the inversion symmetry of the crystal the Fermi surface must approach perpendicular to the Brillouin boundary. As the gate voltage changed to the charge-neutral point, the Fermi surface should go through a "neck" configuration, where the density of states rapidly grows on both sides of the voltage, generating high densities of "electrons" and "holes." Experiments [12] indicate that (a) both electrons and holes are excited in graphene, (b) at zero gate voltage, the electrons are dominant, and (c) the resistivity ρ exhibits a sharp maximum at the electron density $n_1 = 2 \times 10^{11}\ \text{cm}^2$. Feature (b) should arise as follows. The negatively charged electron is close to the positive ions C^+ and the hole is farther away from C^+. Hence, the gain in the Coulomb interaction is greater for the "electron." That is, the "electron" is more easily activated. Thus, the "electrons" are the majority carriers at zero gate voltage as observed. Feature (c) is related to the fact that the conductivity

$$\sigma \equiv \rho^{-1} = \frac{ne^2\tau}{m^*}, \tag{8.16}$$

where n is the carrier density and τ is the relaxation time, must decrease since the effective mass m^* shoots up to ∞ in the small neck limit.

We note that the "neck" Fermi surface was observed in copper, where densities of "electrons" and "holes" are high [13]. Necks in the Fermi surface are known to cause the Hall (Seebeck) coefficient to be negative (positive) [14]. It is interesting to see if graphene shows the same properties. The same easy channels in which the electron runs with a small mass, may be assumed for other hexagonal directions, [011] and [101]. The currents run in three channels $\langle 110 \rangle \equiv [110]$, [011], and [101] and thus the system as a whole does not show anisotropy in transport properties.

The current band theory based on the WS cell model predicts a gapless semiconductor and cannot explain this fact. The WS model is suited for the study of the ground state energy of a crystal. To treat the electron dynamics for a noncubic lattice such as a honeycomb lattice, we must introduce a Cartesian unit cell and use a mass tensor. We present a new theoretical model for electron dynamics. In this model the electron (hole) has a size of a unit carbon hexagon.

We have seen that the "electron" and "hole" have different internal charge distributions, and they are not point particles. Hence, they have different effective masses m_1 and m_2, which are different from the gravitational mass $m = 9.11 \times 10^{-28}$ g. Which carriers are easier to activate or excite? This question can be answered without considering channeling. The electron is near the positive ions and the hole is farther away from the positive ions. Hence, the gain in the Coulomb interaction is greater for the electron. That is, the electrons are more easily activated (or excited). The electrons move in the welcoming potential well channels while the holes do not. This fact also leads to the smaller activation energy for the electrons. We may represent the activation energy difference by [15]

$$\varepsilon_1 < \varepsilon_2 . \tag{8.17}$$

The thermally activated (or excited) electron densities are given by

$$n_j(T) = n_j \exp\left(\frac{-\varepsilon_j}{k_B T}\right) , \tag{8.18}$$

where $j = 1$ and 2 represent the electron and hole, respectively. The prefactor n_j is the density at the high-temperature limit. The "electron" can move easily with a smaller effective mass in the direction $\langle 110 \rangle$ rather than perpendicular to it as we see presently.

8.3 The Full Hamiltonian

Fujita and his group developed a quantum statistical theory of superconductivity in a series of papers [16–20]. We present this theory (see Section 8.4) here again in the present section.

In the ground state there are *no* currents for any system. To describe a supercurrent, we must introduce *moving pairons*, that is, pairons with finite CM momenta. Creation operators for "electron" (1) and "hole" (2) pairons are defined by

$$B_{12}^{(1)\dagger} \equiv B_{k_1 \uparrow k_2 \downarrow}^{(1)\dagger} \equiv c_1^{(1)\dagger} c_2^{(1)\dagger} , \quad B_{34}^{(2)} = c_4^{(2)} c_3^{(2)} . \tag{8.19}$$

We calculate the commutators among B and $B^\dagger$, and obtain

$$\left[B_{12}^{(j)}, B_{34}^{(j)}\right] = 0 , \quad \left[B_{12}^{(j)}\right]^2 = 0 , \tag{8.20}$$

$$\left[B_{12}^{(j)}, B_{34}^{(j)\dagger}\right] = \begin{cases} 1 - n_1^{(j)} - n_2^{(j)} & \text{if } \boldsymbol{k}_1 = \boldsymbol{k}_3 \text{ and } \boldsymbol{k}_2 = \boldsymbol{k}_4 \\ c_2^{(j)} c_4^{(j)\dagger} & \text{if } \boldsymbol{k}_1 = \boldsymbol{k}_3 \text{ and } \boldsymbol{k}_2 \neq \boldsymbol{k}_4 \\ c_1^{(j)} c_3^{(j)\dagger} & \text{if } \boldsymbol{k}_1 \neq \boldsymbol{k}_3 \text{ and } \boldsymbol{k}_2 = \boldsymbol{k}_4 \\ 0 & \text{otherwise} . \end{cases} \tag{8.21}$$

Pairon operators of different types j always commute:

$$\left[B^{(j)}, B^{(i)}\right] = 0 \quad \text{if} \quad j \neq i , \quad \text{and} \tag{8.22}$$

$$n_1^{(j)} \equiv c_{\boldsymbol{k}_1\uparrow}^{(j)\dagger} c_{\boldsymbol{k}_1\uparrow}^{(j)} , \quad n_2^{(j)} \equiv c_{\boldsymbol{k}_2\downarrow}^{(j)\dagger} c_{\boldsymbol{k}_2\downarrow}^{(j)} \tag{8.23}$$

represent the number operators for "electrons" ($j = 1$) and "holes" ($j = 2$).

Let us now introduce the relative and net momenta $(\boldsymbol{k}, \boldsymbol{q})$ such that

$$\boldsymbol{k} \equiv \frac{1}{2}(\boldsymbol{k}_1 - \boldsymbol{k}_2) , \quad \boldsymbol{q} \equiv \boldsymbol{k}_1 + \boldsymbol{k}_2 ; \quad \boldsymbol{k}_1 = \boldsymbol{k} + \frac{1}{2}\boldsymbol{q} , \quad \boldsymbol{k}_2 = -\boldsymbol{k} + \frac{1}{2}\boldsymbol{q} . \tag{8.24}$$

Alternatively we can represent pairon annihilation operators by

$$B_{\boldsymbol{kq}}^{\prime(1)} \equiv B_{\boldsymbol{k}_1\uparrow\boldsymbol{k}_2\downarrow}^{(1)} \equiv c_{-\boldsymbol{k}+\boldsymbol{q}/2\downarrow}^{(1)} c_{\boldsymbol{k}+\boldsymbol{q}/2\uparrow}^{(1)} , \quad B_{\boldsymbol{kq}}^{\prime(2)} = c_{\boldsymbol{k}+\boldsymbol{q}/2\uparrow}^{(2)} c_{-\boldsymbol{k}+\boldsymbol{q}/2\downarrow}^{(2)} . \tag{8.25}$$

The prime on B will be dropped hereafter. In the $\boldsymbol{k}$-$\boldsymbol{q}$ representation the commutation relations are re-expressed as

$$\left[B_{\boldsymbol{kq}}^{(j)}, B_{\boldsymbol{k}'\boldsymbol{q}'}^{(i)}\right] = 0 , \quad \left[B_{\boldsymbol{kq}}^{(j)}\right]^2 = 0 , \tag{8.26}$$

$$\left[B_{\boldsymbol{kq}}^{(j)}, B_{\boldsymbol{k}'\boldsymbol{q}'}^{(i)\dagger}\right] = \begin{cases} \left(1 - n_{\boldsymbol{k}+\boldsymbol{q}/2\uparrow} - n_{-\boldsymbol{k}+\boldsymbol{q}/2\downarrow}\right) \delta_{ji} & \text{if } \boldsymbol{k} = \boldsymbol{k}' \text{ and } \boldsymbol{q} = \boldsymbol{q}' \\ c_{-\boldsymbol{k}+\boldsymbol{q}/2\downarrow}^{(j)} c_{-\boldsymbol{k}'+\boldsymbol{q}'/2\downarrow}^{(j)\dagger} \delta_{ji} & \text{if } \boldsymbol{k} + \frac{\boldsymbol{q}}{2} = \boldsymbol{k}' + \frac{\boldsymbol{q}'}{2} \\ & \text{and } -\boldsymbol{k} + \frac{\boldsymbol{q}}{2} \neq -\boldsymbol{k}' + \frac{\boldsymbol{q}'}{2} \\ c_{\boldsymbol{k}+\boldsymbol{q}/2\uparrow}^{(j)} c_{\boldsymbol{k}'+\boldsymbol{q}'/2\uparrow}^{(j)\dagger} \delta_{ji} & \text{if } \boldsymbol{k} + \frac{\boldsymbol{q}}{2} \neq \boldsymbol{k}' + \frac{\boldsymbol{q}'}{2} \\ & \text{and } -\boldsymbol{k} + \frac{\boldsymbol{q}}{2} = -\boldsymbol{k}' + \frac{\boldsymbol{q}'}{2} \\ 0 & \text{otherwise} . \end{cases} \tag{8.27}$$

Using the new notation, we can rewrite the generalized Bardeen–Cooper–Schrieffer (BCS) Hamiltonian (7.23) as[1])

$$\begin{aligned} \mathcal{H} = & \sum_{\boldsymbol{k},s} \varepsilon_{\boldsymbol{k}}^{(1)} n_{\boldsymbol{k},s}^{(1)} + \sum_{\boldsymbol{k},s} \varepsilon_{\boldsymbol{k}}^{(2)} n_{\boldsymbol{k},s}^{(2)} \\ & - \sum_{\boldsymbol{k}}{}' \sum_{\boldsymbol{q}}{}' \sum_{\boldsymbol{k}'}{}' v_0 \left[B_{\boldsymbol{kq}}^{(1)\dagger} B_{\boldsymbol{k}'\boldsymbol{q}}^{(1)} + B_{\boldsymbol{kq}}^{(1)\dagger} B_{\boldsymbol{k}'\boldsymbol{q}}^{(2)\dagger} + B_{\boldsymbol{kq}}^{(2)} B_{\boldsymbol{k}'\boldsymbol{q}}^{(1)} + B_{\boldsymbol{kq}}^{(2)} B_{\boldsymbol{k}'\boldsymbol{q}}^{(2)\dagger} \right] , \end{aligned} \tag{8.28}$$

1) Here a spherical (circular) Fermi surface is assumed for a 3D (2D) free electron system. Note that (8.28) is reduced to the original BCS Hamiltonian for $\boldsymbol{q} = 0$ (see [23, Equation (24)]).

where v_0 is the pairing interaction strengths per volume among and between "electron" (1) and "hole" (2) pairons. This is the full Hamiltonian for the system of graphene, which can describe moving pairons as well as stationary pairons. Here, the prime on the summations indicates the restriction arising from the phonon exchange, see below.

8.4 Moving Pairons

The phonon-exchange attraction is in action for any pair of electrons near the Fermi surface. In general the bound pair has a net momentum, and hence, it moves. Such a pair is called a *moving pairon*. The energy w_q of a moving pairon for the 2D case can be obtained (see Section 7.6) from

$$w_q a(\boldsymbol{k}, \boldsymbol{q}) = \left[\varepsilon\left(\left|\boldsymbol{k} + \frac{\boldsymbol{q}}{2}\right|\right) + \varepsilon\left(\left|-\boldsymbol{k} + \frac{\boldsymbol{q}}{2}\right|\right)\right] a(\boldsymbol{k}, \boldsymbol{q}) - \frac{v_0}{(2\pi\hbar)^2} \int' \mathrm{d}^2 k' a(\boldsymbol{k}', \boldsymbol{q}) , \tag{8.29}$$

which is *Cooper's equation*, Equation (1) of his 1956 Physical Review Letter [21]. The prime on the k'-integral means the restriction on the integration domain arising from the phonon-exchange attraction, see below. We note that the net momentum $\boldsymbol{q}$ is a constant of motion, which arises from the fact that the phonon exchange is an internal process, and hence cannot change the net momentum. The *pair wavefunctions* $a(\boldsymbol{k}, \boldsymbol{q})$ are coupled with respect to the other variable $\boldsymbol{k}$, meaning that the exact (or energy-eigenstate) pairon wavefunctions are superpositions of the pair wavefunctions $a(\boldsymbol{k}, \boldsymbol{q})$.

Equation (8.29) can be solved as follows. We assume that the energy w_q is negative:

$$w_q < 0 . \tag{8.30}$$

Then,

$$\varepsilon\left(\left|\boldsymbol{k} + \frac{\boldsymbol{q}}{2}\right|\right) + \varepsilon\left(\left|-\boldsymbol{k} + \frac{\boldsymbol{q}}{2}\right|\right) - w_q > 0 .$$

Rearranging the terms in (8.29) and dividing by

$$\varepsilon\left(\left|\boldsymbol{k} + \frac{\boldsymbol{q}}{2}\right|\right) + \varepsilon\left(\left|-\boldsymbol{k} + \frac{\boldsymbol{q}}{2}\right|\right) - w_q ,$$

we obtain

$$a(\boldsymbol{k}, \boldsymbol{q}) = \left[\varepsilon\left(\left|\boldsymbol{k} + \frac{\boldsymbol{q}}{2}\right|\right) + \varepsilon\left(\left|-\boldsymbol{k} + \frac{\boldsymbol{q}}{2}\right|\right) - w_q\right]^{-1} C(\boldsymbol{q}) , \tag{8.31}$$

where

$$C(\boldsymbol{q}) \equiv \frac{v_0}{(2\pi\hbar)^2} \int' \mathrm{d}^2 k' a(\boldsymbol{k}', \boldsymbol{q}) , \tag{8.32}$$

which is k-independent. Introducing (8.31) in (8.29), and dropping the common factor $C(\boldsymbol{q})$, we obtain

$$1 = \frac{v_0}{(2\pi\hbar)^2}\int' \mathrm{d}^2k \left[\varepsilon\left(\left|\boldsymbol{k}+\frac{\boldsymbol{q}}{2}\right|\right)+\varepsilon\left(\left|-\boldsymbol{k}+\frac{\boldsymbol{q}}{2}\right|\right)+|w_q|\right]^{-1} . \tag{8.33}$$

We now assume a free electron model in 3D. The Fermi surface is a sphere of the radius (momentum)

$$k_\mathrm{F} \equiv (2m_1\varepsilon_\mathrm{F})^{1/2} , \tag{8.34}$$

where m_1 represents the effective mass of an electron. The energy $\varepsilon(|\boldsymbol{k}|)$ is given by

$$\varepsilon(|\boldsymbol{k}|) \equiv \varepsilon_k = \frac{|\boldsymbol{k}|^2-|\boldsymbol{k}_\mathrm{F}|^2}{2m_1} . \tag{8.35}$$

The prime on the k-integral in (8.33) means the restriction:

$$0 < \varepsilon\left(\left|\boldsymbol{k}+\frac{\boldsymbol{q}}{2}\right|\right) , \quad \varepsilon\left(\left|-\boldsymbol{k}+\frac{\boldsymbol{q}}{2}\right|\right) < \hbar\omega_\mathrm{D} . \tag{8.36}$$

We may choose the polar axis along $\boldsymbol{q}$ as shown in Figure 8.5. The integration with respect to the azimuthal angle simply yields the factor 2π. The k-integral can then be expressed by (Problem 8.4.1)

$$\frac{(2\pi\hbar)^3}{v_0} = 4\pi\int_0^{\pi/2}\mathrm{d}\theta\,\sin\theta\int_{k_\mathrm{F}+\frac{1}{2}q\cos\theta}^{k_\mathrm{F}+k_\mathrm{D}-\frac{1}{2}q\cos\theta}\frac{k^2\mathrm{d}k}{|w_q|+2\varepsilon_k+(4m_1)^{-1}q^2} , \tag{8.37}$$

where k_D is given by

$$k_\mathrm{D} \equiv m_1\hbar\omega_\mathrm{D}k_\mathrm{F}^{-1} . \tag{8.38}$$

After performing the integration and taking the small-q and small-$(k_\mathrm{D}/k_\mathrm{F})$ limits, we obtain (Problem 8.4.2)

$$w_q = w_0 + \frac{v_\mathrm{F}}{2}q , \tag{8.39}$$

where w_0 is given by (Problem 8.4.3)

$$w_0 = \frac{-2\hbar\omega_\mathrm{D}}{\exp(2/(v_0\mathcal{D}(0)))-1} . \tag{8.40}$$

As expected, the zero-momentum pairon has the lowest energy. The excitation energy is continuous with *no* energy gap. Equation (8.39) was first obtained by Cooper (unpublished) and it is recorded in Schrieffer's book [22, Equation (2.15)]. The energy w_q increases *linearly* with momentum $q(=|\boldsymbol{q}|)$ for small q. This behavior arises from the fact that the density of states is strongly reduced with the

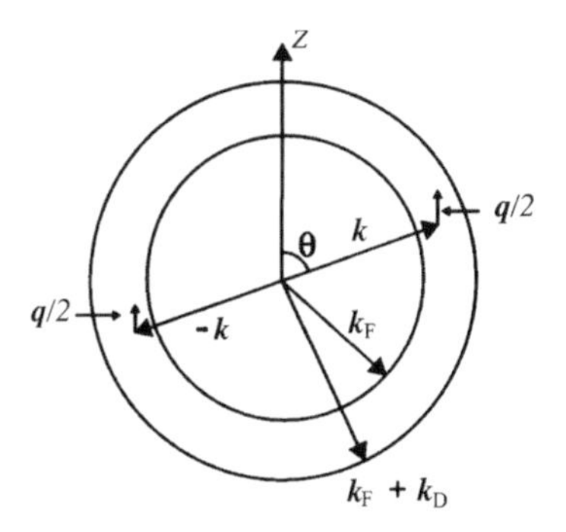

Figure 8.5 The range of the interaction variables (k, θ) is limited to a circular shell of thickness k_D.

increasing momentum q and dominates the q^2 increase of the kinetic energy. The linear dispersion relation means that a *pairon moves like a massless particle with a common speed* $v_F/2$. This relation plays a vital role in the Bose–Einstein condensation (BEC) of pairons (see next section).

Such a linear energy-momentum relation is valid for pairons moving in any dimension. However, the coefficients slightly depend on the dimensions; in fact

$$w_q = w_0 + cq\,, \tag{8.41}$$

where $c = 1/2 v_F$ and $2/\pi v_F$ for three and two dimensions, respectively.

Problem 8.4.1. Verify (8.37).

Problem 8.4.2. Derive an energy-momentum relation (8.39) for the 3D case with the assumption of a Fermi sphere. Hint: Use the diagram in Figure 8.5.

Problem 8.4.3. Show that w_0 for the 3D case is given by (8.40).

8.5 The Bose–Einstein Condensation of Pairons

In Section 8.3, we saw that the pair operators $(B, B^\dagger)$ appearing in the full Hamiltonian $\mathcal{H}$ in (8.28) satisfy rather complicated commutator relations (8.26) and (8.27). In particular part of (8.26)

$$\left[B_{k0}^\dagger\right]^2 \equiv \left[b_k^\dagger\right]^2 = \left[c_{-k\uparrow}^\dagger c_{k\downarrow}^\dagger\right]^2 = 0 \tag{8.42}$$

reflects the fermionic properties of the constituting electrons. Here, $B_{k0}^\dagger \equiv b_k^\dagger$ represents creation operator for *zero-momentum* pairons. Bardeen, Cooper, and Schrieffer [23] studied the ground state of a superconductor, starting with the reduced Hamiltonian $\mathcal{H}_0$, which is obtained from the Hamiltonian $\mathcal{H}$ in (8.28) by retaining the zero-momentum pairons with $\boldsymbol{q} = 0$ written in terms of b by letting $B_{k0}^{(j)} = b_k^{(j)}$, that is,

$$\mathcal{H}_0 = \sum_k 2\varepsilon_k^{(1)} b_k^{(1)\dagger} b_k^{(1)} + \sum_k 2\varepsilon_k^{(2)} b_k^{(2)\dagger} b_k^{(2)}$$
$$- \sum_k{}' \sum_{k'}{}' v_0 \left[b_{k'}^{(1)\dagger} b_k^{(1)} + b_{k'}^{(1)\dagger} b_k^{(2)\dagger} + b_{k'}^{(2)} b_k^{(1)} + b_{k'}^{(2)} b_k^{(2)\dagger} \right] . \tag{8.43}$$

Here, we expressed the "electron" and "hole" kinetic energies in terms of pairon operators. The reduced Hamiltonian $\mathcal{H}_0$ is bilinear in pairon operators $(b, b^\dagger)$, and can be diagonalized exactly. Bardeen, Cooper, and Schrieffer obtained the ground state energy E_0 as

$$E_0 = \hbar\omega_{\mathrm{D}} \mathcal{D}(0) w_0 , \tag{8.44}$$

where $\mathcal{D}(0)$ is the density of states at the Fermi energy. The w_0 is the ground state energy of the pairon, see (8.40). Equation (8.44) means simply that the ground state energy equals the numbers of pairons times the ground state energy of the pairon. In the ground state of any system there is no current. To describe the supercurrent we must introduce moving pairons, that is, nonzero-momentum pairons. We first show that the center of masses of the pairons move as bosons. That is, the number operator of pairons having net momentum q

$$n_q \equiv \sum_k n_{kq} = \sum_k B_{kq}^\dagger B_{kq} \tag{8.45}$$

have the eigenvalues

$$n'_q = 0, 1, 2, \cdots . \tag{8.46}$$

The number operator for the pairons in the state $(\boldsymbol{k}, \boldsymbol{q})$ is

$$n_{kq} \equiv B_{kq}^\dagger B_{kq} = c_{k+q/2}^\dagger c_{-k+q/2}^\dagger c_{-k+q/2} c_{k+q/2} , \tag{8.47}$$

where we omitted the spin indices. Its eigenvalues are limited to *zero* or *one*:

$$n'_{kq} = 0 \text{ or } 1 . \tag{8.48}$$

To explicitly see (8.46), we introduce

$$B_q \equiv \sum_k B_{kq} , \tag{8.49}$$

and obtain

$$[B_q, n_q] = \sum_k \left(1 - n_{k+\frac{1}{2}q} - n_{-k+\frac{1}{2}q}\right) B_{kq} = B_q , \quad \left[n_q, B_q^\dagger\right] = B_q^\dagger . \tag{8.50}$$

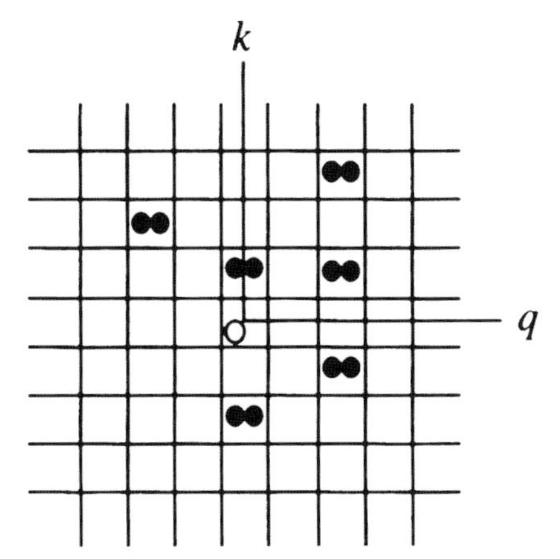

Figure 8.6 The number representation of many electron pairs in the (k, q) space.

Although the occupation number n_q is not connected with B_q as $n_q \neq B_q^\dagger B_q$, the eigenvalues n'_q of n_q satisfying (8.50) can be shown straightforwardly to yield [24] $n'_q = 0, 1, 2, \ldots$ with the eigenstates

$$|0\rangle\ , \quad |1\rangle = B_q^\dagger|0\rangle\ , \quad |2\rangle = B_q^\dagger B_q^\dagger|0\rangle\ , \cdots\ . \tag{8.51}$$

This is important. We illustrate it by taking a one-dimensional motion. The pairon occupation-number states may be represented by drawing quantum cells in the (k, q) space. From (8.48), the number n'_{kq} is limited to 0 or 1, see Figure 8.6. The number of pairons characterized by the net momentum q only, n'_q, is the sum of the numbers of pairs at column q, and clearly it is *zero* or a *positive integer*.

In summary, pairons with both $\boldsymbol{k}$ and $\boldsymbol{q}$ specified are subject to the Pauli exclusion principle, see (8.48). Yet, the occupation numbers n'_q of pairons having a CM momentum $\boldsymbol{q}$ are 0, 1, 2, ...

The most important signature of many bosons is the *Bose–Einstein Condensation*. Earlier we showed that the pairon moves with the linear dispersion relation, see (8.39):

$$w_p = w_0 + \frac{2}{\pi} v_F p \equiv w_0 + cp\ , \tag{8.52}$$

where we designated the pairon net momentum (magnitude) by the more familiar p rather than q.

Let us consider a 2D system of free bosons having a linear dispersion relation: $\varepsilon = cp$, where $c = 2/\pi v_F$. The number of bosons, N, and the Bose distribution function,

$$f_B(\varepsilon; \beta, \mu) \equiv \frac{1}{e^{\beta(\varepsilon-\mu)} - 1} \equiv f_B(\varepsilon) \quad (\geq 0)\ , \tag{8.53}$$

are related by

$$N = \sum_p f_B(\varepsilon_p; \beta, \mu) = N_0 + \sum_{\substack{p \\ \varepsilon_p > 0}}{}' f_B(\varepsilon_p)\ , \tag{8.54}$$

where μ is the chemical potential and

$$N_0 \equiv \left(e^{-\beta\mu} - 1\right)^{-1} \tag{8.55}$$

is the number of *zero*-momentum bosons. Note that $\alpha \equiv \beta\mu$ with $\beta \equiv (k_B T)^{-1}$. The prime on the summation in (8.54) indicates the omission of the zero-momentum state. For notational convenience we write

$$\varepsilon = cp = \frac{2}{\pi} v_F p \quad (> 0) \; . \tag{8.56}$$

We divide (8.54) by the normalization area L^2, and take the bulk limit:

$$N \to \infty \, , \quad L \to \infty \quad \text{while} \quad N L^{-2} \equiv n \; . \tag{8.57}$$

We then obtain

$$n - n_0 \equiv \frac{1}{(2\pi\hbar)^2} \int d^2 p \, f_B(\varepsilon) \, , \tag{8.58}$$

where $n_0 \equiv N_0/L^2$ is the number density of zero-momentum bosons and n the total boson density. After performing the angular integration and changing integration variables, we obtain from (8.58)

$$2\pi\hbar^2 c^2 \beta^2 (n - n_0) = \int_0^\infty dx \frac{x}{\lambda^{-1} e^x - 1} \, , \quad x \equiv \beta\varepsilon \; . \tag{8.59}$$

Here λ is the fugacity defined by

$$\lambda \equiv e^{\beta\mu} \, , \quad (< 1) \; . \tag{8.60}$$

We note that the fugacity λ is less than unity for all temperatures. After expanding the integrand in (8.59) in powers of $\lambda e^{-x} (< 1)$, and carrying out the x-integration, we obtain (Problem 8.5.1)

$$n_x \equiv n - n_0 = \frac{k_B^2 T^2 \phi_2(\lambda)}{2\pi\hbar^2 c^2} \, , \tag{8.61}$$

where $\phi_m(\lambda)$ is given by

$$\phi_m(\lambda) \equiv \sum_{k=1}^{\infty} \frac{\lambda^k}{k^m} \, , \quad 0 \le \lambda \le 1 \; . \tag{8.62}$$

We need $\phi_2(\lambda)$ here, but we introduced ϕ_m for later reference. Equation (8.61) gives a relation among λ, n, and T.

The function $\phi_2(\lambda)$ monotonically increases from zero to the maximum value $\phi_2(1) = 1.645$ as λ is raised from zero to one. In the low-temperature limit, $\lambda = 1$, $\phi_2(\lambda) = \phi_2(1) = 1.645$, and the density of excited bosons, n_x, varies like T^2 as seen in (8.61). This temperature behavior of n_x persists as long as the rhs of (8.61) is smaller than n; the *critical temperature* T_c occurs at $n = k_B^2 T_c^2 \phi_2(1)/2\pi\hbar^2 c^2$. Solving this, we obtain

$$k_B T_c = 1.954\hbar c n^{1/2} \ (= 1.24\hbar v_F n^{1/2}) \ . \tag{8.63}$$

If the temperature is raised beyond T_c, the density of zero-momentum bosons, n_0, becomes vanishingly small, and the fugacity λ can be determined from (Problem 8.5.2)

$$n = \frac{k_B T^2 \phi_2(\lambda)}{2\pi\hbar^2 c^2} \ , \qquad T > T_c \ . \tag{8.64}$$

In summary, the fugacity λ is equal to unity in the condensed region: $T < T_c$, and it becomes smaller than unity for $T > T_c$, where its value is determined from (8.64). Formula (8.63) for the critical temperature T_c is distinct from the famous BCS formula

$$3.53 k_B T_c = 2\Delta_0 \ , \tag{8.65}$$

where Δ_0 is the zero-temperature electron energy gap in the weak coupling limit. The electron energy gap $\Delta(T)$ and the pairon ground state energy w_0 depends on the phonon-exchange coupling energy parameter v_0, which appears in the starting Hamiltonian $\mathcal{H}$ in (8.28). The ground state energy w_0 is negative. Hence, this w_0 cannot be obtained by the perturbation theory, and hence the connection between w_0 and v_0 is very complicated. This makes it difficult to discuss the critical temperature T_c based on the BCS formula (8.65).

Unlike the BCS formula, formula (8.63) is directly connected with the measurable quantities: the pairon density n_0 and the Fermi speed v_F.

We emphasize here that both formulas (8.63) and (8.65) can be derived, starting with the Hamiltonian (8.28) and following statistical mechanical calculations, see more details in reference [25].

Problem 8.5.1. Verify (8.61).

Problem 8.5.2. Derive (8.64).

8.6 Superconductivity in Metallic SWNTs

The pitch in a metallic SWNT contains an *irrational* number of hexagons and there is *no* lattice k-vector along the tube. The unrolled plane may extend along the tube axis (y-axis) and also along the circumference (x-axis) just like a graphene sheet. But there is a significant difference. The true graphene sheet has an inversion symmetry with respect to the plane. The rolled SWNT has an inside and outside. If "holes" are excited on the inside, then they can move along the tube length, providing an extra channel for the transport of charge. "Electrons," which are negatively charged, cannot move in a straight line along the tube since they are attracted by the positively charged wall. We may assume a 1D k-vector along the tube length.

The contribution of the "hole" channel is significant. Transport experiments by Moriyama *et al.* [9] show that the main charge carriers are "holes." In contrast, "electrons" are the normal charge carriers in graphene. Thermopower (Seebeck coefficient) measurements by Kong *et al.* [26] also provide evidence that the charge carriers are "hole"-like.

Superconductivity occurs in a 2D (or 3D) crystal having "electrons" and "holes". The unrolled configuration of a SWNT clearly indicates a possible 2D superconductor. Phonons and electrons share the same Brillouin zone. This affinity is important for the electron–phonon interaction. In fact phonon exchange binds electron pairs, called Cooper pairs (pairons). Since the phonon does not carry a charge, phonon exchange conserves the charge and can, and must, create (or annihilate) a pair of positive (+) and negative (−) pairons simultaneously. Hence, the numbers of ±pairons created are mutually equal. The −pairon, that is, the "electron" pair, moves with a higher speed $2/\pi v_1$ than the +pairon ("hole" pair) since

$$v_1 > v_2 \quad \text{in graphene.} \tag{8.66}$$

Hence, there should be a supercurrent if the pairons undergo a BEC. In the superconducting state ±pairons are condensed at the same momentum p_n. The supercurrent density, calculated by (charge) × (density) × (speed), is

$$j = -2en_0\frac{2}{\pi}(v_1 - v_2)\,, \tag{8.67}$$

where n_0 is the (−pairon) density.

We have shown earlier in Section 8.2 that the "electron" and "hole" have different charge distributions and have therefore different effective masses (m_1, m_2). Then, the Fermi speeds (v_1, v_2) are different, and the supercurrent density is finite as seen in (8.67). It is noteworthy that the supercurrent density j vanishes if we assume that the "hole" is an antielectron with the same mass. The fermion–antifermion symmetry must be broken for superconductivity to occur.

The superconducting state becomes unstable if enough magnetic field is applied. This is so because ± pairon motion (currents) is affected differently by the field. If the condensation (Meissner) energy is overcome by the magnetic field energy, the system reverts to the normal state. There is a critical field. As long as the binding energy is greater than the magnetic energy the system stays in a superconducting state by repelling the magnetic field, which is known as the Meissner effect.

The pairon moves with a linear dispersion relation, see (8.41). The BEC of the pairons occur in a 2D graphene plane with the critical temperature T_c, given by (8.63). We note that no BEC (and no superconductivity) occurs in 1D.

8.7
High-Field Transport in Metallic SWNTs

In 2000, Yao, Kane, and Dekker [27] reported high-field transport in metallic SWNT. In Figure 8.7, we reproduced their I–V curves, after [27, Figure 1]. At low fields (voltage $\sim$ 30 mV), the currents show temperature-dependent dips near the origin, exhibiting non-Ohmic behaviors while at high fields ($\sim$ 5 V) the resistance R versus the bias voltage V shows the relation:

$$R = R_0 + \frac{V}{I_0} \quad (\text{high } V) , \tag{8.68}$$

where R_0 and I_0 are constants. The authors discussed this low-field behavior in terms of a 1D Luttinger liquid model. Many experiments, however, indicate that electrical transport in SWNTs has a two-dimensional (2D) character [6]. In fact, the conductivity in individual nanotubes depends on the circumference and the pitch characterizing a space curve (2D). Hence, nanotube physics requires a 2D theory. In the present work, we present a unified microscopic theory of both low- and high-field conductivities. Carbon nanotubes were discovered by Iijima [6]. The

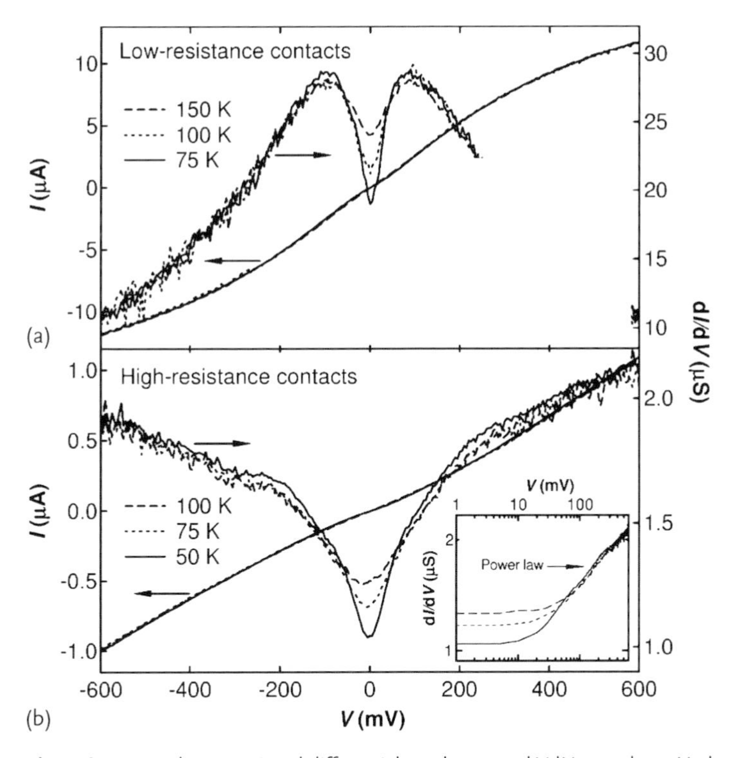

Figure 8.7 Typical current I and differential conductance dI/dV vs. voltage V obtained using (a) low-resistance contacts (LRC) and (b) high-resistance contacts (HRC). The inset in (b) plots dI/dV vs. V on a double-log scale for the HRC sample. After Yao *et al* [27].

important questions are how the electrons or other charged particles traverse the nanotubes and whether these particles are scattered by impurities and phonons or not. To answer these questions, we need the electron energy band structures. Wigner and Seitz [28] developed the WS cell model to study the ground state of a metal. Starting with a given lattice, they obtain a Brillouin zone in the k-space and construct a Fermi surface. This method has been successful for cubic crystals including the face-centered cubic (fcc), the body-centered cubic (bcc), diamond (dia) and zincblende lattices. If we apply the WS cell model to graphene, we then obtain a gapless semiconductor, which is *not* experimentally observed [6].

SWNTs can be produced by rolling graphene sheets into circular cylinders of about 1 nm diameter and microns (μm) in length [1, 2]. Electrical conduction in SWNTs depends on the circumference and pitch as shown in Figure 6.1, Chapter 6, and can be classified into two groups: *semiconducting* or *metallic* [6]. In our previous work [29], we have shown that this division into two groups arises as follows. A SWNT is likely to have an integral number of carbon hexagons around the circumference. If each pitch contains an integral number of hexagons, then the system is periodic along the tube axis, and "holes" (not "electrons") can move along the tube. Such a system is semiconducting and its electrical conductivity increases with temperature, and is characterized by an activation energy ε_3 [30]. The energy ε_3 has a distribution since both pitch and circumference have distributions. The pitch angle is not controlled in the fabrication process. There are numerous other cases where the pitch contains an irrational number of hexagons. In these cases, the system shows a metallic behavior which has been experimentally observed [9].

We primarily deal with metallic SWNTs in the present work. Before dealing with high-field transport, we briefly discuss low-field transport. Tans *et al.* [5] measured the electric currents in metallic SWNTs under bias and gate voltages. Their data from [5, Figure 2], are reproduced in Figure 8.1, where the currents versus the bias voltage are plotted at three gate voltages: A (88.2 mV), B (104.1 mV), and C (120.0 mV) (shown in Figure 8.1).

Significant features are listed in Section 8.1. Tan *et al.* [5] interpreted the flat currents near $V_{\text{bias}} = 0$ in Figure 8.1 in terms of a ballistic electron model [6]. We proposed a different interpretation of the data in Figure 8.1 based on the Cooper pair [21] (pairon) carrier model. Pairons move as bosons, and hence they are produced with no activation energy factor. All features (a)–(e) listed in Section 8.1 can be explained simply with the assumption that the nanotube wall is in a superconducting state (see Section 8.1).

In the currently prevailing theory [6], it is argued that the electron (fermion) motion becomes ballistic at a certain quantum condition. But all fermions are known to be subject to scattering. It is difficult to justify why the ballistic electron is not scattered by impurities and phonons, which naturally exist in nanotubes. Yao, Kane, and Dekker [27] emphasized the importance of phonon scattering effects in their analysis of their data in Figure 8.7. The Cooper pairs [21] in supercurrents, as is known, can run with no resistance (due to impurities and phonons). Clearly the experiments on the currents shown in Figure 8.7 are temperature-dependent, indicating the importance of the electron–phonon scattering effect. If the ballistic electron model is adopted, then the phonon scattering cannot be discussed within the model's framework. We must go beyond the ballistic electron model.

If the SWNT is unrolled, then we have a graphene sheet, which can be superconducting at a finite temperature. We studied, in Section 6.2, the conduction behavior of graphene, starting with the honeycomb lattice and introducing "electrons" and "holes" based on the orthogonal unit cell. Phonons are generated based on the same orthogonal unit cell. In Sections 4.3 and 4.4, we treated phonons and phonon-exchange attraction. In Section 8.3, we constructed a Hamiltonian suitable for the formation of the Cooper pairs and derived the linear dispersion relation for the center of mass motion of the pairons (see Section 8.4). The pairons moving with a linear dispersion relation undergo a BEC in 2D (see Section 8.5). In the next section we discuss the zero-bias anomaly (ZBA) observed in Figure 8.7.

8.8 Zero-Bias Anomaly

The unusual current dip at zero bias in Figure 8.7 is often called the *zero-bias anomaly*. This effect is clearly seen in the low-resistance contacts (LRC) sample. The differential conductance $\mathrm{d}I/\mathrm{d}V$ increases with increasing bias, reaching a maximum at $V \sim 100\,\mathrm{mV}$. With a further bias increase, $\mathrm{d}I/\mathrm{d}V$ drops dramatically. See Figure 8.7a. We will show that the ZBA arises from a breakdown of the superconducting state of the system.

With no bias, the nanotube's wall below $\sim 100\,\mathrm{K}$ is in a superconducting state. If a small bias is applied, then the system is charged, positively or negatively depending on the polarity of the external bias. The applied bias field will not affect the neutral supercurrent but can accelerate the charges at the outer side of the carbon wall. The resulting normal currents carried by conduction electrons are scattered by impurities and phonons. The phonon population changes with temperature, and hence the phonon scattering is temperature-dependent. The normal electric currents along the tube length generate circulating magnetic fields, which eventually destroy the supercurrent running in the wall at a high enough bias. Thus, the current I (μA) versus the voltage V (mV) is nonlinear near the origin because of the supercurrents running in the wall. The differential conductance $\mathrm{d}I/\mathrm{d}V$ is very small and nearly constant (superconducting) for $V < 10\,\mathrm{mV}$ in the high-resistance contacts (HRC) sample, see Figure 8.7b. We stress that if the ballistic electron model [6] is adopted, then the scattering by phonons cannot be discussed. The nonlinear I–V curves below 150 K mean that the carbon wall is superconducting. Thus, the clearly visible temperature effects for both LRC and HRC samples arise from the phonon scattering. We assumed that the system is superconducting below $\sim 150\,\mathrm{K}$. The ZBA arises only from the superconducting state. The superconducting critical temperature T_c must then be higher than 150 K. An experimental check of T_c is highly desirable.

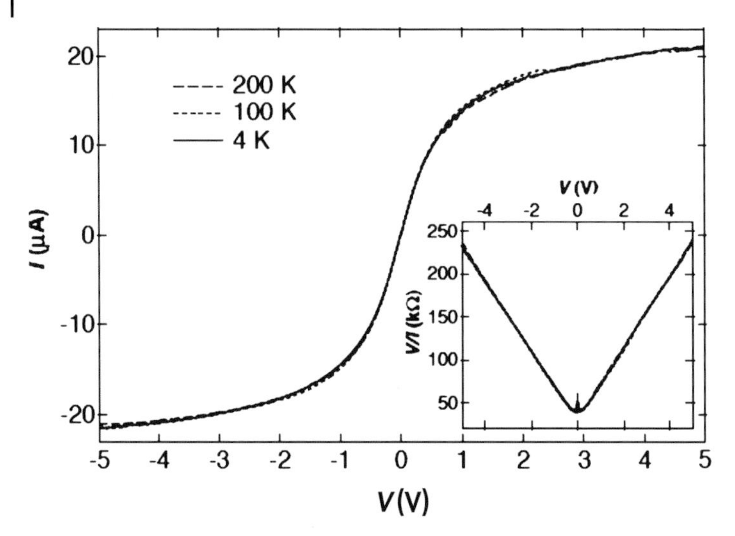

Figure 8.8 Large-bias $I-V$ characteristics at different temperatures using low-resistance contacts for a sample with an electrode spacing of 1 μm. The inset plots $R = V/I$ vs. V. After Yao, *et al.* [31].

8.9
Temperature Behavior and Current Saturation

Yao *et al.* [31] extend the $I-V$ measurements up to 5 V as shown in Figure 8.8. Strikingly, the $I-V$ curves at great bias measured at different temperatures between 4 K and room temperature overlap each other. This temperature behavior is consistent with our picture that the superconductivity state of the metallic SWNT continued throughout the temperature range measured. Thus, the superconductivity temperature T_c must be higher than room temperature.

From the shape of the $I-V$ curves in Figure 8.8, it is clear that the trend of decreasing conductances continues to high bias. Extrapolating the measured $I-V$ curves to higher voltage would lead to a current saturation, that is, a vanishing conductance. The current saturation may arise as follows. When the bias is raised from zero, the system will be charged with "holes." The resulting "hole" currents run along the outer side of the tube. There is an extra contribution to the current I. But these currents must saturate since the maximum "hole" states are limited by the Pauli exclusion principle. The maximum number of "holes" must be much smaller than the number of C's.

8.10
Summary

Various non-Ohmic behaviors observed in metallic SWNTs under bias and gate voltages were discussed based on the bosonic pairon model in this chapter. The ballistic electron model was avoided since this model cannot explain by itself why the ballistic fermionic electrons are not scattered by impurities and phonons. In our microscopic theory we start with a honeycomb lattice for graphene, establish a Fermi surface, identify the charge carriers, and discuss the electrical transport using kinetic theory.

We have uncovered a number of significant facts:

- "Electrons" and "holes" move with different masses in anisotropic environments in graphene (and graphite). In contrast "electrons" and "holes" in germanium (Ge) and silicon (Si) move in isotropic environments.
- For electron dynamics it is important to introduce a rectangular (non-WS) unit cell for the honeycomb lattice. Only then can we establish that "electrons" and "holes" move with different effective masses (m_1, m_2) and different activation energies $(\varepsilon_1, \varepsilon_2)$. Graphene is intrinsically anisotropic.
- A metallic (semiconducting) SWNT contains an irrational (integral) number of carbon hexagons in each pitch. The former case occurs more often.
- The unrolled configuration of the metallic SWNT is a 2D graphene sheet indefinitely extending over the tube length direction *and* the circumference direction. Electrical transport in the wall occurs in 2D (*not* in 1D).
- "Holes" (and not "electrons") can move inside the tube, and this "hole" channel is significant as observed in the thermopower experiments.
- Electrons and phonons move in the same k-space within the (first) Brillouin zone. This affinity makes the electron–phonon interaction important.
- The phonon exchange between electrons binds Cooper pairs (pairons). Positively and negatively charged $(\pm)$pairons are created (or annihilated) in pairs simultaneously. The number of $\pm$pairons are mutually equal at 0 K.
- The pairons move with the linear dispersion relation.
- The CM of the pairons move as bosons.
- The system of the pairons moving in 2D with the linear dispersion relation undergoes a BEC with the critical temperature T_c given in (8.63).
- A superconducting state is generated by the BEC of $\pm$pairons at the CM momentum $p_n = 2\pi\hbar n/L$. The integer n is small and the tube length L is macroscopic.
- The supercurrent density is given by (8.67). The current is finite only if the "electron" and the "hole" have different masses.
- The supercurrent is neutral, and hence it cannot be accelerated by a bias electric field.
- The supercurrent is destroyed if a strong magnetic field is applied. There is a critical field.
- A small magnetic field is repelled by the superconducting carbon tube.
- The superconducting state is destroyed by raising the temperature. It is important to find the critical temperature T_c above which the electrical conduction becomes normal (resistive).

References

1 Iijima, S. and Ishibashi, T. (1993) *Nature*, **363**, 603.
2 Bethune, D.S. Klang, C.H., Vries, M.S. de, Gorman, G., Savoy, R., Vazquez, J., and Beyers, R. (1993) *Nature*, **363**, 605.
3 Ebbesen, T.W., Lezec, H.J., Hiura, H., Bennett, J.W., Ghaemi, H.F., and Thio, T. (1996) *Nature*, **382**, 54.
4 Dai, H., Wong, E.W., and Lieber, C.M. (1996) *Science*, **272**, 523.
5 Tans, S.J., Devoret, M.H., Dai, H., Thess, A., Smalley, R.E., Geerligs, L.J., and Dekker, C. (1997) *Nature*, **386**, 474.
6 Saito, R., Dresselhaus, G., and Dresselhaus, M.S. (1998) *Physical Properties of Carbon Nanotubes*, Imperial College Press, London, pp. 35–39, pp. 139–144, pp. 155–156.
7 Fujita, S., Takato, Y., and Suzuki, A. (2011) *Mod. Phys. Lett. B*, **25**, 223.
8 Tans, S.J., Verschueren, A.R.M., and Dekker, C. (1998) *Nature*, **393**, 49.
9 Moriyama, S., Toratani, K., Tsuya, D., Suzuki, M., Aoyagi, Y., and Ishibashi, K. (2004) *Physica E*, **24**, 46.
10 Bachtold, A., Fuhrer, M.S., Plyasunov, S., Forero, M., Anderson, E.H., Zettl, A., and McEuen, P.L. (2000) *Phys. Rev. Lett.*, **84**, 6082.
11 Jans, S.J., Verschuerew, A.R.M., and Dekker, C. (1998) *Nature*, **393**, 49.
12 Zhang, Y., Tan, Y.W., Stormer, H.L. and Kim, P. (2005) *Nature*, **438**, 201.
13 Ashcroft, N.W. and Mermin, N.D. (1976) *Solid State Physics*, Saunders, Philadelphia, pp. 228–229, pp. 291–293.
14 Fujita, S., Ho, H.-C., and Okamura, Y. (2000) *Int. J. Mod. Phys. B* **14**, 2231.
15 Fujita, S. and Suzuki, A. (2010) *J. Appl. Phys.*, **107**, 013 711.
16 Fujita, S. (1991) *J. Superconduct.*, **4**, 297.
17 Fujita, S. (1992) *J. Superconduct.*, **5**, 83.
18 Fujita, S. and Watanabe, S. (1992) *J. Superconduct.*, **5**, 219.
19 Fujita, S. and Watanabe, S. (1993) *J. Superconduct.*, **6**, 75.
20 Fujita, S. and Godoy, S. (1993) *J. Superconduct.*, **6**, 373.
21 Cooper, L.N. (1956) *Phys. Rev.*, **104**, 1189.
22 Schrieffer, J.R. (1964) *Theory of Superconductivity*, Benjamin, New York, p. 33.
23 Bardeen, J., Cooper, L.N., and Schrieffer, J.R. (1957) *Phys. Rev.* **108**, 1175.
24 Dirac, P.A.M. (1958) *Principle of Quantum Mechanics*, 4th edn, Oxford University Press, London, p. 37, pp. 136–138, p. 211, pp. 253–257.
25 Fujita, S., Ito, K., and Godoy, S. (2009) *Quantum Theory of Conducting Matter: Superconductivity*, Springer, New York, pp. 79–81.
26 Kong, W.J., Lu, L., Zhu, H.W., Wei, B.Q., and Wu, D.H. (2005) *J. Phys.: Condens Matter*, **17**, 1923.
27 Yao, Z., Kane, C.L., and Dekker, C. (2000) *Phys. Rev. Lett.*, **84**, 2941.
28 Iijima, S. (1991) *Nature*, **354**, 56.
29 Wigner, E. and Seitz, F. (1933) *Phys. Rev.*, **43**, 804.
30 Fujita, S., Takato, Y., and Suzuki, A. (2011) *Mod. Phys. Lett. B*, **25**, 223.
31 Yao, Z., Kane, C.L., and Dekker, C. (2000) *Phys. Rev. Lett.*, **84**, 2941.

9
Magnetic Susceptibility

The electron has mass m, charge $-e$, and a half-spin. Hence, it has a spin magnetic moment. Pauli paramagnetism and Landau diamagnetism of the conduction electrons are discussed in this chapter.

9.1
Magnetogyric Ratio

Let us consider a classical electron's motion in a circle in the xy-plane as shown in Figure 9.1. The angular momentum $\boldsymbol{l} \equiv \boldsymbol{r} \times \boldsymbol{p}$ points towards the positive z-axis and its magnitude is given by

$$|\boldsymbol{l}| = mrv \,. \tag{9.1}$$

According to the electromagnetism, a current loop generates a *magnetic moment* $\boldsymbol{\mu}$ (vector) whose magnitude equals the current times the area of the loop, and whose direction is specified by the right-hand screw rule. The magnitude of the moment generated by the electron motion, therefore, is given by

$$\text{Current} \times \text{Area} = \left(\frac{ev}{2\pi r}\right) \times (\pi r^2) = \frac{1}{2} evr \,, \tag{9.2}$$

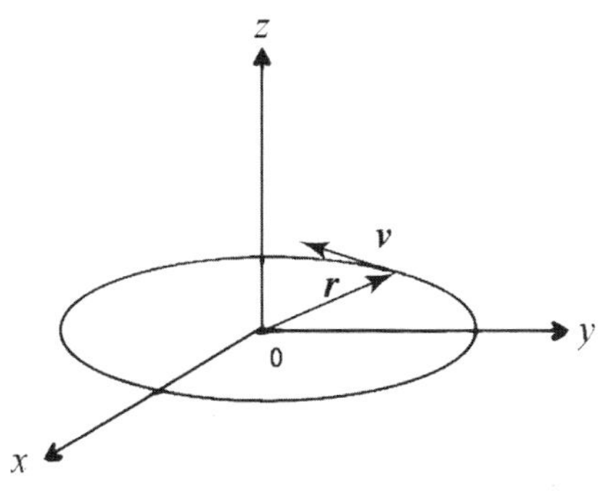

Figure 9.1 An electron with a circular motion generates a magnetic moment $\boldsymbol{\mu}$ proportional to its angular momentum $\boldsymbol{l} = m\boldsymbol{r} \times \boldsymbol{v}$.

Electrical Conduction in Graphene and Nanotubes, First Edition. S. Fujita and A. Suzuki.

and the direction is along the negative z-axis. We observe here that *magnetic moment* $\boldsymbol{\mu}$ *is proportional to the angular momentum* $\boldsymbol{l}$. We may express this relation by

$$\boldsymbol{\mu} = \alpha \boldsymbol{l} . \tag{9.3}$$

This relation, in fact, holds not only for this circular motion but also in general. The proportionality factor

$$\alpha = \frac{-e}{2m} \tag{9.4}$$

is called the *magnetogyric* or *magnetomechanical ratio*. We note that the ratio is proportional to the charge $-e$ and inversely proportional to the (electron) mass m.

We assume that a magnetic field $\boldsymbol{B}$ is applied along the positive z-axis. The potential field energy $\mathcal{V}$ of a magnetic dipole is given by

$$\mathcal{V} = -\mu B \cos\theta = -\mu_z B , \tag{9.5}$$

where θ is the angle between the vectors $\boldsymbol{\mu}$ and $\boldsymbol{B}$.

We may expect that a general relation such as (9.3) holds also in quantum theory. The angular momentum (eigenvalues) is quantized in the units of $\hbar$. The electron has a spin angular momentum $\boldsymbol{s}$ whose z-component can assume either $1/2\hbar$ or $-1/2\hbar$. Let us write

$$s'_z = \frac{1}{2}\hbar\sigma'_z \equiv \frac{1}{2}\hbar\sigma , \quad \sigma \equiv \sigma'_z = \pm 1 . \tag{9.6}$$

Analogous to (9.3) we assume that

$$\mu_z \propto s'_z \propto \sigma . \tag{9.7}$$

We shall write this quantum relation in the form

$$\mu_z = \frac{1}{2} g \mu_{\mathrm{B}} \sigma , \tag{9.8}$$

where

$$\mu_{\mathrm{B}} \equiv \frac{e\hbar}{2mc} = 9.27 \times 10^{-21}\,\mathrm{erg \cdot G^{-1}} \quad \text{(CGS units)} , \tag{9.9}$$

called the *Bohr magneton*,[1] has the dimensions of the magnetic moment. The constant g in (9.8) is a numerical factor of order 1, and is called a *g-factor*. If the magnetic moment of the electron is accounted for by the "spinning" of the charge around the z-axis, then the g-factor should be exactly one. However, the experiments show that this factor is 2. This phenomenon is known as the *spin anomaly*, which is an important indication of the quantum nature of the spin.

1) The Bohr magneton in terms of SI units is defined as $\mu_{\mathrm{B}} \equiv e\hbar/(2m) = 9.27 \times 10^{-24}\,\mathrm{J \cdot T^{-1}}$.

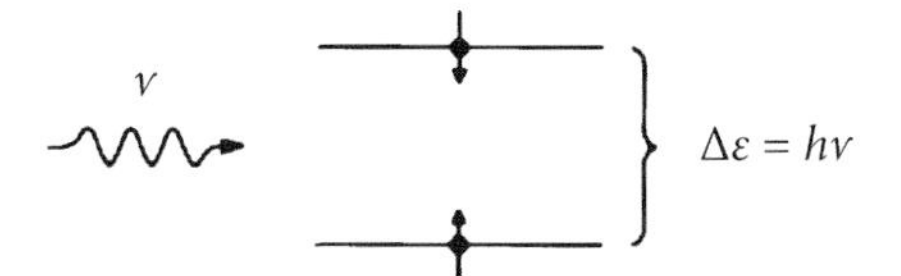

Figure 9.2 Electron spin resonance (ESR). An electron with the up-spin may absorb a photon of the energy $h\nu$ and jump to the upper energy level by flipping its spin if $\Delta\varepsilon = h\nu$.

In the presence of a magnetic field $\boldsymbol{B}$ the electron whose spin is directed along $\boldsymbol{B}$, called the electron with *up-spin*, will have a lower energy than the *down-spin* electron whose spin is directed against $\boldsymbol{B}$. The difference is, according to (9.5) and (9.8),

$$\Delta\varepsilon = \frac{1}{2}g\mu_B(+1) - \frac{1}{2}g\mu_B(-1) = g\mu_B B\,. \tag{9.10}$$

For $B = 7000$ G and $g = 2$, we obtain the numerical estimate $\Delta\varepsilon/k_B \simeq 1$ K.

If an electromagnetic wave with the frequency ν satisfying $h\nu = \Delta\varepsilon$ is applied, then the electron may absorb a photon of the energy $h\nu$ and jump to the upper energy level. Figure 9.2 illustrates this phenomenon, which is known as the *electron spin resonance* (ESR). The frequency corresponding to $\Delta\varepsilon/k_B = 1$ K is

$$\nu = 2.02 \times 10^{10}\ \text{cycles s}^{-1}\,. \tag{9.11}$$

This frequency falls in the microwave region of the electromagnetic radiation spectrum.

9.2
Pauli Paramagnetism

Let us consider an electron moving in free space. The quantum states for the electron can be characterized by momentum $\boldsymbol{p}$ and spin $\sigma(= \pm 1)$. If a weak constant magnetic field $\boldsymbol{B}$ is applied along the positive z-axis, then the energy ε associated with the quantum state $(\boldsymbol{p}, \sigma)$ is given by

$$\varepsilon = \frac{p^2}{2m} - \frac{1}{2}g\mu_B\sigma B \equiv \varepsilon(\boldsymbol{p}, \sigma)\,, \tag{9.12}$$

where the second term arises from the electromagnetic interaction (see (9.5) and (9.8)). Since $g = 2$ for the electron spin, we may simplify (9.12) to

$$\varepsilon = \frac{p^2}{2m} - \mu_B B\sigma\,. \tag{9.13}$$

This expression shows that the electron with up-spin ($\sigma = +1$) has a lower energy than the electron with down-spin ($\sigma = -1$). We say that the spin degeneracy is removed in the presence of a magnetic field.

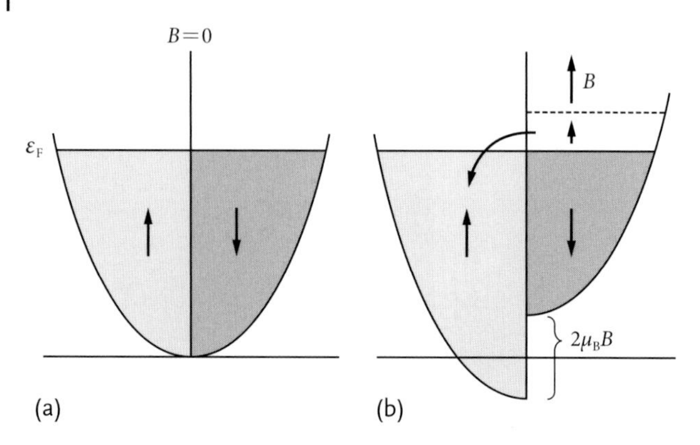

Figure 9.3 The density of states $\mathcal{D}_+$ and $\mathcal{D}_-$, for the free electrons with up (↑) and down (↓) spins, are drawn against the energy ϵ, which is measured upwards. (a) When $B = 0$ the two halves of the Fermi–Dirac distribution are equal, and thus $M = 0$; (b) when a field B is applied, spins in the antiparallel half flip into the parallel half, resulting in a net parallel magnetization.

Let us now consider a collection of free electrons in equilibrium. At the absolute zero temperature, the states with the lowest energies will be occupied by the electrons, the Fermi energy ε_F providing the upper limit. This situation is schematically shown in Figure 9.3, where the density of states, $\mathcal{D}_+(\varepsilon)$ and $\mathcal{D}_-(\varepsilon)$, for electrons with up- and down-spins are drawn against the energy ε. The density of states for free electrons was discussed in Section 3.6. In the absence of the field, both $\mathcal{D}_+(\varepsilon)$ and $\mathcal{D}_-(\varepsilon)$ are the same and are given by one-half of (3.63):

$$\mathcal{D}_0(\varepsilon) \equiv \frac{m^{3/2}}{\sqrt{2}\pi^2\hbar^3}\varepsilon^{1/2} . \tag{9.14}$$

Because of the magnetic energy $-\mu_B B$, the curve for the density of states, $\mathcal{D}_+(\varepsilon)$, for electrons with up-spins will be displaced downward by $\mu_B B$ compared with that for zero field and will be given by

$$\mathcal{D}_+(\varepsilon) = \frac{m^{3/2}}{\sqrt{2}\pi^2\hbar^3}(\varepsilon + \mu_B B)^{1/2} = \mathcal{D}_0(\varepsilon + \mu_B B) , \quad \varepsilon \geq -\mu_B B . \tag{9.15}$$

Similarly the curve for the density of states, $\mathcal{D}_-(\varepsilon)$, for electrons with down-spins is displaced by

$$\mathcal{D}_-(\varepsilon) = \frac{m^{3/2}}{\sqrt{2}\pi^2\hbar^3}(\varepsilon - \mu_B B)^{1/2} = \mathcal{D}_0(\varepsilon - \mu_B B) , \quad \varepsilon \geq \mu_B B . \tag{9.16}$$

From Figure 9.3, the numbers $N_\pm$ of the electrons with up- and down-spins are given by

$$N_\pm = V \int_{\mp\mu_B B}^{\varepsilon_F} d\varepsilon \mathcal{D}_\pm(\varepsilon) = V \int_0^{\varepsilon_F \pm \mu_B B} dx \mathcal{D}_0(x) \quad (x = \varepsilon \pm \mu_B B) . \tag{9.17}$$

The difference $N_+ - N_-$ generates a finite magnetic moment for the system. Each electron with up-spin contributes μ_B, and each electron with down-spin contributes $-\mu_B$. Therefore, the total magnetic moment is $N_+\mu_B - N_-\mu_B$. Dividing this by volume V, we obtain, for the *magnetization*,

$$\begin{aligned} I &\equiv \frac{\mu_B}{V}[N_+ - N_-] \\ &= \mu_B \left(\int_0^{\varepsilon_F + \mu_B B} dx \mathcal{D}_0(x) - \int_0^{\varepsilon_F - \mu_B B} dx \mathcal{D}_0(x) \right) \simeq 2\mu_B^2 B \mathcal{D}_0(\varepsilon_F) , \end{aligned} \tag{9.18}$$

where we retain the term proportional to B only. Using (9.14), we can re-express this as follows:

$$I = \frac{\sqrt{2}\mu_B^2 m^{3/2}}{\pi^2 \hbar^3} \varepsilon_F^{1/2} B > 0 . \tag{9.19}$$

The last expression shows that the magnetization is positive, and proportional to the field B. That is, the system is *paramagnetic*. The susceptibility χ defined through the relation

$$I = \chi B \tag{9.20}$$

is given by

$$\chi = \frac{\sqrt{2}\mu_B^2 m^{3/2}}{\pi^2 \hbar^3} \varepsilon_F^{1/2} . \tag{9.21}$$

By using the relation

$$n = \frac{2}{3} \frac{\sqrt{2} m^{3/2}}{\pi^2 \hbar^3} \varepsilon_F^{3/2} , \tag{9.22}$$

we can rewrite (9.21) as (Problem 9.2.1)

$$\chi_{\text{Pauli}} = \frac{3}{2} \frac{\mu_B^2 n}{\varepsilon_F} . \tag{9.23}$$

This result was first obtained by Pauli [1] and is often referred to as *Pauli paramagnetism*. We can easily extend the theory to a finite temperature case (Problem 9.2.2).

We note that Pauli paramagnetism is weaker than the paramagnetism of isolated atoms approximately by the factor $k_B T/\varepsilon_F$ (if this factor is small).

Problem 9.2.1. Derive (9.23).

Problem 9.2.2. Derive the magnetization at finite temperatures. Hint: The magnetization at a finite temperature is given by

$$I = -\mu_B(\langle n_\uparrow \rangle - \langle n_\downarrow \rangle) = \mu_B \int_0^\infty d\varepsilon \mathcal{D}(\varepsilon)[f_F(\varepsilon - \mu_B B) - f_F(\varepsilon + \mu_B B)] ,$$

where f_F is the Fermi distribution function and $\mathcal{D}(\varepsilon)$ is the density of states in energy space.

9.3
The Landau States and Levels

A classical electron circulates around an applied magnetic field. This is often called *cyclotron motion*. The important quantum effect is the quantization of the cyclotron motion. Let us calculate the energy levels of an electron in a constant magnetic field $\boldsymbol{B}(= \nabla \times \boldsymbol{A})$. We choose the vector potential

$$\boldsymbol{A} = (A_x, A_y, A_z) = (0, Bx, 0) , \tag{9.24}$$

which yields a constant field $\boldsymbol{B}$ in the z-direction (Problem 9.3.1). The Hamiltonian H is then given by (Problem 9.3.2)

$$H = \frac{1}{2m}[\boldsymbol{p} + e\boldsymbol{A}]^2 = \frac{1}{2m}\left[p_x^2 + (p_y + eBx)^2 + p_z^2\right] . \tag{9.25}$$

By replacing the classical momentum $\boldsymbol{p}$ with the corresponding momentum operator $-\mathrm{i}\hbar\nabla$ in (9.25), the Schrödinger equation can now be written down as

$$\mathcal{H}\psi = -\frac{\hbar^2}{2m}\left[\frac{\partial^2}{\partial x^2} + \left(\frac{\partial}{\partial y} + \frac{\mathrm{i}eB}{\hbar}x\right)^2 + \frac{\partial^2}{\partial z^2}\right]\psi = \varepsilon\psi . \tag{9.26}$$

Since the Hamiltonian $\mathcal{H}$ contains neither y nor z explicitly, we assume a wavefunction of the form:

$$\psi(x, y, z) = \phi(x)\exp[-\mathrm{i}(k_y y + k_z z)] . \tag{9.27}$$

Substituting this expression into (9.26) yields the following equation for $\phi(x)$ (Problem 9.3.3):

$$\left[-\frac{\hbar^2}{2m}\frac{\mathrm{d}^2}{\mathrm{d}x^2} + \frac{1}{2}m\omega_\mathrm{c}^2\left(x - \frac{\hbar k_y}{eB}\right)^2\right]\phi(x) = \varepsilon_1\phi(x) , \tag{9.28}$$

$$\varepsilon_1 \equiv \varepsilon - \frac{\hbar^2 k_z^2}{2m} . \tag{9.29}$$

Equation (9.28) is the energy-eigenvalue equation for a harmonic oscillator with the *cyclotron frequency*

$$\omega_\mathrm{c} \equiv \frac{eB}{m} \tag{9.30}$$

and the center of oscillation displaced from the origin by $x_0 = \hbar k_y/(eB)$. The energy eigenvalues are given by

$$\varepsilon_1 = \hbar\omega_\mathrm{c}\left(N_\mathrm{L} + \frac{1}{2}\right) , \quad N_\mathrm{L} = 0, 1, 2, \ldots \tag{9.31}$$

Combining this with (9.29), we obtain

$$\varepsilon \equiv \varepsilon_{N_\mathrm{L},k_z} = \hbar\omega_\mathrm{c}\left(N_\mathrm{L} + \frac{1}{2}\right) + \frac{\hbar^2 k_z^2}{2m} , \quad N_\mathrm{L} = 0, 1, 2, \ldots \tag{9.32}$$

These energy eigenvalues are called the *Landau Levels* (LL) [2]. The corresponding quantum states, called the *Landau states*, are characterized by the quantum number (N_L, k_y, k_z). We note that the energies do not depend on k_y, and they are therefore highly degenerate. The Landau states are quite different from the momentum eigenstates. This has significant consequences on magnetization and galvanomagnetic phenomena. An electron in a Landau state may be pictured as circulating with the angular frequency ω_c around the magnetic field. If radiation having a frequency equal to ω_c is applied, the electron may jump up from one Landau state to another by absorption of photon energy equal to $\hbar\omega_c$. This generates the phenomenon of *cyclotron resonance.*

Problem 9.3.1. Show that the vector potential given by (9.24) generates a magnetic field pointing in the positive z-direction. For a constant magnetic field $\boldsymbol{B}$ we can choose the vector potential $\boldsymbol{A} = \frac{1}{2}\boldsymbol{B} \times \boldsymbol{r}$. Show this by explicitly calculating $\nabla \times \boldsymbol{A}$.

Problem 9.3.2. Derive the Hamiltonian (9.25).

Problem 9.3.3. Derive (9.28) from (9.26) and (9.27). Solving (9.28), obtain its energy eigenvalues (9.31).

9.4 Landau Diamagnetism

The electron always circulates around the magnetic flux so as to reduce the magnetic field. This is called the *motional diamagnetism.* If we calculate this effect classically by considering the system confined to a closed volume, we obtain zero magnetic moments. This is known as *van Leeuwen's theorem.* We first demonstrate this theorem.

Let us take a system of free electrons confined in a volume V. The partition function per electron is

$$Z(B) = \frac{1}{(2\pi\hbar)^3} \int_V \mathrm{d}^3 r \int \mathrm{d}^3 p \exp\left[-\frac{(\boldsymbol{p} + e\boldsymbol{A})^2}{2mk_B T}\right] . \tag{9.33}$$

We introduce kinetic momentum $\boldsymbol{\Pi}$:

$$m\dot{\boldsymbol{r}} \equiv m\boldsymbol{v} = \boldsymbol{p} + e\boldsymbol{A} \equiv \boldsymbol{\Pi} = (\Pi_x, \Pi_y, \Pi_z) . \tag{9.34}$$

After simple calculations, we obtain

$$\mathrm{d}x\mathrm{d}y\mathrm{d}z\mathrm{d}p_x\mathrm{d}p_y\mathrm{d}p_z = \mathrm{d}x\mathrm{d}y\mathrm{d}z\mathrm{d}\Pi_x\mathrm{d}\Pi_y\mathrm{d}\Pi_z . \tag{9.35}$$

Using the last three equations, we can show that $Z(B)$ is equal to the electron partition function with no field;

$$Z(B) = Z(B = 0) . \tag{9.36}$$

In 1930, L.D. Landau [2] showed that quantization of electron circulation yields a *diamagnetic moment*. We shall demonstrate this below.

Electrons obey Fermi–Dirac statistics. Considering a system of free electrons, we define the *free energy* F as (Problems 9.4.1 and 9.4.2)

$$F = N\mu - 2k_B T \sum_i \ln\left[1 + e^{(\mu - \varepsilon_i)/(k_B T)}\right] , \tag{9.37}$$

where the factor 2 arises from the spin degeneracy and ε_i is the Landau energy given by (9.32). The chemical potential μ is determined from the condition:

$$\frac{\partial F}{\partial \mu} = 0 . \tag{9.38}$$

The *total magnetic moment* M for the system can be found from

$$M = -\frac{\partial F}{\partial B} . \tag{9.39}$$

Equation (9.38) is equivalent to the usual condition (Problem 9.4.3) that the total number of electrons, N, can be obtained in terms of the Fermi distribution

$$f(\varepsilon) \equiv \left[\exp\left(\frac{\varepsilon - \mu}{k_B T}\right) + 1\right]^{-1} \tag{9.40}$$

from

$$N = 2\sum_i f(\varepsilon_i) . \tag{9.41}$$

The Landau energy ε_i is characterized by the Landau oscillator quantum number N_L and the z-component momentum $p_z (= \hbar k_z)$. The energy E becomes continuous in the bulk limit. Let us introduce the density of states $\mathcal{D}(\varepsilon)$ $(= \mathrm{d}\mathcal{W}/\mathrm{d}\varepsilon)$ such that

$$\mathcal{D}(\varepsilon)\mathrm{d}\varepsilon = \text{the number of states having an energy between } \varepsilon \text{ and } \varepsilon + \mathrm{d}\varepsilon . \tag{9.42}$$

We now write (9.37) in the form

$$F = N\mu - 2k_B T \int_0^\infty \mathrm{d}\varepsilon \frac{\mathrm{d}\mathcal{W}}{\mathrm{d}\varepsilon} \ln\left[1 + e^{(\mu - \varepsilon)/(k_B T)}\right] . \tag{9.43}$$

The *statistical weight* (number) $\mathcal{W}$ is the total number of states having energy less than

$$\varepsilon = \hbar\omega_c\left(N_L + \frac{1}{2}\right) + \frac{p_z^2}{2m} . \tag{9.44}$$

Conversely, the allowed values of p_z are distributed over the range in which $|p_x|$ does not exceed

$$\left\{2m\left[\varepsilon - \left(N_\mathrm{L} + \frac{1}{2}\right)\hbar\omega_\mathrm{c}\right]\right\}^{1/2} . \tag{9.45}$$

For a fixed pair $(\varepsilon, N_\mathrm{L})$ the increment in the weight, $\mathrm{d}\mathcal{W}$, is given by

$$\begin{aligned}\mathrm{d}\mathcal{W} &= eB\frac{L_1L_2}{(2\pi\hbar)}\int \mathrm{d}p_z\frac{L_3}{(2\pi\hbar)} \\ &= eB\frac{V}{(2\pi\hbar)^2}2\left\{2m\left[\varepsilon - \left(N_\mathrm{L} + \frac{1}{2}\right)\hbar\omega_\mathrm{c}\right]\right\}^{1/2} ,\end{aligned} \tag{9.46}$$

where $V = L_1L_2L_3$ is the volume of the container. After summing (9.46) with respect to N_L, we obtain

$$\mathcal{W}(\varepsilon) = A\frac{(\hbar\omega_\mathrm{c})^{3/2}}{\sqrt{2}\pi}2\sum_{N_\mathrm{L}=0}^{\infty}\sqrt{\varepsilon^* - (2N_\mathrm{L}+1)\pi} , \tag{9.47}$$

where

$$A \equiv V\frac{(2\pi m)^{3/2}}{(2\pi\hbar)^3} , \quad \varepsilon^* \equiv \frac{2\pi\varepsilon}{\hbar\omega_\mathrm{c}} . \tag{9.48}$$

We assume high Fermi degeneracy such that

$$\mu \simeq \varepsilon_\mathrm{F} \gg \hbar\omega_\mathrm{c} \quad \text{for a metal.} \tag{9.49}$$

The sum over N_L in (9.47) converges slowly. We use *Poisson's summation formula* [3, 4] and, after mathematical manipulations, obtain

$$\mathcal{W}(\varepsilon) = \mathcal{W}_0 + \mathcal{W}_\mathrm{L} + \mathcal{W}_\mathrm{osc} , \tag{9.50}$$

where

$$\mathcal{W}_0 = A\frac{4}{3\sqrt{\pi}}\varepsilon^{3/2} , \tag{9.51}$$

$$\mathcal{W}_\mathrm{L} = -A\frac{1}{24\sqrt{\pi}}\frac{(\hbar\omega_\mathrm{c})^2}{\varepsilon^{1/2}} , \tag{9.52}$$

$$\mathcal{W}_\mathrm{osc} = A\frac{(\hbar\omega_\mathrm{c})^{3/2}}{\sqrt{2}\pi^{3/2}}\sum_{\nu=1}^{\infty}\frac{(-1)^\nu}{\nu^{3/2}}\sin\left(\frac{2\pi\nu\varepsilon}{\hbar\omega_\mathrm{c}} - \frac{\pi}{4}\right) . \tag{9.53}$$

The detailed steps leading to (9.50) through (9.53) are given in the Appendix A.5. The term $\mathcal{W}_0$, which is independent of B, gives the weight equal to that for a free electron system with no field. The term $\mathcal{W}_L$ is negative (*diamagnetic*) and can generate a diamagnetic moment. We start with (9.43), integrate by parts, and obtain

$$F = N\mu - 2\int_0^\infty d\varepsilon \mathcal{W}(\varepsilon) f(\varepsilon)$$

$$= N\mu + 2\int_0^\infty d\varepsilon \frac{df}{d\varepsilon} \int_0^\varepsilon d\varepsilon' \mathcal{W}(\varepsilon') . \tag{9.54}$$

The $-df/d\varepsilon$, which can be expressed as (Problem 9.4.2)

$$-\frac{df}{d\varepsilon} = \frac{1}{4k_B T}\text{sech}^2\left(\frac{\varepsilon - \mu}{2k_B T}\right) , \tag{9.55}$$

has a peak near $\varepsilon = \mu$ if $k_B T \ll \mu$, and

$$\int_0^\infty d\varepsilon \frac{df}{d\varepsilon} = -1 . \tag{9.56}$$

For a smoothly changing integrand in the last member of (9.54) $-df/d\varepsilon$ can be regarded as a *Dirac delta function*:

$$-\frac{df}{d\varepsilon} = \delta(\varepsilon - \mu) . \tag{9.57}$$

Using this property and (9.54) and (9.39), we obtain (Problem 9.4.4)

$$F_L = A\frac{1}{6\sqrt{\pi}}(\hbar\omega_c)^2 \varepsilon_F^{1/2} . \tag{9.58}$$

Here we set $\mu = \varepsilon_F$. This is justified since the corrections to $\mu(B, T)$ start with the B^4 term and with a T^2 term. Using (9.58) and (9.49), we obtain

$$\chi_{\text{Landau}} = V^{-1}\frac{\partial M_L}{\partial B} = -V^{-1}\frac{\partial^2 F_L}{\partial B^2} = -\frac{n\mu_B^2}{2\varepsilon_F} , \tag{9.59}$$

where n is the free electron density.

Comparing this result with (9.23), we observe that Landau diamagnetism is one third of the Pauli paramagnetism in magnitude: $\chi_{\text{Landau}} = -\frac{1}{3}\chi_{\text{Pauli}}$. But the calculations in this section are done with the assumption of the free electron model. If the effective mass (m^*) approximation is used, formula (9.59) is corrected by the factor $(m^*/m)^2$ as we can see from (9.58). Hence, the *diamagnetic susceptibility* for a metal is

$$\chi_{\text{Landau}}^{\text{metal}} = \left(\frac{m^*}{m}\right)^2 \chi_{\text{Landau}} . \tag{9.60}$$

We note that the Landau susceptibility is spin-independent. For a metal having a small effective mass, the Landau susceptibility χ_L^{metal} can be greater in magnitude than the Pauli paramagnetic susceptibility χ_{Pauli}. Then the total magnetic susceptibility expressed as

$$\chi = \chi_{\text{Pauli}} + \chi_{\text{Landau}}^{\text{metal}} \tag{9.61}$$

can be negative (diamagnetic). This is observed for GaAs ($m^* = 0.07m$).

The oscillatory term $\mathcal{W}_{\text{osc}}$ in (9.53) yields the *de Haas–van Alphen oscillations*, which will be discussed in Section 10.1.

Problem 9.4.1. The grand partition function Ξ is defined by

$$\Xi \equiv \text{TR}\{\exp(\alpha N - \beta H)\} \equiv \sum_{N=0}^{\infty} \text{e}^{\alpha N} \text{Tr}_{,N}\{\exp(\alpha N - \beta H_N)\} ,$$

where H_N is the Hamiltonian of the N-particle system and Tr stands for the trace (diagonal sum). The symbol TR means a grand ensemble trace. We assume that the internal energy E, the number density n, and the entropy S are respectively given by

$$E = \langle H \rangle \equiv \frac{\text{TR}\{H \exp(\alpha N - \beta H)\}}{\Xi} ,$$

$$n = \frac{\langle N \rangle}{V} = \frac{\text{TR}\{N \exp(\alpha N - \beta H)\}}{V \Xi} ,$$

$$S = k_{\text{B}}(\ln \Xi + \beta E - \alpha \langle N \rangle) .$$

Show that

$$F \equiv E - TS - \mu \langle N \rangle - k_{\text{B}} T \ln \Xi ,$$

where $\mu \equiv k_{\text{B}} T \alpha$ is Gibbs free energy per particle: $G \equiv E - TS + PV = \mu \langle N \rangle$.

Problem 9.4.2.

1. Evaluate the grand partition function Ξ for a free electron system characterized by $H = \sum_{j=1}^{N} p_j^2/(2m)$.
2. Show that

$$N = \frac{\partial}{\partial \alpha} \ln \Xi = 2 \sum_p f(\varepsilon_p) , \quad E = -\frac{\partial}{\partial \beta} \ln \Xi = 2 \sum_p \varepsilon_p f(\varepsilon_p) ,$$

where 2 is the spin degeneracy factor, ε_p and $f(\varepsilon)$ are respectively given by

$$\varepsilon_p = \frac{\boldsymbol{p}^2}{2m} , \quad f(\varepsilon) = (\text{e}^{\beta(\varepsilon - \mu)} - 1)^{-1} , \quad \beta \equiv \frac{1}{k_{\text{B}} T} .$$

Problem 9.4.3. Using the free energy F in (9.37), obtain (9.40) and (9.41).

Problem 9.4.4. Verify (9.58).

References

1 Pauli, W. (1927) *Z. Phys.*, **41**, 81.
2 L.D. Landau (1930) *Z. Phys.*, **64**, 629.
3 Morse, P.M. and Feshbach, H. (1953) *Methods of Theoretical Physics*, McGraw-Hill, New York, pp. 466–467.
4 Courant, R. and Hilbert, D. (1953) *Methods of Mathematical Physics*, vol. 1, Wiley-Interscience, New York, pp. 76–77.

10
Magnetic Oscillations

The de Haas–van Alphen oscillations in susceptibility are often analyzed using Onsager's formula, which is derived. The statistical mechanical theory of the oscillations for the quasifree electron moving in 3D and 2D is developed in this chapter.

10.1
Onsager's Formula

A metal is a system in which the conduction electrons, either *"electrons"* or *"holes,"* move without much resistance. "Electrons" ("holes") are fermions having negative (positive) charge $q = -(+)e$ and positive effective masses m^*, which respond to the Lorentz force $\boldsymbol{F} = q(\boldsymbol{E} + \boldsymbol{v} \times \boldsymbol{B})$. These conduction electrons are generated in a metal, depending on the curvature sign of the metal's Fermi surface. Their existence can be checked by the linear heat capacity at the lowest temperatures. As we saw in Chapter 3, each metal has a Fermi surface, often quite complicated.

The most frequently used means to probe the Fermi surface is to observe de Haas–van Alphen (dHvA) oscillations and analyze the data using Onsager's formula (10.2). A magnetic field of the order $1\,\mathrm{T} = 10^4\,\mathrm{G}$ is applied in a special lattice direction. The experiments are normally carried out using pure samples at the liquid helium temperatures to reduce the impurity and phonon scattering effects.

The *magnetic susceptibility* χ is defined by

$$I = \chi B\,, \tag{10.1}$$

where I is the *magnetization* (magnetic moment per unit volume) and B the applied magnetic field (magnitude). When the experiments are done on pure samples and at very low temperatures, the susceptibility χ in some metals is found to exhibit oscillations with varying magnetic field strength B. This phenomenon was first discovered in 1930 by de Haas and van Alphen [1, 2] in their study of bismuth (Bi), and it is called *de Haas–van Alphen oscillations*. Magnetic oscillations in magnetoresistance (MR) similar to the dHvA oscillations are called Shubnikov–de Haas (SdH) oscillations, which is discussed in Section 10.5. As we shall show, these oscillations have a quantum mechanical origin. Currently, analyses of dHvA oscillations are

Electrical Conduction in Graphene and Nanotubes, First Edition. S. Fujita and A. Suzuki.

done routinely in terms of *Onsager's formula* (10.2). According to Onsager's theory [3], the nth maximum (counted from $1/B = 0$) occurs for a field B given by the relation

$$n + \gamma = \frac{1}{2\pi\hbar e}\frac{A}{B} \equiv \frac{1}{(2\pi\hbar)^2}\Phi_0\frac{A}{B} , \tag{10.2}$$

where A is any extremal area of intersections between the Fermi surface and the family of planes $\boldsymbol{B} \cdot \boldsymbol{p} \equiv \boldsymbol{B} \cdot (\hbar \boldsymbol{k}) =$ constant, and γ is a *phase* (number) less than unity. The constant

$$\Phi_0 \equiv \frac{2\pi\hbar}{e} \equiv \frac{h}{e} = 4.135 \times 10^{-1}\,\mathrm{G\,cm}^2 \tag{10.3}$$

is called the electron *flux quantum*.

As an example, consider an *ellipsoidal Fermi surface*, as shown in Figure 10.1a:

$$\varepsilon_{\mathrm{F}} = \frac{p_1^2}{2m_1} + \frac{p_2^2}{2m_2} + \frac{p_3^2}{2m_3} , \qquad m_1, m_2, m_3 > 0 . \tag{10.4}$$

The subscript "F" on ε will be omitted hereafter in this section. Assume that the field $\boldsymbol{B}$ is applied along the p_3-axis. All the intersections are ellipses represented by

$$\varepsilon - \frac{p_3^2}{2m_3} = \frac{p_1^2}{2m_1} + \frac{p_2^2}{2m_2} \quad (p_3 = \text{constant}) . \tag{10.5}$$

The maximum area A of the intersection occurs at $p_3 = 0$ (the "*belly*"; see Figure 10.1a), and its area A is

$$A = \pi(2m_1\varepsilon)^{1/2}(2m_2\varepsilon)^{1/2} = 2\pi(m_1 m_2)^{1/2}\varepsilon . \tag{10.6}$$

Using (10.2) and solving for ε, we obtain

$$\varepsilon = eB(m_1 m_2)^{-1/2}(n + \gamma)\hbar , \tag{10.7}$$

which indicates that the energy ε is quantized as the energy of the simple harmonic oscillator with the angular frequency

$$\omega_0 \equiv \frac{eB}{(m_1 m_2)^{1/2}} . \tag{10.8}$$

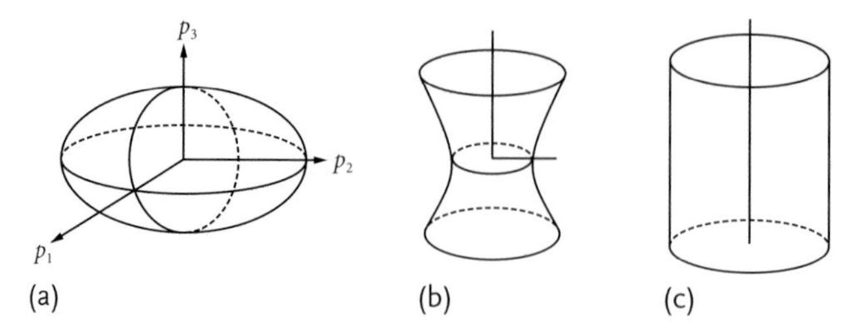

Figure 10.1 Fermi surfaces: (a) ellipsoid, (b) hyperboloid (neck), (c) cylinder.

As a second example, we take a *hyperboloidal Fermi surface* that can be represented by (10.4) with $m_1, m_2 > 0$ and $m_3 < 0$. Assume the same orientation of $\boldsymbol{B}$. Equations (10.5) and (10.6) then hold, where the area A represents the minimal area of the intersection at $p_3 = 0$ (the "*neck*"; see Figure 10.1b). As a third example, assume that $m_3 = \infty$ in (10.4), which represents a *Fermi cylinder* (see Figure 10.1c). In this case the area A is given by (10.6) for the same orientation of $\boldsymbol{B}$.

All three geometrical shapes are discussed by Onsager in his correspondence [3]. At the time of writing in 1952, only the ellipsoidal Fermi surface was known in experiments. Today we know that all three cases occur in reality. When tested by experiments, the agreements between theory and experiment are excellent. The cases of ellipsoidal and hyperboloidal surfaces were found in noble metals such as copper (Cu), silver (Ag), and gold (Au). The dHvA oscillations in Ag are shown in Figure 10.2, where the susceptibility χ is plotted against B^{-1} in arbitrary units, after Schönberg and Gold [4, 5]. The magnetic field is along a $\langle 111 \rangle$ direction. The two distinct periods are due to the "neck" and "belly" orbits indicated, the high-frequency oscillation coming from the larger belly orbit. By counting the number of high-frequency periods in a single low-frequency period, for example, between the two arrows, we can deduce directly that $A_{111}(\text{belly})/A_{111}(\text{neck}) = 51$, which is most remarkable.

Onsager's derivation of (10.2) in his original paper [3] is quite illuminating. Let us follow his arguments. For any closed k-orbit, there should be a closed orbit in

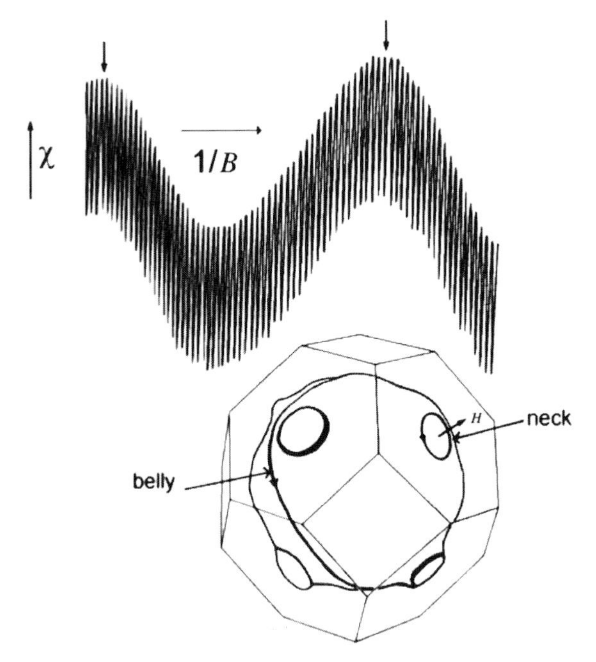

Figure 10.2 The dHvA oscillations in silver with the magnetic field along a $\langle 111 \rangle$ direction, after Schönberg and Gold [4, 5]. The two distinct periods are due to the neck and belly orbits indicated in the inset.

the position space, called a closed r-orbit. The periodic component of the motion, which involves the components of $\boldsymbol{p} \equiv \hbar \boldsymbol{k}$ and $\boldsymbol{r}$ perpendicular to $\boldsymbol{B}$, is quantized. We apply the *Bohr–Sommerfeld quantization rule*

$$\oint \boldsymbol{p} \cdot \mathrm{d}\boldsymbol{r} = (n + \gamma)2\pi\hbar \tag{10.9}$$

to the r-orbit. The magnetic moment $\boldsymbol{\mu}$ is proportional to the angular momentum $\boldsymbol{j}$:

$$\boldsymbol{\mu} = \alpha \boldsymbol{j} \,, \quad \alpha = \text{constant} \,. \tag{10.10}$$

The cross section Ω of the r-orbit is determined such that the enclosed magnetic flux Φ, given by $B\Omega$, equals $(n + \gamma)$ times the flux quantum $\Phi_0 \equiv h/e$:

$$\Phi = (n + \gamma)\Phi_0 = B\Omega \,. \tag{10.11}$$

This is *Onsager's magnetic flux quantization.*

For a free electron the closed circular path in the k-space perpendicular to the field becomes a similar path in the r-space, turned through a right angle, and with the linear dimension changed in the ratio $(eB)^{-1}$ (Problem 10.1.1). This may hold for a nearly circular closed orbit. If we assume this relation for the Bloch electron, the area enclosed by the closed k-orbit, A, is proportional to that enclosed by the closed r-orbit, Ω:

$$A = (eB)^2\Omega \,. \tag{10.12}$$

Combining the last two equations, we obtain (10.2). □

The most remarkable argument advanced by Onsager is that an electron in a closed r-orbit may move, keeping a finite number n of flux quanta, each carrying $\Phi_0 \equiv h/e$, within the orbit. This comes from the physical principle that the magnetic field $\boldsymbol{B}$ does not work on the electron and, therefore, the field does not change the electron's kinetic energy. This property should hold for any charged particle. The flux quantization for the Cooper pair in a superconductor was observed in 1961 by Deaver and Fairbank [6] and Doll and Näbauer [7]. Because the Cooper pair has charge (magnitude) $2e$, the observed flux quantum is found to be $-h/(2e)$, that is, *half the electron flux quantum* Φ_0 defined in (10.3). The phase γ in (10.9) can be set equal to 1/2. This can be deduced by taking the case of a free electron for which quantum calculations are carried out exactly. The quantum number n can arbitrarily be large. Hence, Onsager's formula can be applied for *any* strength of field (Problem 10.1.1).

However, (10.2) turns out to contain a limitation. The curvatures along the closed k-orbit must either be entirely positive or negative. The k-orbit cannot have a mixture of a positive-curvature section and a negative-curvature section.

Problem 10.1.1. Consider a free electron having mass m and charge q, subject to a constant magnetic field $\boldsymbol{B}$.

1. Write down Newton's equation of motion.
2. Show thatthe magnetic force $q\boldsymbol{v} \times \boldsymbol{B}$ does not work on the electron; that is, the kinetic energy $E \equiv mv^2/2$ does not change with time.
3. Show that the component of $\boldsymbol{v}$ parallel to $\boldsymbol{B}$ is a constant.
4. Show that the electron spirals about the field $\boldsymbol{B}$ with the angular frequency $\omega = eB/m$.
5. Show that the orbit projected on a plane perpendicular to the field $\boldsymbol{B}$ is a circle of radius $R = p_\perp/\omega = mv_\perp/eB$, where $v_\perp$ represents the speed of the circular motion. Find the maximum radius $R_{\max}$.
6. Define the *kinetic momentum* $\boldsymbol{\Pi} \equiv m\boldsymbol{v}$ and express the energy ε in terms of Π_j.
7. Choose the $x_3(=z)$-axis along $\boldsymbol{B}$. Show that the curve represented by $E(p_1, p_2, 0) = \varepsilon$ is a circle of radius $(2m\varepsilon)^{1/2} = \Pi$.
8. Show that the areas of the circles obtained in parts 5 and 7 differ by the factor $(eB)^2$.

10.2 Statistical Mechanical Calculations: 3D

Susceptibility is an equilibrium property and, therefore, can be calculated by using standard statistical mechanics. Here, we demonstrate Onsager's formula (10.2) using a free electron model.

The *free energy* F is, from (9.43),

$$F = N\mu - 2k_B T \int_0^\infty d\varepsilon \frac{dW}{d\varepsilon} \ln\left[1 - e^{(\mu-\varepsilon)/(k_B T)}\right] . \tag{10.13}$$

The oscillatory statistical weight W_{osc} is, from (9.53),

$$W_{\text{osc}} = A \frac{(\hbar\omega_c)^{3/2}}{\sqrt{2}\pi^{3/2}} \sum_{\nu=1}^{\infty} \frac{(-1)^\nu}{\nu^{3/2}} \sin\left(\frac{2\pi\nu\varepsilon}{\hbar\omega_c} - \frac{\pi}{4}\right) . \tag{10.14}$$

We note that W_{osc} oscillates with alternating signs. In fact the relevant energy ε is of the order of the Fermi energy ε_F, which is much greater than the cyclotron frequency ω_c times the Planck constant $\hbar$. Hence, if there are many oscillations within the width of $df/d\varepsilon$ of the order $k_B T$, then the contribution to F must vanish. This condition is shown in Figure 10.3. Let us study this behavior in detail. Using (9.54), we obtain

$$F = N\mu + 2\int_0^\infty d\varepsilon \frac{df}{d\varepsilon} \int_0^\varepsilon d\varepsilon' W_{\text{osc}}(\varepsilon') . \tag{10.15}$$

The critical temperature T_c below which the oscillations can be observed is

$$k_B T_c \sim \hbar\omega_c . \tag{10.16}$$

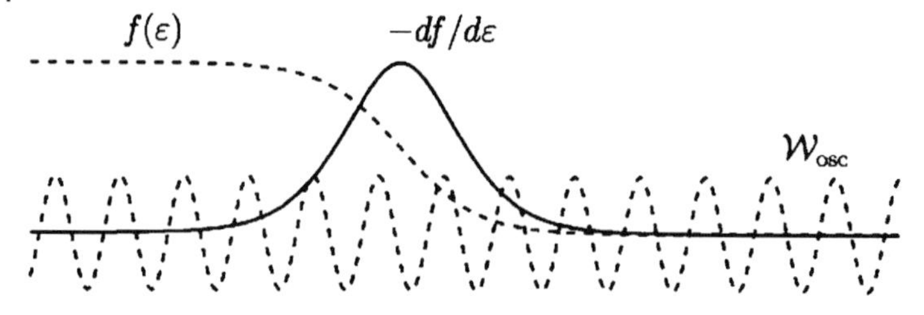

Figure 10.3 Numerous oscillations in W_{osc} within the width of $-\mathrm{d}f/\mathrm{d}\varepsilon$ cancel out the contribution to the free energy F.

Below this critical temperature ($T < T_{\mathrm{c}}$), we cannot replace $-\mathrm{d}f/\mathrm{d}\varepsilon$ by $\delta(\varepsilon-\mu)$ since the integrand varies violently. The integral in (10.15),

$$\int_0^\infty \mathrm{d}\varepsilon \frac{\mathrm{d}f}{\mathrm{d}\varepsilon} \int_0^\varepsilon \mathrm{d}\varepsilon' \mathcal{W}_{\text{osc}}(\varepsilon') = \int_0^\infty \mathrm{d}\varepsilon \mathcal{W}_{\text{osc}}(\varepsilon) f(\varepsilon) , \tag{10.17}$$

must be calculated with care. We introduce a new variable $\zeta = \beta(\varepsilon-\mu)$ and extend the lower limit to $-\infty(\beta\mu \to \infty)$ and obtain

$$\int_0^\infty \mathrm{d}\varepsilon \cdots \frac{1}{\mathrm{e}^{\beta(\varepsilon-\mu)}+1} = \beta^{-1} \int_{-\mu\beta}^\infty \mathrm{d}\zeta \cdots \frac{1}{\mathrm{e}^\zeta+1}$$
$$\to \beta^{-1} \int_{-\infty}^\infty \mathrm{d}\zeta \cdots \frac{1}{\mathrm{e}^\zeta+1} . \tag{10.18}$$

Using $\sin(A+B) = \sin A \cos B + \cos A \sin B$ and

$$\int_{-\infty}^\infty \mathrm{d}\zeta \mathrm{e}^{\mathrm{i}\alpha\zeta} \frac{1}{\mathrm{e}^\zeta+1} = \frac{\pi}{\mathrm{i}\sinh \pi\alpha} , \tag{10.19}$$

(Problem 10.2.1), we obtain from (10.15) (Problem 10.2.2)

$$F_{\text{osc}} = A\sqrt{\frac{2}{\pi}}(\hbar\omega_{\mathrm{c}})^{3/2} k_{\mathrm{B}} T \sum_{\nu=1}^\infty \frac{(-1)^\nu}{\nu^{3/2}} \frac{\cos(2\pi\nu\varepsilon_{\mathrm{F}}/(\hbar\omega_{\mathrm{c}})-\pi/4)}{\sinh(2\pi^2\nu k_{\mathrm{B}} T/(\hbar\omega_{\mathrm{c}}))} . \tag{10.20}$$

Although (10.20) contains an infinite sum with respect to ν just as the infinite sum in (9.47), its summation character is quite different. Only the first term with $\nu = 1$ in the sum is important in practice because $(\sinh(2\pi^2\nu k_{\mathrm{B}} T/(\hbar\omega_{\mathrm{c}})))^{-1} \ll 1$. Thus, we obtain

$$F_{\text{osc}} = -A\sqrt{\frac{2}{\pi}}(\hbar\omega_{\mathrm{c}})^{3/2} k_{\mathrm{B}} T \frac{\cos(2\pi\varepsilon_{\mathrm{F}}/(\hbar\omega_{\mathrm{c}})-\pi/4)}{\sinh(2\pi^2 k_{\mathrm{B}} T/(\hbar\omega_{\mathrm{c}}))} . \tag{10.21}$$

Using this equation we calculate the magnetization $I = -V^{-1}\partial F/\partial B$ and obtain

$$I_{\text{osc}} = \frac{1}{\sqrt{2}} n\mu_{\mathrm{B}} \frac{k_{\mathrm{B}} T}{\varepsilon_{\mathrm{F}}} \left(\frac{\hbar\omega_{\mathrm{c}}}{\varepsilon_{\mathrm{F}}}\right)^{1/2} \frac{\cos(2\pi\varepsilon_{\mathrm{F}}/(\hbar\omega_{\mathrm{c}})-\pi/4)}{\sinh(2\pi^2 k_{\mathrm{B}} T/(\hbar\omega_{\mathrm{c}}))} . \tag{10.22}$$

The neglected terms are exponentially smaller than those in (10.22) since $\exp(k_B T/(\hbar\omega_c)) \gg 1$. In the low field limit, the oscillation number in the range $k_B T$ becomes great, and hence, the contribution of the sinusoidal oscillations to the free energy must cancel out. This effect is represented by the factor

$$\frac{\pi k_B T}{\sinh(2\pi^2 K_B T/(\hbar\omega_c))} .$$

We define the *susceptibility* χ by

$$\chi = \frac{I}{B} . \tag{10.23}$$

Note that the magnetization I is not necessarily proportional to the field B. Using Equations (9.23), (9.58), (10.22), and (10.23), we obtain (Problem 10.2.3)

$$\chi = \frac{1}{2}\frac{n\mu_B^2}{\varepsilon_F}\left[3 - \left(\frac{m^*}{m}\right)^2 + \phi(T, B)\right] , \tag{10.24}$$

where

$$\phi(T, B) = 2\sqrt{2}\frac{k_B T}{(\hbar\omega_c\varepsilon_F)^{1/2}} \cos\left(2\pi\frac{\varepsilon_F}{\hbar\omega_c} - \frac{\pi}{4}\right) e^{-2\pi^2 k_B T/(\hbar\omega_c)} . \tag{10.25}$$

Our calculations indicate that

(a) The oscillation period is $\varepsilon_F/(\hbar\omega_c)$. This result confirms Onsager's formula (10.2). In fact the maximum area of πp_F^2 occurs at $p_z = 0$. Hence,

$$\frac{1}{(2\pi\hbar)^2}\frac{2\pi\hbar}{e}\frac{\pi p_F^2}{B} = \frac{\varepsilon_F}{\hbar\omega_c} ,$$

if the quadratic dispersion relation $\varepsilon = p^2/(2m^*)$ holds. We note that *all* electrons participate in the *cyclotronic motion* with the same frequency ω_c, and the signal is substantial.

(b) The *envelope of the oscillations* exponentially decreases in B^{-1} as

$$\exp\left(\frac{-2\pi^2 k_B T}{\hbar\omega_c}\right) = \exp\left(-\frac{\delta}{B}\right) , \quad \delta \equiv \frac{2\pi^2 k_B T m^* \hbar}{e} . \tag{10.26}$$

Thus, if the "*decay rate*" δ in B^{-1} is measured carefully, the effective mass m^* may be obtained directly through

$$m^* = \frac{e\delta}{2\pi^2\hbar k_B T} . \tag{10.27}$$

The calculations in this section were carried out by assuming a quasifree electron model. The actual physical condition in solids is more complicated. We cannot use the quasifree particle model alone to explain the experimental data.

Problem 10.2.1. Verify (10.19). Hint: Consider an integral on the real axis

$$I(\alpha, R) = \int_{-R}^{R} \mathrm{d}x \frac{\mathrm{e}^{\mathrm{i}\alpha(x+\mathrm{i}y)}}{\mathrm{e}^{x}+1}\,, \quad z = x + \mathrm{i}y\,, \quad \alpha, R > 0\,.$$

We add an integral over a semicircle of the radius R in the upper z-plane to form an integral over a closed contour. We then take the limit as $R \to \infty$. Note that the integral over the semicircle vanishes in this limit if $\alpha > 0$. The integral on the real axis, $I(\alpha, \infty)$, becomes the desired integral in (10.19). Evaluate the integral over the closed contour by using the residue theorem.

Problem 10.2.2. Derive (10.20).

Problem 10.2.3. Verify (10.24).

10.3 Statistical Mechanical Calculations: 2D

The dHvA oscillations occur in 2D and 3D. The 2D system is intrinsically paramagnetic since Landau's diamagnetism is absent, which is shown here.

Let us take a dilute system of quasifree electrons moving in a plane. Applying a magnetic field $\boldsymbol{B}$ perpendicular to the plane, each electron will be in the Landau state with the energy

$$\varepsilon = \hbar\omega_{\mathrm{c}}\left(\mathrm{N_L} + \frac{1}{2}\right)\,, \quad \omega_{\mathrm{c}} \equiv \frac{eB}{m^*}\,, \quad \mathrm{N_L} = 0, 1, 2, \ldots\,, \tag{10.28}$$

where m^* is the cyclotron effective mass. We introduce kinetic momenta:

$$\Pi_x = p_x + eA_x\,, \quad \Pi_y = p_y + eA_y\,. \tag{10.29}$$

The Hamiltonian $\mathcal{H}$ for the quasifree electron is then

$$\mathcal{H} = \frac{1}{2m^*}\left(\Pi_x^2 + \Pi_y^2\right) \equiv \frac{1}{2m^*}\Pi^2\,. \tag{10.30}$$

After simple calculations, we obtain

$$\mathrm{d}x\mathrm{d}\Pi_x\mathrm{d}y\mathrm{d}\Pi_y = \mathrm{d}x\mathrm{d}p_x\mathrm{d}y\mathrm{d}p_y\,. \tag{10.31}$$

We can then represent quantum states by the small quasiphase space elements $\mathrm{d}x\mathrm{d}\Pi_x\mathrm{d}y\mathrm{d}\Pi_y$. The Hamiltonian $\mathcal{H}$ in (10.30) does not depend on the position (x, y). Assuming large normalization lengths (L_x, L_y), we can represent the Landau states by the concentric shells having the statistical weight

$$2\pi\Pi\Delta\Pi\frac{L_x L_y}{(2\pi\hbar)^2} = \frac{eBA}{2\pi\hbar}\,, \tag{10.32}$$

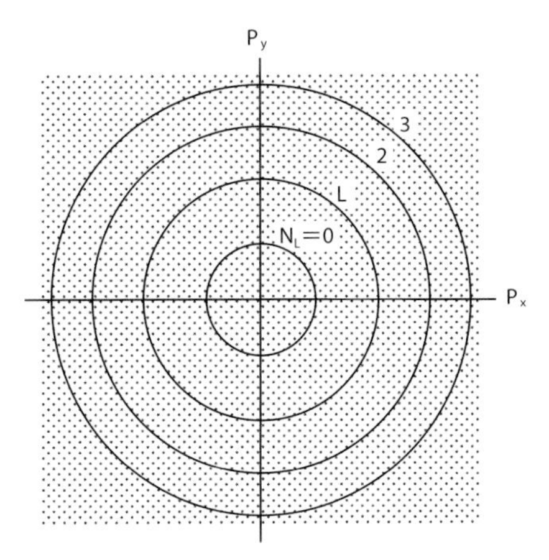

Figure 10.4 Quantization scheme for free electrons: without magnetic field (dots) and in a magnetic field (circles) perpendicular to the paper.

with $A = L_x L_y$ and $\hbar\omega_c = \Delta(\Pi^2/(2m^*)) = \Pi\Delta\Pi/m^*$ in the $\Pi_x\Pi_y$-space as shown in Figure 10.4. As the field B is raised the separation $\hbar\omega_c$ grows, and the quantum states are collected or bunched together. As a result of the bunching, the density of states, $\mathcal{D}(\varepsilon)$, should change periodically since the Landau levels (LLs) are equally spaced. The statistical weight $\mathcal{W}$ is the total number of states having energies less than

$$\varepsilon = \hbar\omega_c\left(N_L + \frac{1}{2}\right) .$$

From Figure 10.4, this $\mathcal{W}$ is given by

$$\mathcal{W} = \frac{L_x L_y}{(2\pi\hbar)^2} 2\pi\Pi\Delta\Pi \cdot 2 \sum_{N_L=0}^{\infty} \Theta\left[\varepsilon - \hbar\omega_c\left(N_L + \frac{1}{2}\right)\right] , \tag{10.33}$$

where $\Theta(x)$ is the *Heaviside step function*:

$$\Theta(x) = \begin{cases} 1 & \text{if } x > 0 \\ 0 & \text{if } x < 0 . \end{cases} \tag{10.34}$$

We introduce the dimensionless variable $\varepsilon^* \equiv 2\pi\varepsilon/(\hbar\omega_c)$, and rewrite $\mathcal{W}$ as

$$\mathcal{W}(\varepsilon) = 2C(\hbar\omega_c) \sum_{N_L=0}^{\infty} \Theta[\varepsilon^* - (2N_L + 1)\pi] , \quad C = 2\pi m^* A(2\pi\hbar)^{-2} . \tag{10.35}$$

We assume a high Fermi degeneracy such that

$$\mu \simeq \varepsilon_F \gg \hbar\omega_c . \tag{10.36}$$

The sum in (10.35) can be computed by using *Poisson's summation formula* [8]:

$$\sum_{n=-\infty}^{\infty} f(2\pi n) = \frac{1}{2\pi} \sum_{n=-\infty}^{\infty} \int_{-\infty}^{\infty} d\tau\, f(\tau) e^{-i\omega\tau} . \tag{10.37}$$

After the mathematical steps detailed in Appendix A.5, we obtain [9]

$$\mathcal{W}(E) = \mathcal{W}_0 + \mathcal{W}_{osc} , \tag{10.38}$$

$$\mathcal{W}_0 = C\hbar\omega_c \left(\frac{\varepsilon}{\pi}\right) = A\left(\frac{m^*}{\pi\hbar^2}\right)\varepsilon , \tag{10.39}$$

$$\mathcal{W}_{osc} = C\hbar\omega_c \frac{2}{\pi} \sum_{\nu=1}^{\infty} \frac{(-1)^\nu}{\nu} \sin\left(\frac{2\pi\nu\varepsilon}{\hbar\omega_c}\right) . \tag{10.40}$$

The B-independent term W_0 is the statistical weight for the system with no fields. The oscillatory term $\mathcal{W}_{osc}$ contains an infinite sum with respect to ν, but only the first term $\nu = 1$ is important in practice. The term $\mathcal{W}_{osc}$ can generate magnetic oscillations. There is no term proportional to B^2 generating the Landau diamagnetism.

We calculate the free energy F in (10.15) using the statistical weight $\mathcal{W}$ in (10.38) through (10.40), and obtain (Problem 10.3.1)

$$F = N\mu + A\frac{2m^*}{\pi\hbar^2}\varepsilon_F + A\frac{2e}{\pi\hbar}Bk_B T \sum_{\nu=1} \frac{(-1)^\nu}{\nu} \frac{\cos(2\pi\nu\varepsilon_F/(\hbar\omega_c))}{\sinh(2\pi^2\nu k_B T m^*/(\hbar e))} , \tag{10.41}$$

where we used the integration formulas in (10.20). We took the low-temperature limit except for the oscillatory terms. The *magnetization* I, the total magnetic moment per unit area, can be obtained from

$$I = -\frac{1}{A}\frac{\partial F}{\partial B} . \tag{10.42}$$

Thus far, we did not consider the Pauli magnetization I_{Pauli} due to the electron spin (see (9.18)):

$$I_{Pauli} = -\frac{2\mu_B^2}{A} N_0(\varepsilon_F) = 2n\mu_B \frac{\mu_B B}{\varepsilon_F} . \tag{10.43}$$

Using (10.41) through (10.43), we obtain the total magnetization:

$$\begin{aligned} I_{total} &= I_{Pauli} + I \\ &= 2n\mu_B \frac{\mu_B B}{\varepsilon_F}\left[1 - \left(\frac{\varepsilon_F}{\mu_B B}\right)\frac{k_B T}{\varepsilon_F}\left(\frac{m}{m^*}\right)\frac{\cos(2\pi\varepsilon_F/(\hbar\omega_c))}{\sinh(2\pi^2 m^* k_B T/(\hbar e B))}\right] . \end{aligned} \tag{10.44}$$

In this calculation, we neglected the spurious contribution of the B-derivatives of the quantities inside the $\cos(2\pi\varepsilon_F/(\hbar\omega_c))$. This condition is absent when we calculate the magnetization I through (10.42) directly. The magnetic susceptibility χ is defined by the ratio

$$\chi = \frac{I}{B} \,. \tag{10.45}$$

Only the first oscillatory term, $\nu = 1$, is important and kept in (10.44) since $\sinh(2\pi^2 m^* k_B T/(\hbar e B)) \gg 1$. The negative sign indicates a diamagnetic nature.

Figure 10.4 clearly shows that all electrons, not just those excited electrons near the Fermi surface, are subject to the magnetic field and all are in the Landau states. This is reflected by the fact that the Pauli magnetization I_{Pauli} is proportional to the electron density n, as seen in (10.43). The oscillatory magnetization is also proportional to the density n. At a finite temperature T, $-\mathrm{d}f/\mathrm{d}\varepsilon$ has a width. In this range of the order of $k_B T$ many oscillations occur if the field B is lowered. Assuming this condition, we obtained (10.44), and hence, this equation is valid for any finite T. At $T = 0$, the width vanishes and the oscillatory terms also vanish.

When the system is subjected to an external electric field, all electrons will respond, and hence, the magnetoconductivity σ should be proportional to the electron density n. The oscillatory statistical weight generates a SdH oscillation [1, 2] with the factor smaller by the factor $k_B T/\varepsilon_F$ in σ. Hence, the dHvA and SdH oscillations should be similar.

Störmer *et al.* [9] measured the SdH oscillations at 1.5 K in GaAs/AlGaAs, by stacking 4000 layers equivalent to the area of 240 cm^2. (Without stacking the signal is too small to observe.) Their data are shown in Figure 10.5. We see here that

(a) the oscillation periods $\varepsilon_F/\hbar\omega_c$ match,
(b) the magnetization $(-I)$ rather than the susceptibility χ behaves more similarly to the diagonal resistance ρ_{xx},
(c) the central line of the oscillation (background) is roughly independent of the field, and
(d) the envelopes for $-I$ and ρ_{xx} are similar.

Feature (a) means that both oscillations arise from the same cause, the periodic oscillation of the statistical weight $\mathcal{W}$. Feature (b) simply comes from the same field dependence (the B-independence) of the background (central line) of the oscillations. The density of states for a 2D quasifree electron system with no field is independent of the energy. The behavior (c) means that GaAs/AlGaAs is described adequately by this quasifree electron model. Feature (d) requires further discussion. Our formula (10.44) indicates that the period of the oscillations, $\varepsilon_F/(\hbar\omega_c) = m^*\varepsilon_F/(\hbar e B)$, and the exponential decay rate, $\pi k_B T/(\hbar\omega_c) = \pi k_B T m^*/(\hbar e)$, of the envelope $(\sinh(2\pi^2 k_B T m^*/(\hbar e B)))^{-1}$ are both controlled by the effective mass m^*. This feature is found not to be supported by the observed experiments. We shall discuss this point later in Chapter 11.

In conclusion, the 2D quasifree electron system is intrinsically paramagnetic, since there is no Landau diamagnetism. However, there are magnetic oscillations. The GaAs/AlGaAs heterostructure is often used for the study of the Quantum Hall Effect (QHE). The parental 3D GaAs is diamagnetic, and hence, the magnetic behavior is greatly different in 2D and 3D.

Problem 10.3.1. Verify (10.41).

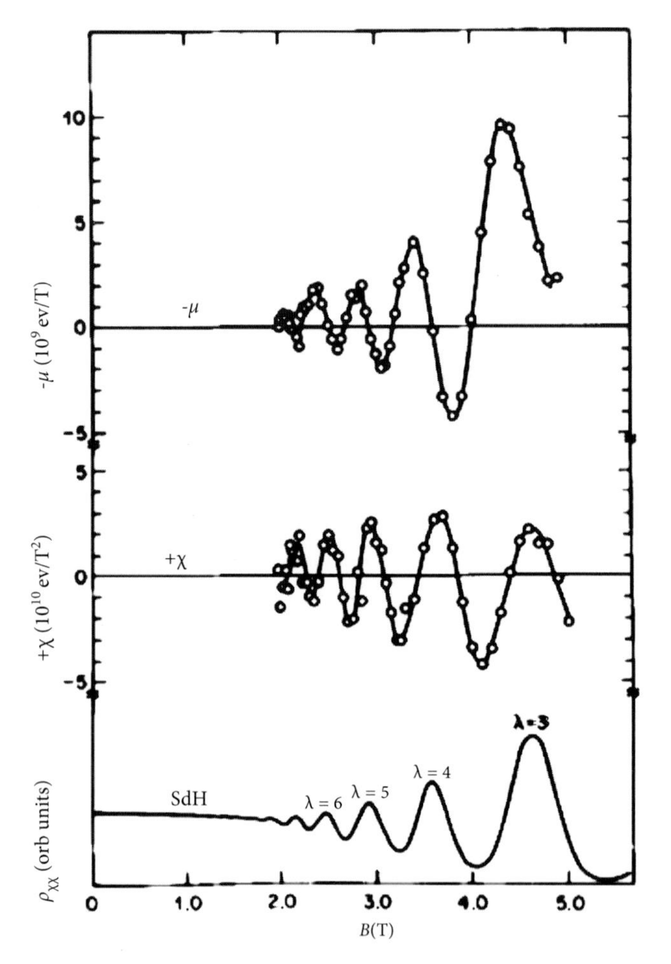

Figure 10.5 Experimental results on a stack of 4000 layers of 2D electron systems equivalent to an area of 240 cm^2 after Störmer *et al.* [9]. μ is the magnetic moment, and χ is the susceptibility. The trace denoted SdH is Shubnikov–de Haas data on a separate specimen of the same sample.

10.4 Anisotropic Magnetoresistance in Copper

10.4.1 Introduction

If the Fermi surface is nonspherical, the MR becomes anisotropic. Cu has open orbits in k-space as represented by trace b in Figure 10.6 [10]. This open orbit contains positive and negative curvatures along the contour of equal energy. No physical electron can move along the orbit as we see presently. An "electron" ("hole") is an elementary excitation which is generated on the positive (negative) side of the Fermi surface with the convention that the positive side contains the positive normal vector at the surface point, pointing in the energy-increasing direction. Thus, the "electron" ("hole") has an energy higher (lower) than the Fermi energy and circulates counterclockwise (clockwise) when viewed from the tip of the applied magnetic field vector (a standard definition). Since the static magnetic field cannot supply energy, no physical electron can travel *electron*-like in one section of the energy contour and *hole*-like in another. Klauder and Kunzler [11] observed a striking anisotropic MR as reproduced in Figure 10.7. The MR is over 400 times the zero-field resistance in some directions. We study an isotropic MR, applying kinetic theory and using the Fermi surface.

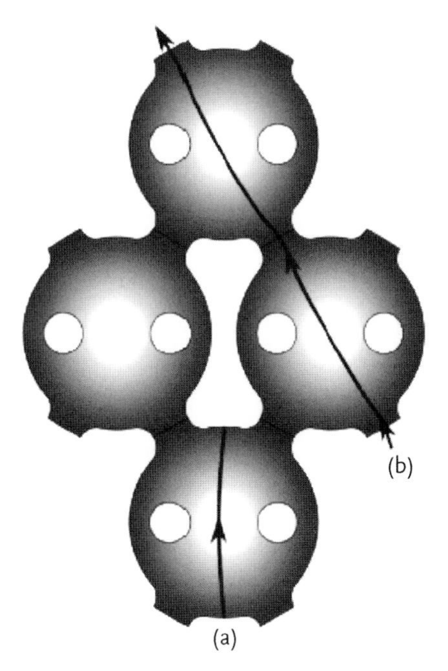

Figure 10.6 The figure shows a closed orbit "a" in k-space that can be traced by the electron, and an open orbit "b" that extends over the two Brillouin zones and that cannot be traveled by an electron.

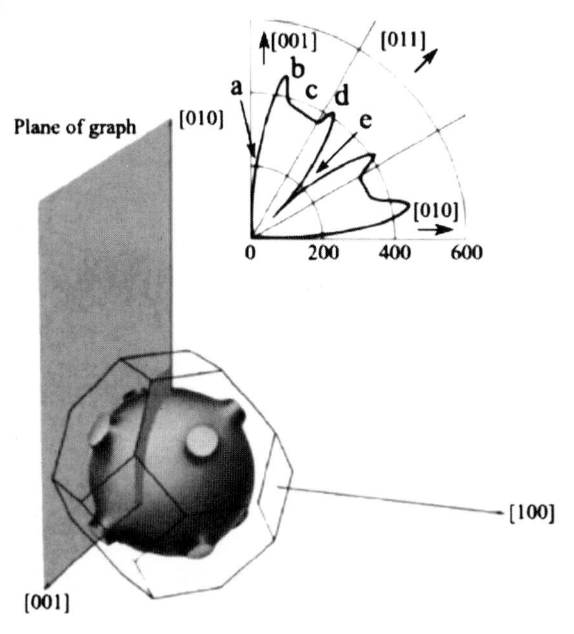

Figure 10.7 The striking anisotropy of the MR in Cu after Klauder and Kunzler [11]. The [001] and [010] directions of the Cu crystal are shown. The current flows in the [100] direction. The magnetic field is in the plane (100); its magnitude is fixed at 1.8 T~(=18 kG), and its direction varies continuously from [001] to [010] as shown in the inset.

Electron transport has traditionally been dealt with using kinetic theory or the Boltzmann equation method. In the presence of a static magnetic field, the classical electron orbit is curved. Then, the basic kinetic theoretical model in which the electron moves on a straight line, breaks down. Furthermore, the collision term of the Boltzmann equation containing the scattering cross section cannot be written down. Fortunately, quantum theory can save the situation. If the magnetic field is applied, then the classical electron continuously changes from straight line motion at zero field to curved motion at a finite magnetic field. When the magnetic field is gradually applied, the energy of the electron does not change, but the resulting spiral motion always acts so as to reduce the magnetic fields. Hence, the total energy of the electron with its surrounding fields is less than the sum of the electron energy and the unperturbed field energy. The electron *dressed* with the fields is in a bound (negative energy) state, and it can resist the break-up. The guiding center of the circulation can move in all directions in the absence of the electric field. If a weak electric field is applied in a direction, the dressed electron whose position is the guiding center, preferentially jumps in the field direction, and generates a current. We can apply kinetic theory to the guiding center motion, and obtain a formula for the electrical conductivity [12, 13]:

$$\sigma = \frac{n_c e^2 \tau}{M^*} , \qquad (10.46)$$

where n_c is the density of the dressed electrons, e the charge, M^* the magnetotransport (effective) mass, and τ the relaxation time. The magnetotransport mass M^* is distinct from the cyclotron mass m^*. Equation (10.46) can also be obtained by the Boltzmann equation method as shown earlier by Fujita *et al.* [12, 13]. In [12, 13], the dressed electron is identical to the composite fermion [14–17] used in the theory of the QHE [18, 19]. Briefly, the electron circulates around a finite number of flux quanta (fluxons) that are intact according to Onsager's flux quantization hypothesis [20]. Applying relativity, we may see that the fluxons move around the electron. From this viewpoint, the dressed electron is considered to carry a number of fluxons. Thus, the dressed electron is composed of an electron and fluxons. The composite particle moves as a fermion (boson) if it carries an even (odd) number of fluxons [21–23]. The free-energy minimum consideration favors a population dominance of c-fermions, each with two fluxons, over c-bosons, each with one fluxon, in the experimental conditions at liquid helium temperatures. The entropy is much higher for the c-fermions than for the c-bosons. The magnetic oscillation, which occurs only with Fermionic carriers, is observed in Cu. This experimental fact also supports the idea that the carriers in the magnetotransport are c-fermions.

Pippard in his book, *Magnetoresistance in Metal*, [24], argued that the MR for the quasifree particle system vanishes after using the relaxation time approximation in the Boltzmann equation method. The MR in actual experimental conditions is found to be always finite. Equation (10.46), in fact, contains the magnetotransport mass M^* distinct from the electron mass m^*. This fact alone makes the MR nonzero.

The MR is defined by

$$\mathrm{MR} = \frac{\Delta\rho}{\rho_0} = \frac{\rho(B) - \rho_0}{\rho_0}\,, \tag{10.47}$$

where $\rho(B)$ is the magnetoresistivity at the field magnitude B and $\rho_0 \equiv \rho(0)$, the resistivity at zero field.

First, we regard the small necks on the Fermi surface as *singular points* (see Figure 10.7). There are eight singular points in total on it. If the magnetic field $\boldsymbol{B}$ is along the direction [001], then there are two planes (parallel to the plane that the two vectors [010] and [100] make) containing four singular points each. The same condition also holds when the field $\boldsymbol{B}$ is along the direction [010]. These conditions correspond to the major minima of MR in Figure 10.7. Next, we consider the case in which the field $\boldsymbol{B}$ is along [011]. There are three planes perpendicular to [011] which contain two, four, and two singular points. This case corresponds to the second deepest minimum of MR. Lastly, the broad minima in data of MR correspond to the case where the field $\boldsymbol{B}$ is such that there are four planes perpendicular to $\boldsymbol{B}$, each containing two singular points. In this case, there is a range of angles in which this condition holds. Hence, these minima should be broad. This singular-points model can explain the presence of these minima in MR. We propose a more realistic model in the following section.

10.4.2
Theory

We shall introduce the following theoretical model:

(i) We assume that the magnetoconductivity σ can be calculated based on (10.46). The effective mass M^* and the relaxation time τ are unlikely to depend on the direction of the field $\boldsymbol{B}$. Only the conduction electron density n_c depends on the $\boldsymbol{B}$-direction relative to the lattice.
(ii) We assume that each neck (bad point) is represented by a sphere of radius a centered at the eight singular points on the ideal Fermi surface. When the magnetic field $\boldsymbol{B}$ is applied, the electron, then, circulates perpendicular to $\boldsymbol{B}$ in the k-space. If it hits the bad sphere, then it cannot complete the orbit, and cannot contribute to the conduction.

In Figure 10.7, we see the following five main features as the magnetic field $\boldsymbol{B}$ is rotated in the (100) plane from [001] to [011]. The MR has a deepest minimum (labeled a in Figure 10.7), a greatest maximum (b), a broad and flat minimum (c), a second greatest maximum (d), and a second deepest minimum (e). These features are repeated in the reversed order as the field is rotated from [011] to [010] due to the symmetry of the Fermi surface. We note that the three minima, a, c, and e were qualitatively explained earlier based on the singular-points model. There are eight bad spheres located on the Fermi surface in the direction $\langle 111 \rangle$ from the center O.

We first consider case c. In Figure 10.8, the Fermi surface viewed from [100] is shown. The four dark parts represent the nonconducting k-space volumes (bad volumes). The bad volume contains balls. There are four bad volumes here, and all the centers of the balls in the projected plane perpendicular to [100] lie on the circle of the radius R. This result is connected with the Fermi momentum $p_F \equiv \hbar k$ by

$$R = \sqrt{\frac{2}{3}}k \, . \tag{10.48}$$

Each bad volume can be calculated by using the integration formulas (Problem 10.4.1):

$$\begin{aligned} I(x_0) &= \int_{x_0-a}^{x_0+a} \pi(k^2 - x^2)\mathrm{d}x \\ &= \pi\left[2ak^2 - \frac{1}{3}(x_0+a)^3 + \frac{1}{3}(x_0-a)^3\right] \\ &= 2\pi ak^2 - \frac{2}{3}\pi\left(3x_0^2 a + a^3\right) , \end{aligned} \tag{10.49}$$

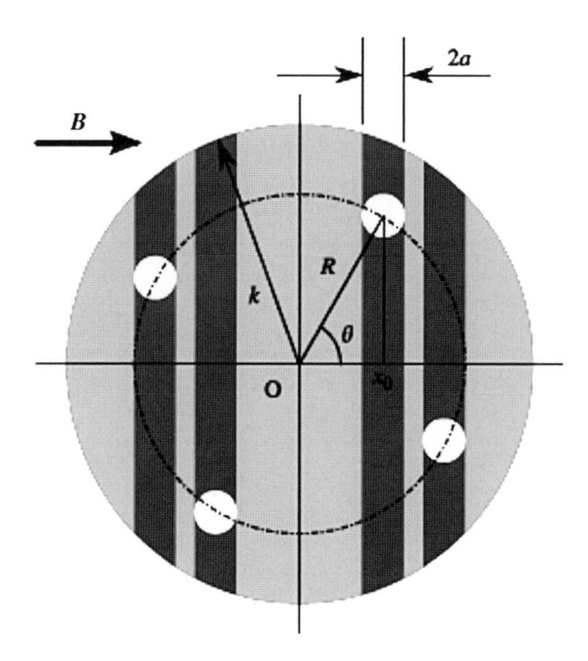

Figure 10.8 The electron circulates perpendicular to the magnetic field $\boldsymbol{B}$ in the k-space. If the electron hits the bad ball of radius a, it will not complete circulation, and it does not contribute to the conduction. The four dark slices viewed from [100], each a width of $2a$, contain nonconducting electrons.

where $x_0 = R\cos\theta$ is the x-component of the ball center, and θ is the angle shown in Figure 10.8. The four centers of the balls lie on the circle, separated by $\pi/2$ in angle. We consider the sum of a pair of two bad volumes associated with the centers at θ and $\theta + \pi/2$. Using (10.49), $\cos(\theta + \pi/2) = -\sin\theta$, and $\sin^2\theta + \cos^2\theta = 1$, we obtain

$$4\pi a k^2 - \frac{4}{3}\pi a^3 - 2\pi a R^2 \,. \tag{10.50}$$

The volume of the other pair with the centers at $\theta + \pi$ and $\theta + 3\pi/2$ contributes the same amount. Therefore, using (10.48), the total bad volume is

$$V_c = \frac{16}{3}\pi a k^2 - \frac{8}{3}\pi a^3 \,. \tag{10.51}$$

The volume V_c does not depend on the angle θ, supporting the broad minimum c observed in the experimental data.

Similarly, we can calculate the bad volumes for the other cases, a and c, and the resultant volumes are

$$V_a = \frac{1}{2} V_c \,, \tag{10.52}$$

$$V_e = \frac{5}{8} V_c \,. \tag{10.53}$$

Going from c to e, the four-slice volume monotonically changes to the three-slice volume. This means that the MR changes smoothly without taking a maximum. From c to a, the four-slice volume changes monotonically to the two-slice volume. Thus, the present model generates no MR maxima. We shall give an explanation for the observed MR maxima in the next section.

The conducting (good) volume is equal to the total ideal Fermi sphere volume $(\frac{4}{3})\pi k^3$ subtracted by the bad volume. The conduction electron density n_c is given by the ideal density n multiplied by the ratio of the good volume over the ideal Fermi sphere volume:

$$n_c = n \frac{(4/3)\pi k^3 - V}{(4/3)\pi k^3} . \tag{10.54}$$

We call the inverse of the magnetoresistivity $\rho(\boldsymbol{B})$ the magnetoconductivity $\sigma(\boldsymbol{B})$. Rewriting $\Delta\rho/\rho_0$ in terms of σ, we obtain

$$\mathrm{MR} \equiv \frac{\Delta\rho}{\rho_0} = \frac{1/\sigma - 1/\sigma_0}{1/\sigma_0} = \frac{\sigma_0}{\sigma} - 1 , \tag{10.55}$$

where $\sigma_0 \equiv 1/\rho_0$ is the zero-field conductivity, which can be calculated with the following formula:

$$\sigma_0 = \frac{n_0 e^2 \tau_0}{m^*} , \tag{10.56}$$

where m^* is the cyclotron mass, and the suffix 0 denotes the zero-field quantities.

Since in the experiments, the MR is very large, ~ 100, compared with unity, we may ignore -1 in (10.55). Using (10.46) and (10.56), we obtain

$$\frac{\Delta\rho}{\rho_0} = \frac{M^*}{m^*} \frac{\tau_0}{\tau} \frac{n_0}{n} . \tag{10.57}$$

This indicates that the lower the magneto conduction electron density n_c, the higher becomes the MR.

10.4.3 Discussion

In Figure 10.7, we observe a MR maximum near [011]. The calculation using our model shows a monotonic change of MR from the four-slice configuration c to the three-slice configuration e. We propose the following explanation. In going from c to e, an overlap of the bad volumes must occur, and the overlapping should not be line-sharp, as assumed in our model. Then the fluctuations, which must occur, generate dissipation. This results in a resistivity maximum in d. In going from the four-slice configuration c to the two-slice configuration a, an overlapping must also occur. Since, in this case, the two overlaps take place simultaneously, the fluctuations and the resultant resistivity increase should be greater, generating a MR maximum higher in b than that in d.

In Figure 10.7, we observe that the MR minimum varies from 80 to 340 (from e to c). This arises from the change in the conduction electron density n_c. Now, we estimate the ratio of the bad ball diameter a over the Fermi momentum k, using the two MR values. With (10.51), (10.53), (10.54), and (10.57) we obtain

$$\frac{(4/3)\pi k^3 - (5/8) V_c}{(4/3)\pi k^3 - V_c} = \frac{340}{80} . \tag{10.58}$$

This yields the ratio:

$$\frac{a}{k} = 0.23 , \tag{10.59}$$

which is reasonable.

We see in Figure 10.7 that the MR rises quadratically with the field angle away from the minimum at e. Our model explains this behavior as follows.

The center of one of the balls lies at $\theta = \pi/2$ for case e. We introduce a small deviation angle ϕ such that

$$\phi = \theta - \frac{\pi}{2} . \tag{10.60}$$

Then, we have

$$x_0 = R\cos\left(\frac{\phi + \pi}{2}\right) = R\sin\phi \approx R\phi . \tag{10.61}$$

Using this and (10.49), we calculate the bad volume near e and obtain

$$V = \frac{5}{8} V_c + 2\pi a R^2 \phi^2 . \tag{10.62}$$

Thus, this shows that the MR rises quadratically in the deviation angle ϕ on the positive and negative sides. This is in agreement with the MR data as shown in Figure 10.7 . This quadratic behavior holds true for the regions between a and b.

In our model a spherical ball was used for a bad volume. We may consider an ellipsoidal (two parameters) model for the improvement.

Other noble metals such as Ag and Au are known to have the Fermi surface with necks. If the bad balls are greater in relative size (a/k), our theory predicts more prominent MR. Experimental confirmation of this behavior is highly desirable. We suggest that the experiments be done below 1 K, where the phonon scattering is negligible and the MR minima become more visible. Only the minima, and not the maxima, contain important information about the Fermi surface.

In conclusion, the spectacular angular dependence of the magnetoresistance in Cu can be explained by using the Drude formula based on the "neck" Fermi surface. The resistance minima can be used to estimate the "neck" size. The resistance (dissipation) maxima arise from the density fluctuations.

Problem 10.4.1. Verify (10.49).

10.5 Shubnikov–de Haas Oscillations

Oscillations in magnetoresistance, similar to the dHvA oscillation in the magnetic susceptibility, were first observed by Shubnikov and de Haas in 1930 [25]. These oscillations are often called the *Shubnikov–de Haas oscillations*. The susceptibility is an equilibrium property and can therefore be calculated by standard statistical mechanical methods. The MR is a nonequilibrium property, and its treatment requires a kinetic theory. The magnetic oscillations in both cases arise from the periodically varying density of states. We shall see that the observation of the oscillations gives a direct measurement of the magnetotransport mass M^*. The observation also gives quantitative information on the cyclotron mass m^*.

Let us take a dilute system of electrons moving in a plane. Applying a magnetic field $\boldsymbol{B}$ perpendicular to the plane, each electron will be in the Landau state with the energy (see Section 9.3):

$$\varepsilon = \left(n + \frac{1}{2}\right)\hbar\omega_{\mathrm{c}}\,, \qquad \omega_{\mathrm{c}} \equiv \frac{eB}{m^*}\,, \qquad N_{\mathrm{L}} = 0, 1, 2, \ldots\,, \tag{10.63}$$

where m^* is the cyclotron mass. The degeneracy of the LL is given by (Problem 10.5.1)

$$\frac{eBA}{2\pi\hbar}\,, \qquad A = \text{sample area}\,. \tag{10.64}$$

The weaker is the field, the more LLs, separated by $\hbar\omega_{\mathrm{c}}$, are occupied by the electrons. In this Landau state the electron can be viewed as circulating around a guiding center. The radius of circulation $l \equiv (\hbar/(eB))^{1/2}$ for the Landau ground state is about 250 Å at a field $B = 1.0\,\mathrm{T}$. If we now apply a weak electric field $\boldsymbol{E}$ in the x-direction, then the guiding center jumps and generates a current.

Let us first consider the case with no magnetic field. We assume a uniform distribution of impurities with the density n_{I}. Solving the Boltzmann equation, we obtain the conductivity (Problem 10.5.2):

$$\sigma = \frac{2e^2}{m^*(2\pi\hbar)^2}\int \mathrm{d}^2 p \frac{\varepsilon}{\Gamma}\left(-\frac{\mathrm{d}f}{\mathrm{d}\varepsilon}\right)\,, \qquad \varepsilon = \frac{p^2}{2m^*}\,, \tag{10.65}$$

where Γ is the energy (ε)-dependent relaxation rate,

$$\Gamma(\varepsilon) = n_{\mathrm{I}}\int \mathrm{d}\Omega\left(\frac{p}{m^*}\right) I(p,\theta)(1-\cos\theta)\,, \tag{10.66}$$

where θ = scattering angle and $I(p,\theta)$ = scattering cross section, and the Fermi distribution function

$$f(\varepsilon) \equiv \frac{1}{\mathrm{e}^{\beta(\varepsilon-\mu)}+1} \tag{10.67}$$

with $\beta \equiv (k_{\mathrm{B}}T)^{-1}$ and μ = chemical potential is normalized such that

$$n = \frac{2}{(2\pi\hbar)^2}\int \mathrm{d}^2 p\, f(\varepsilon)\,, \tag{10.68}$$

where the factor 2 is due to the spin degeneracy. We introduce the density of states, $\mathcal{D}(\varepsilon)$, such that

$$\frac{2}{(2\pi\hbar)^2}\int d^2p\cdots = \int d\varepsilon\mathcal{D}(\varepsilon)\cdots . \tag{10.69}$$

We can then rewrite (10.65) as

$$\sigma = \frac{e^2}{m^*}\int_0^\infty d\varepsilon\mathcal{D}(\varepsilon)\frac{\varepsilon}{\Gamma}\left(-\frac{df}{d\varepsilon}\right) . \tag{10.70}$$

The Fermi distribution function $f(\varepsilon)$ drops steeply near $\varepsilon = \mu$ at low temperatures:

$$k_B T \ll \varepsilon_F . \tag{10.71}$$

The density of states, $\mathcal{D}(\varepsilon)$, is a slowly varying function of the energy ε. For a 2D quasifree electron system, the density of states is independent of the energy ε. Then the Dirac *delta-function replacement formula*

$$-\frac{df}{d\varepsilon} = \delta(\varepsilon - \mu) \tag{10.72}$$

can be used. Assuming this formula, using

$$\int_0^\infty d\varepsilon\mathcal{D}(\varepsilon)\varepsilon\left(-\frac{df}{d\varepsilon}\right) = \int_0^\infty d\varepsilon\mathcal{D}(\varepsilon)f(\varepsilon) , \tag{10.73}$$

and comparing (10.65) and (10.70), we obtain

$$\tau = \int_0^\infty d\varepsilon\mathcal{D}(\varepsilon)\frac{1}{\Gamma(\varepsilon)}f(\varepsilon) . \tag{10.74}$$

Note that the temperature dependence of the relaxation time τ is introduced through the Fermi distribution function $f(\varepsilon)$.

Next we consider the case with a magnetic field. We assume that a dressed electron is a fermion with magnetotransport mass M^* and charge e. Applying kinetic theory to the dressed electrons, we obtain the standard formula for the conductivity: $\sigma = ne^2\tau/M^*$. As discussed earlier, the dressed electrons move in all directions (isotropically) in the absence of the electric field.

We introduce *kinetic momenta*:

$$\Pi_x \equiv p_x + eA_x , \quad \Pi_y \equiv p_y + eA_y . \tag{10.75}$$

The quasifree electron Hamiltonian $\mathcal{H}$ is

$$\mathcal{H} = \frac{1}{2m^*}\left(\Pi_x^2 + \Pi_y^2\right) \equiv \frac{1}{2m^*}\Pi^2 . \tag{10.76}$$

The variables (Π_x, Π_y) are the same kinetic momenta introduced earlier in (9.34). Only we are dealing with a 2D system here. After simple calculations, we obtain

$$\mathrm{d}x\mathrm{d}\Pi_x\mathrm{d}y\mathrm{d}\Pi_y \equiv \mathrm{d}x\mathrm{d}p_x\mathrm{d}y\mathrm{d}p_y \,. \tag{10.77}$$

We can represent the quantum states by quasiphase space element $\mathrm{d}x\mathrm{d}\Pi_x\mathrm{d}y\mathrm{d}\Pi_y$. The Hamiltonian $\mathcal{H}$ in (10.76) does not depend on the position (x, y). Assuming large normalization lengths (L_x, L_y), $A = L_x L_y$, we can then represent the Landau states by the concentric shells in the $\Pi_x\Pi_y$-space (see Figure 10.9), having the statistical weight

$$\frac{2\pi L_x L_y}{(2\pi\hbar)^2}\Pi\Delta\Pi = \frac{A}{2\pi\hbar}\omega_c m^* = \frac{eAB}{2\pi\hbar} \tag{10.78}$$

with the energy separation $\hbar\omega_c = \Delta(\Pi^2/(2m^*)) = \Pi\Delta\Pi/m^*$. Equation (10.78) confirms that the LL degeneracy is $eBA/(2\pi\hbar)$, as stated in (10.64).

Let us consider the motion of the field-dressed electrons (guiding center). We assume that the dressed electron is a fermion with magnetotransport mass M^* and charge e. The kinetic energy is represented by

$$\mathcal{H}_K = \frac{1}{2M^*}\left(\Pi_x^2 + \Pi_y^2\right) \equiv \frac{1}{2M^*}\Pi^2 \,. \tag{10.79}$$

According to Onsager's flux quantization represented by (10.11), the magnetic fluxes can be counted in units of $\Phi_0 = e/h$. The dressed electron is composed of an electron and two elementary fluxes (fluxons). A further explanation of the present model will be given later.

Let us introduce a *distribution function* $\varphi(\Pi, t)$ in the $\Pi_x\Pi_y$-space normalized such that

$$\frac{2}{(2\pi\hbar)^2}\int \mathrm{d}^2\Pi\varphi(\Pi_x, \Pi_y, t) = \frac{N}{A} = n \,. \tag{10.80}$$

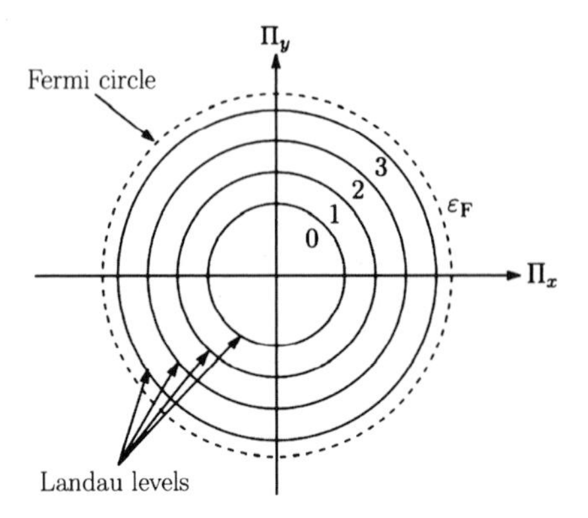

Figure 10.9 The circulation part of the Landau states is represented by the circular shell in $\Pi_x\Pi_y$-space.

The Boltzmann equation for a homogeneous stationary system is

$$e(\boldsymbol{E} + \boldsymbol{v} \times \boldsymbol{B}) \cdot \frac{\partial \varphi}{\partial \Pi} = \int \mathrm{d}\Omega \, \frac{\Pi}{M^*} n_{\mathrm{I}} I(\Pi, \theta)[\varphi(\Pi') - \varphi(\Pi)] \,, \tag{10.81}$$

where θ is the angle of deflection, that is, the angle between the initial and final kinetic momenta (Π, Π'). In the actual experimental condition the magnetic force term can be neglected. Assuming this condition for now, we obtain the Boltzmann equation for a field-free system.[1] Hence, we obtain the same conductivity formula (10.65) with m^* replaced by M^*, yielding (10.46).

As the field B is raised, the separation $\hbar\omega_{\mathrm{c}}$ becomes greater and the quantum states are bunched together. The density of states should contain an oscillatory part:

$$\sin\left(\frac{2\pi\varepsilon'}{\hbar\omega_{\mathrm{c}}} + \phi_0\right) , \quad \varepsilon' = \frac{\Pi'^2}{2m^*} \,, \tag{10.82}$$

where ϕ_0 is a phase. Since

$$\frac{\varepsilon_{\mathrm{F}}}{\hbar\omega_{\mathrm{c}}} \gg 1 \quad \text{(weak field)} \,, \tag{10.83}$$

the phase ϕ_0 will be dropped hereafter. Physically, the sinusoidal variations in (10.82) arise as follows. From the Heisenberg uncertainty principle (phase space consideration) and the Pauli exclusion principle, the Fermi energy ε_{F} remains approximately constant as the field B varies. The density of states is high when ε_{F} matches the N_{L}th level, while it is small when ε_{F} falls between neighboring LLs.

If the density of states, $\mathcal{D}(\varepsilon)$, oscillates violently in the drop of the Fermi distribution function $f(\varepsilon) \equiv (\mathrm{e}^{\beta(\varepsilon-\mu)} + 1)^{-1}$, one cannot use the delta-function replacement formula. The use of (10.72) is limited to the case in which the integrand is a smooth function near $\varepsilon = \mu$. The width of $\mathrm{d}f/\mathrm{d}\varepsilon$ is of the order $k_{\mathrm{B}}T$. The critical temperature T_{c} below which the oscillations can be observed is

$$k_{\mathrm{B}}T \sim \hbar\omega_{\mathrm{c}} \,. \tag{10.84}$$

Below the critical temperature $T < T_{\mathrm{c}}$, we may proceed as follows. Let us consider the integral

$$I = \int_0^\infty \mathrm{d}\varepsilon \, f(\varepsilon) \sin\left(\frac{2\pi\varepsilon'}{\hbar\omega_{\mathrm{c}}}\right) \quad \varepsilon \equiv \frac{\Pi^2}{2M^*} \,. \tag{10.85}$$

For the temperature satisfying $\beta\varepsilon = \varepsilon/(k_{\mathrm{B}}T) \gg 1$, we can use the same mathematical steps as going from (10.17) to (10.20) and obtain

$$I = \pi k_{\mathrm{B}} T \frac{\cos(2\pi\varepsilon_{\mathrm{F}}/(\hbar\omega_{\mathrm{c}}))}{\sinh(2\pi^2 M^* k_{\mathrm{B}} T/(\hbar e B))} \,. \tag{10.86}$$

1) The Boltzmann equation for a magnetic field-free system, where a small constant electric field $\boldsymbol{E}$ is applied to the positive x-axis, is given by

$$-eE\frac{\partial\varphi(\boldsymbol{p})}{\partial p_x} = n_{\mathrm{I}} \int \mathrm{d}\Omega \, \frac{p}{m} \, I[\varphi(\boldsymbol{p}') - \varphi(\boldsymbol{p})] \,.$$

Here we used

$$M^*\mu(T=0) = m^*\varepsilon_F = \frac{1}{2}p_F^2 \,, \tag{10.87}$$

which follows from the fact that the Fermi momentum p_F is the same for both dressed and undressed electrons.

In summary, (i) the SdH oscillation period is $\varepsilon_F/(\hbar\omega_c)$, which is the same for the dHvA oscillations. This arises from the bunching of the quantum states. (ii) The envelope of the oscillations exponentially decreases like $[\sinh(2\pi^2 M^* k_B T/(\hbar e B))]^{-1}$. Thus, if the "decay rate" δ defined through

$$\sinh\left(\frac{\delta}{B}\right) \equiv \sinh\left(\frac{2\pi^2 M^* k_B T}{\hbar e B}\right) \tag{10.88}$$

is measured carefully, the magnetotransport mass M^* can be obtained *directly* through $M^* = e\hbar\delta/(2\pi^2 k_B T)$. This finding is quite remarkable. For example, the relaxation rate τ^{-1} can now be obtained through (10.46) with the measured magnetoconductivity. All electrons, not just those excited electrons near the Fermi surface, are subject to the $\boldsymbol{E}$-field. Hence, the carrier density n_c appearing in (10.46) is the *total* density of the dressed electrons. This n_c also appears in the Hall resistivity expression:

$$\rho_H \equiv \frac{E_H}{j} = \frac{v_d B}{e n_c v_d} = \frac{B}{e n_c} \,, \tag{10.89}$$

where the Hall effect condition $E_H = v_d B$, $v_d =$ drift velocity, was used.

In *cyclotron motion* the electron with the effective mass m^* circulates around the magnetic fluxes. Hence, the cyclotron frequency ω_c is given by eB/m^*. The guiding center (dressed electron) moves with the magnetotransport mass M^*, and therefore, this M^* appears in the hyperbolic sine term in (10.86).

In 1952 Dingle [26] developed a theory for the dHvA oscillations. He proposed to explain the envelope behavior in terms of a *Dingle temperature* T_D such that the exponential decay factor be

$$\exp\left(\frac{-\lambda(T+T_D)}{B}\right) \,, \quad \lambda = \text{constant} \,. \tag{10.90}$$

Instead of the modification in the temperature, we introduced the magnetotransport mass M^* to explain the envelope behavior. The susceptibility ξ is an equilibrium property, and hence, ξ should be calculated without considering the relaxation mechanism. In our theory, the envelope of the oscillations is obtained by taking the average of the sinusoidal density of states with the Fermi distribution of the dressed electrons. There is no place where the impurities come into play. The theory may be checked by varying the impurity density. Our theory predicts little change in the clearly defined envelope. The scattering will change the *relaxation rate* τ^{-1} and the *magnetoconductivity* for the center of the oscillations though (10.46).

In the present theory, the two masses m^* and M^* were introduced corresponding to two physical processes: the cyclotron motion of the electron and the magnetotransport motion of the dressed electron. The dressed electrons are present whether the system is probed in equilibrium or in nonequilibrium as long as the system is subjected to a magnetic field. The presence of the c-particles can be checked by measuring the susceptibility or the heat capacity of the system. All dressed electrons are subject to the magnetic field, and hence, the magnetic susceptibility χ is proportional to the carrier density n_c, although the χ depends critically on the Fermi surface. This explains why the magnetic oscillations in the conductivity and the susceptibility are similar.

Problem 10.5.1. The Landau levels are highly degenerate. Derive that the degeneracy, that is, the number of electrons that can occupy each Landau level, is given by $eBA/(2\pi\hbar)$, where A is the sample area perpendicular to the magnetic field $\boldsymbol{B}$.

Problem 10.5.2. Derive the conductivity formula (10.65).

References

1 Haas, W.J. de and Alphen, P.M. van (1930) *Leiden Commun.*, 208d, 212a.
2 Haas, W.J. de and Alphen, P.M. van (1932), *Leiden Commun.*, 220d.
3 Onsager, L. (1952) *Philos. Mag.*, **43**, 1006.
4 Halse, M.R. (1969) *Philos. Trans. R. Soc. A*, **265**, 507.
5 Schönberg, D. and Gold, A.V. (1969) Physics of Metals 1: *Electrons*, ed. Ziman, J.M., Cambridge University Press, Cambridge, UK, p. 112.
6 Deaver, B.S. and Fairbank, W.M. (1961) *Phys. Rev. Lett.*, **7**, 51.
7 Doll, R. and Näbauer, M. (1961) *Phys. Rev. Lett.*, **7**, 51.
8 Shubnikov, L. and Haas, W.J. de (1930) *Leiden Commun.*, 207a, 207c, 207d, 210a.
9 Störmer, H.L., Haavasoja, T., Narayanamurti, V., Gossard, A.C., and Wiegmann, W. (1983) *J. Vac. Sci. Technol. B*, **1**, 423.
10 Ashcroft, N.W. and Mermin, N.D. (1976) *Solid State Physics*, Saunders, Philadelphia, pp. 291–293.
11 J.R. Klauder and Kunzler, J.E. (1960) in: Harrison, W.A. and Webb, M.B. (eds), *The Fermi Surface*, Wiley, New York.
12 Fujita, S., Horie, S., Suzuki, A., Morabito, D.I. (2006) *Indian J. Pure Appl. Phys.*, **44**, 850.
13 Fujita, S., Ito, K., Kumek, Y., Okamura, Y. (2004) *Phys. Rev. B*, **70**, 075 304.
14 Zhang, S.C., Hansson, T.H., and Klvelson, S. (1989) *Phys. Rev. Lett.*, **62**, 82.
15 Jain, J.K. (1989) *Phys. Rev. Lett.*, **63**, 199.
16 Jain, J.K. (1989) *Phys. Rev. B*, **40**, 8079.
17 Jain, J.K. (1990) *Phys. Rev. B* **41**, 7653.
18 Ezawa, Z.F. (2000) *Quantum Hall Effect*, World Scientific, Singapore.
19 Prange, R.E. and Girvin, S.M. (eds) (1990) *Quantum Hall Effect*, Springer, New York.
20 Onsager, L. (1952) *Philos. Mag.* **43**, 1006.
21 Bethe, H.A. and Jackiw, R.J. (1968) *Intermediate Quantum Mechanics*, 2nd edn, Benjamin, New York, p. 23;
22 Ehrenfest, P. and Oppenheimer, J.R. (1931) *Phys. Rev.* **37**, 311.
23 Fujita, S. and Morabito, D.L. (1998) *Mod. Phys. Lett. B*, **12** 1061.
24 Pippard, A.R. (1989) *Magnetoresistance in Metals*, Cambridge University Press, Cambridge, UK, pp. 3–5.
25 Shubnikov, L. and Haas, W. de (1930) *Leiden Commun.* **207a**, 3.
26 Dingle, R.B. (1952) *Proc. R. Soc. A*, **211**, 500.

11
Quantum Hall Effect

Major experimental facts associated with the Quantum Hall Effect (QHE) and a brief theoretical survey are given in this chapter. A theory of treating integer and fractional QHE is developed in a unified manner based on the model, in which the phonon exchange between the electron and the elementary magnetic flux (fluxon, half-spin fermion) binds an electron–fluxon composite. The center of mass of the composit boson moves with the linear dispersion relation: $\varepsilon = (2/\pi)v_F q$, v_F = Fermi speed, q = momentum. The 2D system of free composite bosons undergoes a Bose–Einstein condensation at $k_B T_c = 1.241\,\hbar v_F n_0^{1/2}$, where n_0 is the density. The QHE state at the Landau-level occupation ratio $\nu = P/Q$, integer P, odd integer Q, is shown to be the superconducting state with an energy gap generated by the condensed composite (c-)bosons, each containing one electron and Q fluxons and carrying the fractional charge e/Q. This state is formed by phonon exchange from the c-fermions, each with $Q-1$ fluxons, occupying the lowest P Landau states. The c-boson density n_0 is one Pth the electron density n_e, making the QHE state less stable with increasing P. In Jain's theory of fractional hierarchy, the effective magnetic field for the c-fermion vanishes at the sermonic fraction P/Q, even Q. This idea is extended to the c-particle (boson, fermion) which moves field-free in 2D at $\nu = P/Q$. The plateau value $(Q/P)(h/e^2)$ at the fractional ratio $\nu = P/Q$, odd Q, directly indicates the fractional charge (magnitude) e/Q for the composite with Q fluxons.

11.1
Experimental Facts

In 1980 von Klitzing *et al.* [1] reported the discovery of the *integer Quantum Hall Effect* (QHE). In 1982 Tsui *et al.* [2] discovered the *fractional QHE*. The GaAs/AlGaAs heterojunction is shown in Figure 11.1. The electrons are trapped in GaAs near the interface within a distance of the order of 100 Å. The QHE measurements are normally carried out in the geometry shown in Figure 11.2. The *Hall field* E_H is the Hall voltage V_H divided by the sample width W. The *Hall resistivity* ρ_H is defined

Electrical Conduction in Graphene and Nanotubes, First Edition. S. Fujita and A. Suzuki.

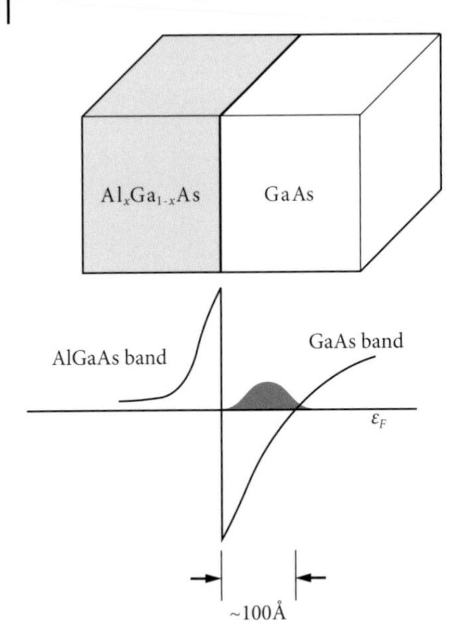

Figure 11.1 The GaAs/AlGaAs heterojunction. The electrons are localized within less than 100 Å of the interface. The shaded area represents the electron distribution.

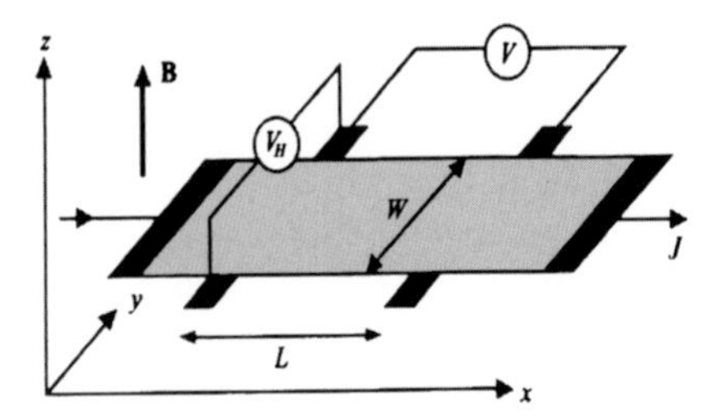

Figure 11.2 Hall effect measurements. The Hall voltage V_H is generated perpendicular to the current J and the magnetic field $\boldsymbol{B}$.

by the ratio:

$$\rho_H \equiv \frac{E_H}{j}, \quad E_H \equiv \frac{V_H}{W}, \tag{11.1}$$

where

$$j \equiv \frac{J}{L} = \frac{E}{\rho}, \quad E = \text{external electric field} \equiv \frac{V}{L} \tag{11.2}$$

is the *current density* defined as the current J over the sample length L. The external electric field E is defined as the voltage V over the length L. Equation (11.2) defines the *resistivity* ρ according to *Ohm's law*. Figure 11.3 represents the data reported by Tsui [3] for the Hall resistivity ρ_H and the resistivity ρ in GaAs/AlGaAs at 60 mK.

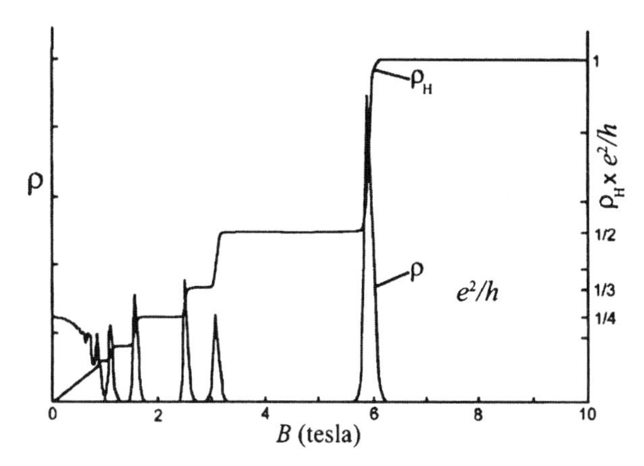

Figure 11.3 Observed QHE in the GaAs/AlGaAs heterojunction at 60 mK, after Tsui [3]. The Hall resistivity ρ_H and the resistance ρ are shown as a function of the magnetic field B in tesla.

The Quantum Hall (QH) states at the *Landau-Level (LL) occupation ratio*, also called the *filling factor*, $\nu = 1, 2, \ldots, 6$ are visible in Figure 11.3. Clearly, each QH state with a *Hall resistivity plateau* (horizontal stretch) is *superconducting* (zero resistance). In the normal metal the Hall resistivity ρ_H is linear in the applied magnetic field B (Problem 11.1.1). The *Hall coefficient* R_H is defined as the ratio E_H/jB:

$$R_H \equiv \frac{E_H}{jB} = \frac{\rho_H}{B} . \tag{11.3}$$

The plateau stability arises from the Meissner effect, which expels the excess weak magnetic field measured relative to the center position (field) of the Hall resistivity plateau from the superconducting body. The stability against the electric field arises from the *gap* in the elementary excitation energy spectrum, which will be explained later. The superconducting state with an energy gap is sometimes referred to as the *incompressible quantum fluid state* in the literature [4–9]. We avoid this characterization in the present text since no pressure is applied to the system. Besides, some research groups have investigated the pressure dependence of the QH state. In such a case the term incompressible quantum fluid becomes confusing. The *superconducting state* (no resistance) characterizes the state more directly. The plateau heights are *quantized* in units of h/e^2, $h =$ planck constant, $e =$ electron charge (magnitude), see Figure 11.3. Each plateau height is material- and shape-independent, indicating the fundamental quantum nature, the boundary-condition independence, and the stability of the QH state.

Figure 11.4 represents the QHE data reproduced after Willett *et al.* [10]. The data indicate a remarkable similarity among and between the integer QHE at $\nu = P$, $P = 1, 2, 3, \ldots$, *and* the fractional QHE at $\nu = P/Q$, positive integer (numerator) P, $P = 1, 2, \ldots$ and odd integer (denominator) Q, $Q = 1, 3, \ldots$ Compare in particular the QH state at $\nu = 2$ and $2/3$, suggesting a single cause for both integer and fractional QHE. The QHE is temperature-dependent, including the superconducting transition at a critical temperature T_c, see below.

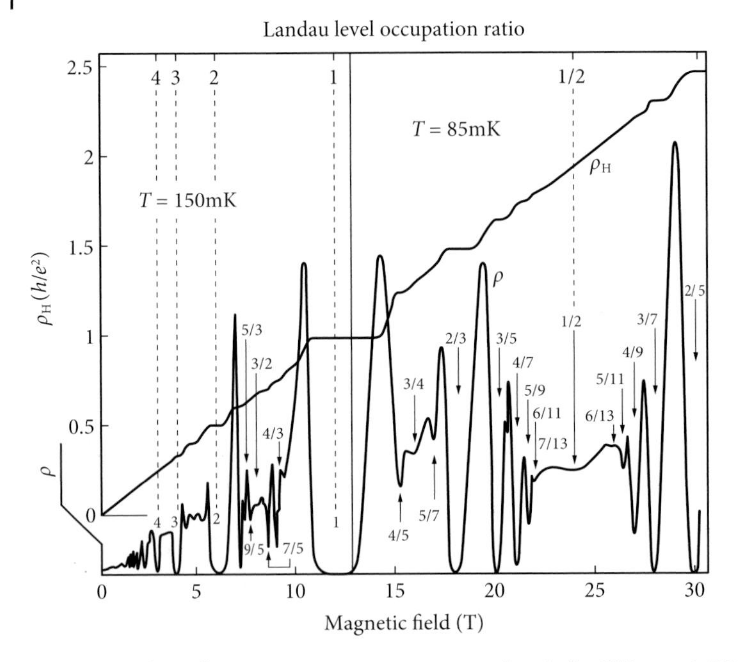

Figure 11.4 The Hall resistivity $\rho_H (\equiv \rho_{xy})$ in h/e^2 and the resistivity $\rho(\equiv \rho_{xx})$ in GaAs/AlGaAs at 85, 150 mK are plotted as a function of the magnetic field B in tesla, reproduced after Willett *et al.* [10]. The high-field ρ trace is reduced in amplitude by a factor of 2.5. The upper numbers indicate the Landau-level occupation ratio ν.

The resistivity data under fields of up to 50 T from Leadley *et al.* [11] are shown in Figure 11.5. The QH state below T_c is recognized by the Hall resistivity plateau and zero resistivity. We see that the critical temperature T_c at $\nu = 1/3$ is higher than 0.59 K and lower than 1.40 K. The T_c at $\nu = 2/7$ is lower than 0.59 since the resistivity ρ is finite.

In 2007 Novoselov *et al.* [12, 13] discovered room-temperature QHE in graphene, $T = 300$ K, $B = 29$ T. It is a theoretical challenge to explain the superconductivity and QHE at such high temperatures.

Problem 11.1.1. Prove that ρ_H in (11.1) is linear in B.

11.2
Theoretical Developments

The 1985 Nobel Prize was awarded to von Klitzing, the discoverer of the integer QHE. The 1998 Nobel Prizes were shared to Tsui, Störmer (experimental discovery) and Laughlin (theory) for their contribution to the fractional QHE.

The prevalent theories [4, 14, 18, 20–23] based on the *Laughlin wavefunction* in the Schrödinger picture deal with the QHE at 0 K and immediately above [18, 19]. The

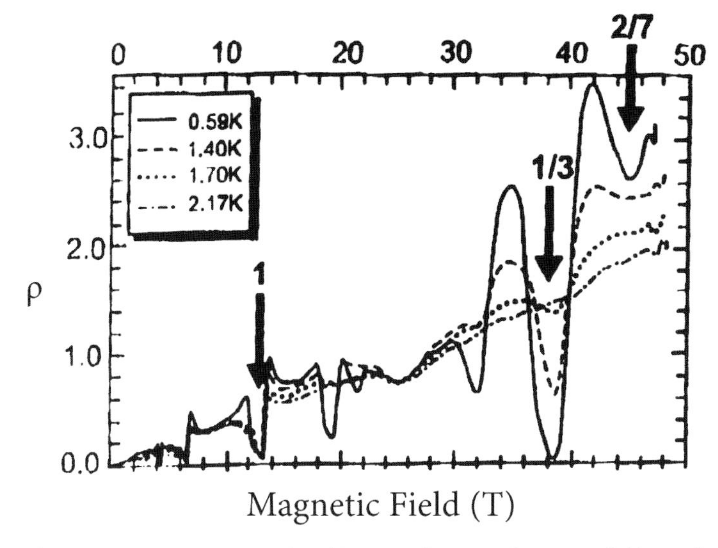

Figure 11.5 Resistivity ρ in GaAs/AlGaAs as a function of magnetic field B in the units of tesla, after Leadley *et al.* [11].

system ground state, however, does not carry a current. To interpret the experimental data it is convenient to introduce *composite* (c-) *particles* (bosons, fermions). The c-bosons (fermions), each containing an electron *and* an odd (even) number of *fluxons* were introduced by Zhang *et al.* [20] and others (Jain [14–17]) for the description of the fractional QHE (Fermi liquid). All QH states with distinctive plateaus in ρ_{H} are observed below around 0.5 K. It is desirable to treat the QHE below and above T_c in a unified manner. The extreme accuracy (precision $\sim 10^{-8}$), in which each plateau is observed, means that the current density j must be computed exactly with no averaging. In the prevalent theories [4–9] the electron–electron interaction and Pauli's exclusion principle are regarded as the cause for the QHE. Both are essentially repulsive and cannot account for the fact that the c-particles are bound, that is, they are in negative-energy states. Besides, the prevalent theories have limitations:

- The zero temperature limit is taken at the outset. Then the question why QHE is observed below 0.5 K and not at higher temperatures in GaAs/AlGaAs cannot be answered. We should have a theory for all temperatures. We should find the critical temperature T_c, below which the superconducting state appears.
- The high field limit is taken at the outset. The integer QHE is observed for small integer P only. The question why the QHE is not observed for large P (weak field) cannot be answered. We better describe the phenomena for all fields.
- The ρ_{H} value $(Q/P)(h/e^2)$ is obtained in a single stroke. To obtain the Hall resistivity ρ_{H} we need two separate measurements of the Hall field E_{H} and the current density j. We must calculate (E_{H}, j) and take the ratio: $\rho_{\mathrm{H}} = E_{\mathrm{H}}/j$.

There is a remarkable similarity between the QHE *and* the High-Temperature Superconductivity (HTSC), both occurring in 2D systems. The major superconducting properties observed in the HTSC are (a) zero resistance, (b) a sharp phase change, (c) the energy gap below T_c, (d) the flux quantization, (e) Meissner effect, and (f) Josephson effects. All of these have been observed in the GaAs/AlGaAs interface. We regard *phonon-exchange attraction* as the cause of both QHE and superconductivity. Starting with a reasonable Hamiltonian, we calculate everything using standard statistical mechanics. We develop a unified theory of the integer and fractional quantum Hall effects in Sections 11.3.3 and 11.3.4.

11.3 Theory of the Quantum Hall Effect

11.3.1 Introduction

Experimental data by Willett *et al.* [10], reproduced in Figure 11.4, show that the Hall resistivity $\rho_H \equiv E_H/j$ (E_H: Hall field, j: current density) in the GaAs/AlGaAs heterojunction at extremely low temperatures (85–150 mK) has plateaus at various fractional LL occupation ratios (*filling factor*) $\nu = P/Q$ with positive integer P and odd integer Q, where the resistivity $\rho \equiv E/j$ (E: applied field) vanishes. In particular at $\nu = 1/3$, the plateau in ρ_H and the drop in ρ (not shown) are as distinctive as the *integer QHE plateau* and *zero resistivity* at $\nu = 1$, indicating the superconducting state with an energy gap. The plateau heights are quantized in units of h/e^2 (h: Plank constant, e: electron charge {magnitude}). The ground state of the GaAs/AlGaAs heterojunction at $\nu = 1/Q$, can be described in terms of the Laughlin wavefunction [4, 18]. Laughlin [18, 19] pointed out a remarkable similarity between the QHE and the cuprate superconductivity, both occurring in 2D. Zhang *et al.* [20] discussed the fractional QHE in terms of the c-bosons, each made up of an electron *and* an odd number of fluxons. The fermionic nature of the fluxon will be discussed below.

The data shown in Figure 11.4 indicate that the Hall resistivity ρ_H is linear in B at $\nu = 1/2$, exhibiting a *Fermi-liquid state.* Jain [14–17] discussed this state and the QHE in terms of the *c-fermions,* each made up of an electron and two fluxons. The c-particle appearing at 0 K must have a lower energy than the constituents (electrons, fluxons), which requires an *attractive* interaction Hamiltonian. Following the Bardeen–Cooper–Schriefer (BCS) theory of superconductivity [24], where the Cooper pair [25] is formed by the phonon-exchange attraction, we regard *phonon-exchange attraction* as the cause of the QHE. The most remarkable feature of Laughlin's theory is his introduction of the *fractional charge* for the elementary excitation. The particle-number nonconserving processes, such as the phonon exchange and the formation of the c-particle, can best be treated using the second-quantization (field-theoretical) formulation [26]. The Fermi statistics of the electrons concisely in terms of the anticommutation rules for the creation and annihilation operators.

All QHE states with distinctive plateaus in ρ_H in GaAs/AlGaAs are observed below around 0.5 K [18, 19]. We regard this temperature as the critical temperature T_c for the QHE. The T_c can be recognized by the Hall resistivity plateau. It is desirable to treat both quantum and classical Hall effects below and above T_c in a unified manner. We shall develop a finite temperature theory, starting with a reasonable model Hamiltonian and calculating everything using standard quantum statistical methods. The extreme accuracy (precision $\sim 10^{-8}$), in which each plateau value is observed, means that the current density j must be computed without averaging. We accomplish this using simple kinetic theory and calculating j in terms of the condensed c-bosons.

The countability concept of the fluxons, known as the *flux quantization*:

$$B = \frac{N}{A}\frac{h}{e} \equiv \frac{N}{A}\Phi_0 \,, \tag{11.4}$$

where A = sample area, N = integer, $\Phi_0 \equiv h/e$ = flux quantum, was originally due to Onsager [27]. This idea and Onsager's formula are routinely used in the analysis of the de Haas–van Alphen effect and the determination of the Fermi surface.

We now discuss the quantum statistics of the fluxons. The magnetic (electric) field is an axial (polar) vector and the associated fluxon (photon) is a half-spin fermion (full-spin boson), which is in line with Dirac's theory [26] that *every quantum particle having the position of an observable is a half-spin fermion.* The magnetic (electric) field is an axial (polar) vector and the associated fluxon (photon) is a half-spin fermion (full-spin boson). The magnetic (electric) flux line cannot (can) terminate at a sink, which also supports the fermionic (bosonic) nature of the associated fluxon (photon). No half-spin fermion can annihilate by itself because of angular momentum conservation. The electron spin originates in the relativistic quantum equation (Dirac's theory of electron) [26]. The discrete (two) quantum numbers ($\sigma_z = \pm 1$) cannot change in the continuous nonrelativistic limit, and hence the spin must be conserved. The countability and statistics of the fluxon are the fundamental particle properties, and hence they cannot be derived starting with a Hamiltonian. We postulate that *the fluxon is a half-spin fermion with zero mass and zero charge.* Only half-spin fermions can form composites of definite quantum statistics.

We assume that the magnetic field $\boldsymbol{B}$ is applied perpendicular to the interface. The 2D Landau level energy:

$$\varepsilon = \hbar\omega_c\left(N_L + \frac{1}{2}\right) \,, \quad \omega_c \equiv \frac{eB}{m^*} \,, \tag{11.5}$$

with the states (N_L, k_y), $N_L = 0, 1, 2, \ldots$, have a great degeneracy (hidden variable k_y).

We shall develop a microscopic theory of the QHE in analogy with the theory of the cuprate superconductivity [28].

11.3.2
The Model

The center of mass (CM) of *any* composite particle (c-particle) moves as a fermion (boson). The eigenvalues of the CM momentum are limited to 0 or 1 (unlimited) if it contains an odd (even) number of elementary fermions. This rule is known as the *Ehrenfest–Oppenheimer–Bethe* (EOB) *rule* [29, 30]. This rule states that a c-particle with respect to the center of mass motion moves as a fermion (boson) if it contains an odd (even) number of elementary fermions. Hence, the CM motion of the composite containing an electron and Q fluxons is bosonic (fermionic) if Q is odd (even). The system of c-bosons condenses below some critical temperature T_c and exhibits a superconducting state while the system of c-fermions shows a Fermi liquid behavior.

A longitudinal phonon, acoustic or optical, generates a density wave, which affects electron (fluxon) motion through charge displacement (current). The exchange of a phonon between electron and fluxon generates an *attractive* transition. The phonon exchange between two electrons generates a transition in the electron states with the *effective (attractive) interaction*, see (4.94):

$$|V_q|^2 \frac{\hbar\omega_q}{(\varepsilon_{|k+q|} - \varepsilon_k)^2 - \hbar^2\omega_q^2}\,, \tag{11.6}$$

where ε_k is the electron energy, $\hbar\omega_q$ the phonon energy, and V_q the *electron–phonon* interaction strength. Let us assume an interface formed in (001). The planar arrays of Ga^{3+}(A) and As^{3-}(B) are located alternately in equilibrium along $\langle 100\rangle$ as ABA′B′AB. A longitudinal phonon, acoustic or optical, running in $\langle 100\rangle$ can generate a density wave which affects the electron (fluxon) motion by the ionic charge displacement (current). The same condition also holds in $\langle 010\rangle$. The lattice wave proceeding in the (001) plane can be regarded as a superposition of the waves proceeding in $\langle 100\rangle$ and $\langle 010\rangle$, and hence the associated phonon can generate a *2D charge-density wave*, and *the electron (fluxon)–phonon interaction.*

The exchange of a phonon between an *electron* and a *fluxon* also generates a transition in the electron states with the effective (attractive) interaction:

$$|V_q V_q'| \frac{\hbar\omega_q}{(\varepsilon_{|k+q|} - \varepsilon_k)^2 - \hbar^2\omega_q^2}\,, \tag{11.7}$$

where V_q' (V_q) is the fluxon–phonon (electron–phonon) interaction strength. The Landau oscillator quantum number N_L is omitted; the bold $\boldsymbol{k}$ denotes the momentum (k_y) and the italic $k(=|\boldsymbol{k}|)$ the magnitude. There are two processes, one with the *absorption* of a phonon with momentum $\boldsymbol{q}$ and the other with the *emission* of a phonon with momentum $-\boldsymbol{q}$, see Figure 11.6a and b, which contribute to the effective interaction with the energy denominators $(\varepsilon_{|k+q|} - \varepsilon_k - \hbar\omega_q)^{-1}$ and $(\varepsilon_{|k+q|} - \varepsilon_k + \hbar\omega_q)^{-1}$, generating (11.7). The interaction is attractive (negative) and most effective when the states before and after the exchange have the same energy ($\varepsilon_{|k+q|} - \varepsilon_k = 0$) as in the degenerate LL.

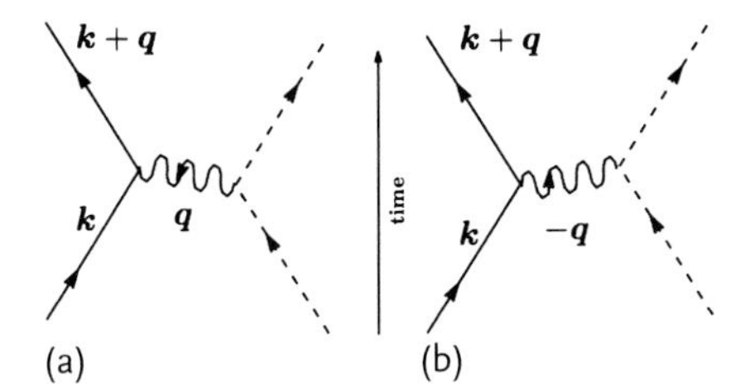

Figure 11.6 (a,b) Two phonon (wavy line) exchange processes, which contribute to the effective interaction in (11.7).

BCS [24] assumed the existence of Cooper pairs [25] in a superconductor, and wrote down a Hamiltonian containing the "electron" and "hole" kinetic energies and the pairing interaction Hamiltonian with the phonon variables eliminated, see Section 7.4. We start with a BCS-like Hamiltonian $\mathcal{H}$ for the present model system:

$$\begin{aligned}\mathcal{H} = &\sum_k{}' \sum_s \varepsilon_k^{(1)} n_{ks}^{(1)} + \sum_k{}' \sum_s \varepsilon_k^{(2)} n_{ks}^{(2)} \\ &+ \sum_k{}' \sum_s \varepsilon_k^{(3)} n_{ks}^{(3)} - \sum_q{}' \sum_k{}' \sum_{k'}{}' \sum_s v_0 \\ &\times \left[B_{k'qs}^{(1)\dagger} B_{kqs}^{(1)} + B_{k'qs}^{(1)\dagger} B_{kqs}^{(2)\dagger} + B_{k'qs}^{(2)} B_{kqs}^{(1)} + B_{k'qs}^{(2)} B_{kqs}^{(2)\dagger} \right] ,\end{aligned} \quad (11.8)$$

where $n_{ks}^{(j)} = c_{ks}^{(j)\dagger} c_{ks}^{(j)}$ is the number operator for the "electron" (1) ("hole" (2), fluxon(3)) at momentum $\boldsymbol{k}$ and spin s with the energy $\varepsilon_{k,s}^{(j)}$, with annihilation (creation) operators c ($c^\dagger$) satisfying the Fermi anticommutation rules:

$$\left\{ c_{ks}^{(i)}, c_{k's'}^{(j)\dagger} \right\} \equiv c_{ks}^{(i)} c_{k's'}^{(j)\dagger} + c_{k's'}^{(j)\dagger} c_{ks}^{(i)} = \delta_{k,k'} \delta_{s,s'} \delta_{i,j} , \quad \left\{ c_{ks}^{(i)}, c_{k's'}^{(j)} \right\} = 0 . \quad (11.9)$$

The fluxon number operator $n_{ks}^{(3)}$ is represented by $a_{ks}^\dagger a_{ks}$ with a ($a^\dagger$) satisfying the anticommutation rules:

$$\left\{ a_{ks}, a_{k's'}^\dagger \right\} = \delta_{k,k'} \delta_{s,s'} , \quad \{ a_{ks}, a_{k's'} \} = 0 . \quad (11.10)$$

The phonon exchange can create electron–fluxon composites, bosonic or fermionic, depending on the number of fluxons. The center of mass of any composite moves as a fermion (boson) if it contains odd (even) numbers of elementary fermions. We call the conduction-electron composite with an odd (even) number of fluxons the *composite (c-) boson* (c-fermion). The electron (hole)-type c-particles carry negative (positive) charge. We expect that electron (hole)-type Cooper-pair-like c-bosons are generated by the phonon-exchange attraction from a pair of electron (hole)-type c-fermions. The pair operators B are defined by

$$B_{kq,s}^{(1)\dagger} \equiv c_{k+q/2,s}^{(1)\dagger} c_{-k+q/2,-s}^{(1)\dagger} , \quad B_{kq,s}^{(2)} \equiv c_{-k+q/2,-s}^{(2)} c_{k+q/2,s}^{(2)} . \quad (11.11)$$

The prime on the summation in (11.8) means the restriction: $0 < \varepsilon_{ks}^{(j)} < \hbar\omega_{\mathrm{D}}$, ω_{D} = Debye frequency. The pairing interaction terms in (11.8) conserve the charge. The term $-v_0 B^{(1)\dagger}_{k'qs} B^{(1)}_{kqs}$, where $v_0 \equiv |V_q V'_q|(\hbar\omega_0 A)^{-1}$, $A =$ sample area, is the pairing strength, generates a transition in the electron-type c-fermion states, see Figure 7.16a. Similarly, the exchange of a phonon generates a transition between the hole-type c-fermion states, see Figure 7.16b, represented by $-v_0 B^{(2)\dagger}_{k'qs} B^{(2)\dagger}_{kqs}$. The phonon exchange can also pair-create (pair-annihilate) electron (hole)-type c-boson pairs, and the effects of these processes are represented by $-v_0 B^{(1)\dagger}_{k'qs} B^{(2)\dagger}_{kqs}$ $\left(-v_0 B^{(1)}_{kqs} B^{(2)}_{kqs}\right)$, see Figure 7.18a,b.

The pairing interaction terms in (11.8) are formally identical to those in (7.23). Only we deal here with c-fermions instead of conduction electrons. We denote creation and annihilation operators by the same symbols c.

We now extend our theory to include elementary fermions (electron, fluxon) as members of the c-fermion set. The Cooper pair (electron, electron) is regarded as the c-boson. We can then treat the superconductivity and the QHE in a unified manner. The c-boson containing one electron and one fluxon can be used to describe the integer QHE as we see later in this chapter.

We shall use the Hamiltonian $\mathcal{H}$ in (11.8) and discuss both integer and fractional QHE. We also use the same Hamiltonian to describe the QHE in graphene in the following chapter.

11.3.3
The Integer QHE

We start with the Hamiltonian $\mathcal{H}$ in (11.8). We regard conduction electrons and fluxons as c-fermions (building blocks). The pair operators,

$$B^{(j)\dagger}_{kq} \equiv c^{(j)}_{k+q/2} a_{-k+q/2} , \quad B^{(j)}_{kq} \equiv a_{-k+q/2} c^{(j)}_{k+q/2} , \tag{11.12}$$

are introduced, where we drop the spin indices. The upper indices $j = 1(2)$ mean "electron" ("hole"). The interaction Hamiltonians

$$-v_0 B^{(j)\dagger}_{kq} B^{(k)\dagger}_{kq} \left(-v_0 B^{(j)}_{kq} B^{(k)}_{kq}\right) , \quad j \neq k ,$$

pair-create (pair-annihilate) electron-type c-bosons and hole-type c-bosons. The c-bosons containing *one electron* and *one fluxon*, formed at $\nu = 1$, will be called the *fundamental* (f) *c-bosons*. See Figure 7.18a,b. The c-bosons can be bound and stabilized by the interaction Hamiltonians $-v_0 B^{(j)\dagger}_{k'q} B^{(j)}_{kq}$. The fc-bosons (fundamental c-bosons) can undergo the Bose–Einstein condensation (BEC) below the critical temperature T_c. The fc-bosons condensed at a momentum along the sample length are shown in Figure 11.7a. Above T_c, they can move in all directions in the plane as shown in Figure 11.7b.

First take the −fc-boson. The ground state energy w_0 can be calculated by solving the Cooper-like equation [25]:

$$w_0 \Psi(\boldsymbol{k}) = \varepsilon_{\boldsymbol{k}} \Psi(\boldsymbol{k}) - \frac{v_0}{(2\pi\hbar)^2} \int' \mathrm{d}^2 k' \Psi(\boldsymbol{k}') , \tag{11.13}$$

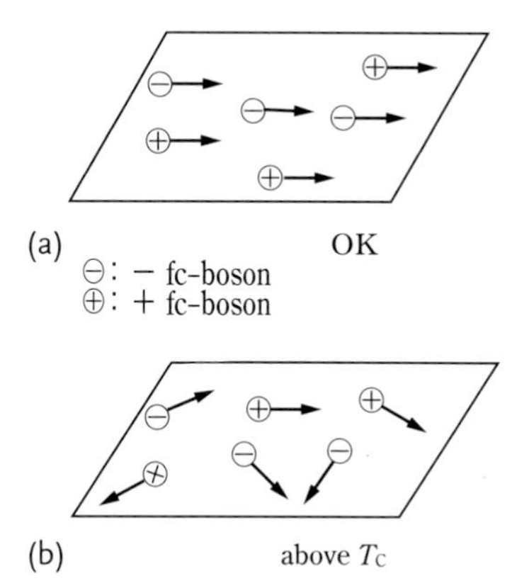

Figure 11.7 ±fc-bosons at 0 K (or below T_c) (a) and noncondensed fc-bosons above T_c (b).

where Ψ is the reduced wavefunction for the stationary fc-bosons; we neglected the fluxon energy. We obtain after simple calculations

$$w_0 = \frac{-\hbar\omega_D}{\exp((v_0 \mathcal{D}_0)^{-1}) - 1} < 0 \,, \tag{11.14}$$

where $\mathcal{D}_0 \equiv \mathcal{D}(\varepsilon_F)$ is the density of states per spin. Note that the binding energy $|w_0|$ does *not* depend on the "electron" mass. Hence, the ±fc-bosons have the same energy w_0.

At 0 K only *stationary* fc-bosons are generated. The ground state energy W_0 of the system of fc-bosons is

$$W_0 = 2N_0 w_0 \,, \tag{11.15}$$

where N_0 is the − (or +) fc-boson number.

At a finite T there are moving (*non*condensed) fc-bosons, whose energies $w_q^{(j)}$ are obtained from [25, 32]

$$w_q^{(j)} \Psi(\boldsymbol{k}, \boldsymbol{q}) = \varepsilon_{|\boldsymbol{k}+\boldsymbol{q}|}^{(j)} \Psi(\boldsymbol{k}, \boldsymbol{q}) - \frac{v_0}{(2\pi\hbar)^2} \int' \mathrm{d}^2 k' \Psi(\boldsymbol{k}', \boldsymbol{q}) \,, \tag{11.16}$$

which is reduced to (11.12) in the small momentum (magnitude) q limit (Problem 11.3.1). For small q, we obtain

$$w_q^{(j)} = w_0 + \frac{2}{\pi} v_F^{(j)} |\boldsymbol{q}| \,, \tag{11.17}$$

where $v_F^{(j)} \equiv (2\varepsilon_F / m_j)^{1/2}$ is the Fermi speed. The brief derivation of the formulas (11.16) and (11.17) is given in Appendix A.6. The energy $w_q^{(j)}$ depends *linearly* on the momentum q. The linear dispersion relation for the 3D Cooper pair was obtained by Cooper (unpublished), and was recorded in Schrieffer's book [31]. By the way, the same linear dispersion relation (11.22) also holds for the 2D Cooper pair in high T_c cuprate superconductors [28]. This relation was clearly observed in the lowest energy region by Lanzara *et al.* [33] with Angle-Resolved Photoemission (ARPS) Spectroscopy. Formula (11.19) for the critical temperature, see below, also holds for the cuprates [28].

The system of *free massless bosons* undergoes a BEC in 2D at the critical temperature [28]:

$$k_B T_c = 1.954 \hbar c n^{1/2} , \tag{11.18}$$

where c is the boson speed, and n the density. The brief derivation is given in Appendix A.6. This is not a violation of *Hohenberg's theorem* that there be no long-range order in 2D, which is obtained using an f-sum rule (mass conservation). The theorem does not hold for massless bosons. Substituting $c = 2/\pi v_F$ in (11.18), we obtain [32] (Problem 11.3.2)

$$k_B T_c = 1.24 \hbar v_F n_0^{1/2} , \quad n_0 \equiv \frac{N_0}{A} . \tag{11.19}$$

The interboson distance $R_0 \equiv 1/\sqrt{n_0}$ calculated from this equation is $1.24\hbar v_F/(k_B T_c)$. The boson size r_0 calculated from (11.18), using the uncertainty relation ($q_{max} r_0 \sim \hbar$) and $|w_0| \sim k_B T_c$, is $r_0 = (2/\pi)\hbar v_F (k_B T_c)^{-1}$, which is a few times smaller than R_0. Thus, the bosons do *not* overlap in space, and the *free boson model* is justified.

Let us take GaAs/AlGaAs. We assume $m^* = 0.067 m_e$, m_e = electron mass. For the electron density 10^{11} cm^{-2}, we have $v_F = 1.36 \times 10^6$ cm s^{-1}. Not all electrons are bound with fluxons since the simultaneous generation of $\pm$ fc-bosons is required. If we assume $n_0 = 10^{10}$ cm^{-2}, we obtain $T_c = 1.29$ K which is reasonable. The precise measurement of T_c may be made in a sample of constricted geometry. The plateau width should vanish at T_c since $\varepsilon_g = 0$.

In the presence of the Bose condensate below T_c the unfluxed electron carries the energy $E_k^{(j)} = \sqrt{\varepsilon_k^{(j)2} + \Delta^2}$, where the quasielectron energy gap Δ is the solution of

$$1 = v_0 \mathcal{D}_0 \int_0^{\hbar\omega_D} d\varepsilon \frac{1}{(\varepsilon^2 + \Delta^2)^{1/2}} \left[1 + \exp\left(-\beta\left(\varepsilon^2 + \Delta^2\right)^{1/2}\right)\right]^{-1} , \tag{11.20}$$

where $\beta \equiv (k_B T)^{-1}$ and $\mathcal{D}_0 \equiv \mathcal{D}(\varepsilon_F)$ is the density of states per spin. Note that the gap Δ depends on T. At T_c there is *no* condensate and hence Δ vanishes. The *moving* fc-boson below T_c with the condensate background has the energy $\tilde{w}_q$, obtained from

$$\tilde{w}_q^{(j)} \Psi(\boldsymbol{k}, \boldsymbol{q}) = E_{|\boldsymbol{k}+\boldsymbol{q}|}^{(j)} \psi(\boldsymbol{k}, \boldsymbol{q}) - \frac{v_0}{(2\pi\hbar)^2} \int' d^2k' \Psi(\boldsymbol{k}', \boldsymbol{q}) , \tag{11.21}$$

where $E^{(j)}$ replaced $\varepsilon^{(j)}$ in (11.16). We obtain

$$\tilde{w}_q^{(j)} = \tilde{w}_0 + \frac{2}{\pi} v_F^{(j)} |\boldsymbol{q}| \equiv w_0 + \varepsilon_g + \frac{2}{\pi} v_F^{(j)} q , \tag{11.22}$$

where $\tilde{w}_0(T)$ is determined from

$$1 = \mathcal{D}_0 v_0 \int_0^{\hbar\omega_D} \frac{d\varepsilon}{|\tilde{w}_0| + (\varepsilon^2 + \Delta^2)^{1/2}} \,. \tag{11.23}$$

The energy difference,

$$\tilde{w}_0(T) - w_0 \equiv \varepsilon_g(T) > 0 \,, \tag{11.24}$$

represents the *T-dependent gap* between the moving and stationary fc-bosons. The energy $\tilde{w}_q$ is negative. Otherwise, the fc-boson should break up. This limits $\varepsilon_g(T)$ to be less than $|w_0|$. Hence, the energy gap $\varepsilon_g(T)$ is $|w_0|$ at 0 K. It declines to zero as the temperature approaches T_c from below.

The fc-boson, having the linear dispersion relation (11.17) or (11.22), can move in all directions in the plane with the constant speed $(2/\pi)v_F^{(j)}$. The supercurrent is generated by $\mp$fc-bosons monochromatically condensed, running along the sample length, see Figure 11.7a. The supercurrent density (magnitude) j, calculated by the rule: j = (charge) × (carrier density) × (drift velocity), is given by

$$j \equiv e^* n_0 v_d = e^* n_0 \frac{2}{\pi} \left| v_F^{(1)} - v_F^{(2)} \right| \,, \tag{11.25}$$

where e^* is the *effective* charge. The induced Hall field (magnitude) E_H equals $v_d B$. The magnetic flux is quantized:

$$B = n_\phi \Phi_0 \,, n_\phi = \text{fluxon density} \,,$$
$$\Phi_0 \equiv (h/e) = \text{magnetic flux (fluxon)} \,. \tag{11.26}$$

Hence we obtain

$$\rho_H \equiv \frac{E_H}{j} = \frac{v_d B}{e^* n_0 v_d} = \frac{n_\phi \Phi_0}{n_0 e^*} = \frac{1}{e^* n_0} n_\phi \left(\frac{h}{e}\right) \,. \tag{11.27}$$

For $e^* = e$, $n_\phi = n_0$, $\rho_H = h/e^2$, explaining the plateau value observed.

The supercurrent generated by equal numbers of $\mp$fc-bosons condensed monochromatically is neutral. This is reflected in the calculations in (11.25). The supercondensate whose motion generates the supercurrent must be neutral. If it has a charge, it would be accelerated indefinitely by the external field because the impurities and phonons cannot stop the supercurrent growing. That is, the circuit containing a superconducting sample and a battery must be burnt out if the supercondensate is not neutral. In the calculation of ρ_H in (11.27), we used the *unaveraged* drift velocity difference $(2/\pi)|v_F^{(1)} - v_F^{(2)}|$, which is significant. *Only* the unaveraged drift velocity cancels out exactly from numerator/denominator, leading to an exceedingly accurate plateau value. Thus, we explained why the precise plateau value in h/e^2 can be observed in experiment.

The conduction electrons are scattered by phonons and impurities. The drift velocity v_d depends on the scattering rate, and ρ_H is proportional to the field B

while ρ is finite. Immediately above T_c the noncondensed fc-bosons moving in all directions, see Figure 11.7b, dominate the magnetotransport, and should show a non-Fermi liquid behavior.

The resistivity ρ at $\nu = 1$ exponentially rises on both sides, see Figure 11.4. This can be explained as follows. The excited fc-boson has an energy gap ε_g. Hence, its density has an *Arrhenius-law Boltzmann factor*

$$\exp\left[-\frac{\varepsilon_g(|B_c|, T)}{(k_B T)}\right], \quad B_c \equiv B - B_1 , \tag{11.28}$$

where $\varepsilon_g(|B_c|, T)$ approaches zero at the super-to-normal boundary. After the field difference from the center $B_1 \equiv n_c \Phi_0 = n_c(h/e)$, B_c, passes the boundary, ρ returns to normal since the carrier density loses the Boltzmann factor.

In summary *the moving fc-boson below T_c has an energy gap ε_g, generating the stable plateau. The fc-bosons condensed at a finite momentum account for the supercurrent with no resistance and the plateau value* equal to h/e^2.

The theory developed in Section 11.3 can simply be extended to the integer QHE. The field magnitude is smaller at $\nu = P = 2, 3, \ldots$ The LL degeneracy is proportional to B, and hence P LLs must be considered. First consider the case $P = 2$. Without the phonon-exchange attraction the electrons occupy the lowest two LLs with spin. The electrons at each level form fc-bosons. See Figure 11.8a and b. The fc-boson density n_0 at each LL is one-half the density at $\nu = 1$, which is equal to the electron density n_e fixed for the sample. Extending the theory to a general integer, we have

$$n_0 = \frac{n_e}{P} . \tag{11.29}$$

This means that both $T_c(\propto n_0^{1/2})$ and ε_g are smaller, making the plateau width (a measure of ε_g) smaller in agreement with experiments, see Figure 11.4. The fc-boson has a lower energy than the conduction electron. Hence, at the extreme low temperatures the supercurrent due to the condensed fc-bosons dominates the normal current due to the conduction electrons and noncondensed fc-bosons, giving rise to the dip in ρ. All dips should reach zero if the temperature is further lowered,

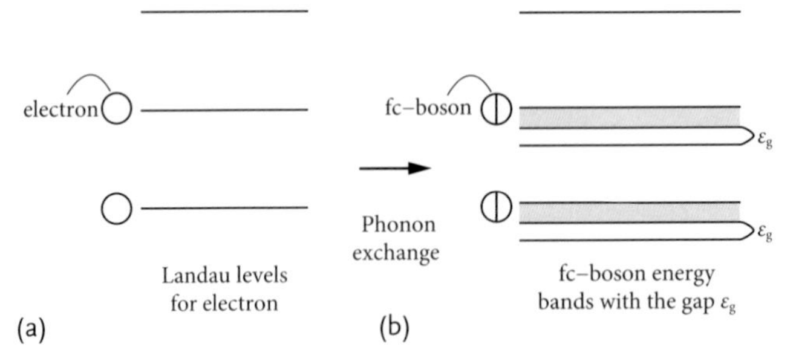

Figure 11.8 (a) Landau levels for electron; (b) fc-boson energy bands. The c-fermions which fill up the lowest two LLs, form the QH state at $\nu = 2$ after the phonon-exchange attraction and the BEC of the fc-bosons.

see Figure 11.3, where the data at 60 mK reported by Tsui [3] are shown. The QHE states with small integers ($\nu = 1, 2, \ldots, 6$) are clearly visible. Other QHE states with greater P are overshadowed by the neighboring QHE states with small P.

Problem 11.3.1. Verify that (11.16) is reduced to (11.13) in the small q-limit.

Problem 11.3.2. Derive (11.19).

11.3.4
The Fractional QHE

Let us consider a general case $\nu = P/Q$, odd Q. Assume that there are P sets of c-fermions with $Q - 1$ fluxons, which occupy the lowest P LLs, see Figure 11.9a, where we choose $Q = 3$, $P = 2$. The c-fermions subject to the available B-field form c-bosons with Q fluxons, see Figure 11.9b. Note the similarity between Figures 11.8 and 11.9. In this configuration the c-boson density n_0 is given by (11.29) and the fluxon density n_ϕ is given by

$$n_\phi = \frac{n_0}{P} . \tag{11.30}$$

Using (11.27), (11.29), and (11.30), we obtain

$$\rho_{\mathrm{H}} \equiv \frac{E_{\mathrm{H}}}{J} = \frac{v_{\mathrm{d}}}{e^* n_0 v_{\mathrm{H}}} n_\phi \frac{h}{e} = \frac{Q}{P}\frac{h}{e^2} , \tag{11.31}$$

as observed. We see that the integer Q indicates the number of fluxons in the c-boson and the integer P the number of LLs occupied by the parental c-fermions, each with $Q - 1$ fluxons. To derive the last equality in (11.31), $\rho_{\mathrm{H}} = (Q/P)(h/e^2)$, we assumed the fractional charge,

$$e^* = \frac{e}{Q} , \tag{11.32}$$

following Laughlin [18, 19] and Haldane [34]. Equation (11.25) along with (11.29) and (11.30) yield that T_{c} decreases as $P^{-1/2}$ with increasing P. Jain's hierarchy of fractionals can be obtained by examining the fractional around 1/2, using (11.29) and (11.30).

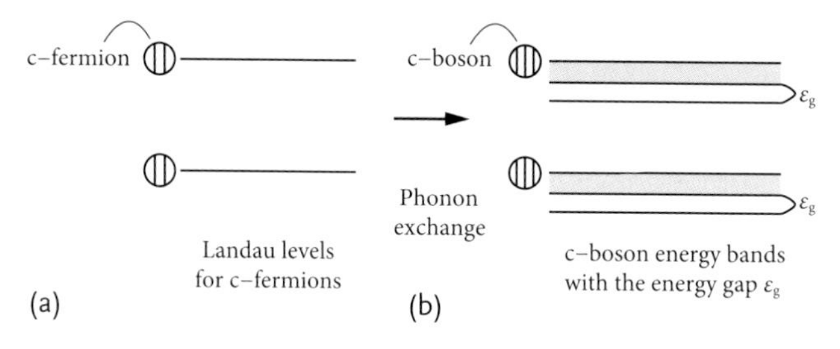

Figure 11.9 The QHE state at $\nu = 2/3$. (a) The c-fermions, each with two fluxons, occupy fully the lowest two LLs; (b) With the phonon exchange, c-bosons, each with three fluxons, are created and occupy the energy bands with the energy gap ε_g.

11.4 Discussion

Jain's unification scheme suggests that the fermion nature of electrons and c-fermions must be considered in the formulation of the c-bosons, whose condensation was thought to generate the QHE. We have incorporated this feature in our *c-boson formation model*, where the lowest P LLs are filled by c-fermions with $Q-1$ fluxons without the phonon exchange, while c-bosons with Q fluxons are formed with it. In the course of the derivation of (11.31), we found (11.29), which allows us to discuss the stability of the QHE states quantitatively. For example, the plateau width is almost the same at $\nu = 3/7$ and $3/5$ in Figure 11.4. The small difference can arise from the v_F difference due to the effective mass difference associated with the parental c-fermions with $Q = 6$ and 4. Equations (11.6) and (11.19) indicate that both T_c and ε_g decrease as $P^{-1/2}$ with increasing P. This trend is in agreement with all of the plateaus and dips identified in Figure 11.4. Equation (11.6) also means that no QHE is realized for very large P since T_c becomes less than the observation temperature. We predict that many features appearing in Figure 11.4 are greatly reduced at the extremely low temperatures, say 10 mK, so that the only remaining features are the broad plateaus in ρ_H and the accompanying zero resistivity at $\nu = 1/Q$, odd Q, separated by the drops in ρ_H and the sharply peaked ρ at $\nu = 1/Q$, even Q, similar to the behavior observed in Figure 11.4.

In summary the QHE state at $\nu = P/Q$, odd Q, is shown to be the superconducting state with an energy gap generated by the condensed c-bosons, each with Q fluxons carrying the fractional charge e/Q. In the present theory the effective field B^* is defined in the form due to Jain [14–17]:

$$B^* \equiv B - B_\nu = B - \frac{1}{\nu} n_e \frac{h}{e} , \qquad n_e = \text{electron density} \tag{11.33}$$

with ν being fermionic (bosonic) fractions P/Q, even (odd) Q. This means that the c-particles move field-free at the exact fraction. For odd Q, the c-bosons, if condensed, can move undisturbed due to the Meissner effect under a small effective B^* field and generate a stable plateau in ρ_H due to the energy gap. The energy gap is difficult to explain based on the c-fermion model alone.

Note added in proof In this Chapter 11 the fractional quantum Hall effect is treated by using Laughlin's theory and results about the fractional charges carried by composite bosons. It is found that the quantum statistical theory in terms of composite particles is more comprehensive. All phenomena about the fractional quantum Hall effect can be described within the frame work of our quantum statistical theory as given in our paper [35].

References

1 Klitzing, K. von, Dorda, G., and Pepper, M. (1980) *Phys. Rev. Lett.*, **45**, 494.
2 Tsui, D.C., Störmer, H.L., and Gossard, A.C. (1982) *Phys. Rev. Lett.*, **48**, 1559.
3 Tsui, D.C. (1989) *4. DFG-Rundgespräch über den Quanten Hall Effekt*, Schleching, Germany.
4 Prange, R.E. and Girvin, S.M. (eds) (1990) *Quantum Hall Effect*, Springer, New York.
5 Stone, M. (ed.) (1992) *Quantum Hall Effect*, World Scientific, Singapore.
6 Das Sarma, S. and Pinczuk, A. (1997) *Perspectives in Quantum Hall Effects*, John Wiley, New York.
7 Chakraboty, T. and Pietiläinen, P. (1998) *Quantum Hall Effect*, Springer, Berlin.
8 Ezawa, Z.F. (2000) *Quantum Hall Effects*, World Scientific, Singapore.
9 Halperin, B.I., Lee, P.A., and Read, H. (1993) *Phys. Rev. B*, **47**, 7312.
10 Willett, R., Eisenstein, J.P., Störmer, H.L., Tsui, D.C., Gossard, A.C., and English, J.H. (1987) *Phys. Rev. Lett.*, **59**, 1776.
11 Leadley, D.R., Burgt, M. van der, Nicholas, R.J., Foxon, G.T., and Harris, J.J. (1996) *Phys. Rev. B*, **53**, 2057.
12 Novoselov, K.S., Gaim, A.K., Morozov, S.V., Jiang, D., Katsnelson, M.I., Grigorieva, I.V., Dubonos, S.V., and Firsov, A.A. (2005) *Nature*, **438**, 197.
13 Novoselov, K.S., Jiang, Z., Zhang, Y., Morozov, S.V., Stormer, H.I., Zeitler, U., Maan, J.C., Boebinger, G.S., Kim, P., and Gaim, A.K. (2007) *Science*, **315**, 1379.
14 Jain, J.K. (1989) *Phys. Rev. Lett.*, **63**, 199.
15 Jain, J.K. (1989) *Phys. Rev. B*, **40**, 8079.
16 Jain, J.K. (1990) *ibid.*, **41**, 7653.
17 Jain, J.K. (1992) *Surf. Sci.*, **263**, 65.
18 Laughlin, R.B.(1983) *Phys. Rev. Lett.*, **50**, 1395.
19 Laughlin, R.B. (1988) *Science*, **242**, 525.
20 Zhang, S.C., Hansson, T.H., and Kevelson, S. (1989) *Phys. Rev. Lett.*, **62**, 82.
21 Girvin, S.M. and MacDonald, A.H. (1987) *Phys. Rev. Lett.*, **58**, 1252.
22 Read, N. (1989) *Phys. Rev. Lett.*, **62**, 86.
23 Shankar, R. and Murthy, G. (1997) *Phys. Rev. Lett.*, **79**, 4437.
24 Bardeen, J., Cooper, L.N. and Schriefler, J.R. (1957) *Phys. Rev.*, **108**, 1175.
25 Cooper, L.N. (1956) *Phys. Rev.*, **104**, 1189.
26 Dirac, P.A.M. (1958) *Principles of Quantum Mechanics*, 4th edn, Oxford Univ. Press., Oxford, pp. 248–252, p. 267.
27 Onsager, L. (1952) *Philos. Mag.*, **43**, 1006.
28 Fujita, S. and Godoy, S. (1996) *Quantum Statistical Theory of Superconductivity*, Plenum, New York, pp. 164–167, pp. 184–185, pp. 247–250.
29 Bethe, H.A. and Jackiw, R. (1968) *Intermediate Quantum Mechanics*, 2nd edn, Benjamin, New York, p. 23.
30 Fujita, S., Gau, S-P, and Suzuki, A. (2001) *JKPS*, **38**, 456.
31 Schrieffer, J.R. (1971) *Theory of Superconductivity*, Second printing, Addison-Wesley Publ. Com., Reading, Massachusetts.
32 Fujita, S., Tamura, Y., and Suzuki, A. (2001) *Mod. Phys. Lett. B*, **15**, 817.
33 Lanzara, A., Bogdanov, P.V., Zhou, X.J., Kellar, S.A., Feng, D.L., Lu, E.D., Yoshida, T., Elsaki, E., Fujimori, A., Klishio, K., Shimoyama, J.I., Noda, T., Uchida, S., Hussain, Z. and Shen, Z.-X. (2001) *Nature*, **412**, 510.
34 Haldane, F.D. (1983) *Phys. Rev. Lett.*, **51**, 605.
35 Fujita, S., Suzuki, A., and Ho, H.C. (2013) ArXiv: 1304.7631v1 [cond-mat,mes-hall].

12
Quantum Hall Effect in Graphene

The unusual Quantum Hall Effect (QHE) in graphene is often discussed in terms of Dirac fermions moving with a linear dispersion relation. The same phenomenon will be explained in terms of the more traditional composite bosons, which move with a linear dispersion relation. The "electron" (wavepacket) moves more easily in the direction [110c-axis] $\equiv$ [110] of the honeycomb lattice than perpendicular to it, while the "hole" moves more easily in [001]. Since "electrons" and "holes" move in different channels, the number densities can be high especially when the Fermi surface has "necks." The strong QHE at filling factor $\nu = 2$ arises from the phonon-exchange attraction in the neighborhood of the "neck" Fermi surfaces. The plateau observed for the Hall conductivity is due to the Bose–Einstein condensation of the composite bosons, a pair of c-ferimons, each having one electron and two fluxons.

12.1
Introduction

Experiments [1–3] indicate that there are two kinds of oscillations for the magnetoresistivity ρ in graphene when plotted as a function of the external magnetic field (magnitude) B. Shubnikov–de Haas (SdH) oscillations appear on the low-field side while Quantum Hall Effect (QHE) oscillations appear on the high-field side. We present a microscopic theory. We start with the graphene honeycomb crystal, construct a two-dimensional Fermi surface for the electron dynamics, develop a Bardeen–Cooper–Schrieffer (BCS)-like theory [4] based on the phonon-exchange attraction between the electron and the fluxon [5], and describe the QHE [6]. The 2D Landau Levels (LL) generate an oscillatory density of states. If multiple oscillations occur within the drop of the Fermi distribution function, then the SdH oscillation emerges for the magnetoconductivity. This means that the carriers in the SdH oscillations must be fermions. The QHE arises from the condensed composite bosons.

In 2005 Novoselov *et al.* [1] discovered a QHE in graphene. The gate field effects in graphene are reproduced in Figure 12.1 after [1, Figure 1]. The conductivity σ as a function of gate voltage V_g is shown in Figure 12.1a while the Hall coefficient R_H measured at magnetic field $B = 2\,T$ is shown in Figure 12.1b, where

Electrical Conduction in Graphene and Nanotubes, First Edition. S. Fujita and A. Suzuki.

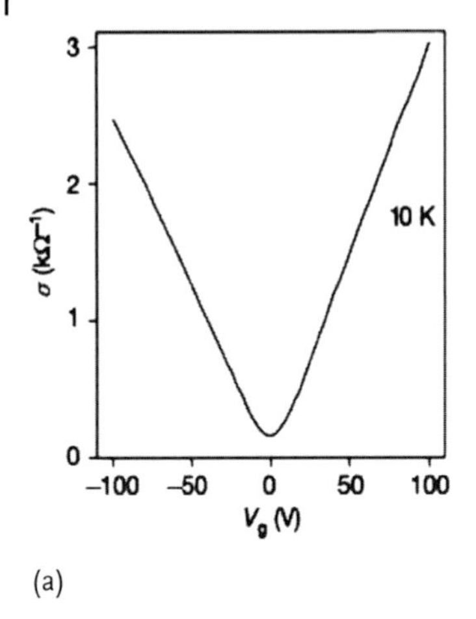

(a)

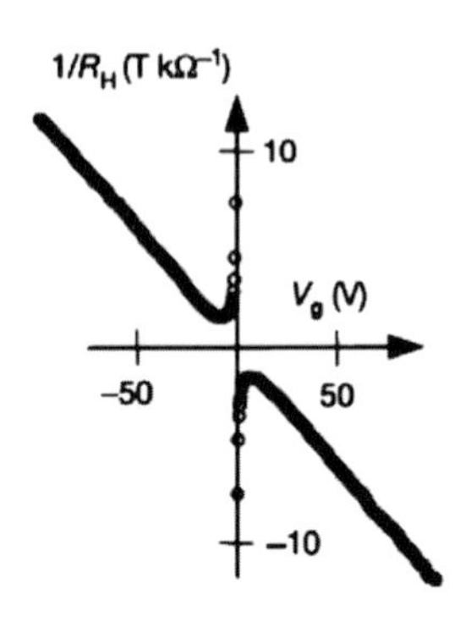

(b)

Figure 12.1 The gate field effects in graphene. Graphene's conductivity in (a) σ (a) and Hall coefficient R_H (b) as a function of gate voltage V_g. R_H was measured in magnetic fields B of 2 T. $R_H = 1/ne$ is inverted to emphasize the linear dependence $n \propto V_g$. The Hall coefficient changes its sign near $V_g = 0$. After [1, Figure 1].

$R_H = 1/ne$, with n being the conduction electron density, is inverted to show the n-linear dependence:

$$\frac{1}{R_H} = ne \quad \text{for} \quad V_g > 20\,\text{V}\,. \tag{12.1}$$

The $1/R_H$ diverges near $V_g = 0$. The linear relation fit (12.1) yields

$$n = (7.3 \times 10^{10}\,\text{cm}^{-2}\,\text{V}^{-1})V_g\,. \tag{12.2}$$

The conductivity σ (a) rises linearly on both sides at high V_g:

$$\sigma \propto V_g\,, \quad V_g > 10\,\text{V}\,. \tag{12.3}$$

This behavior is similar to the case of a metallic single-wall nanotube (SWNT), see Figure 8.8 (inset), where the σ–V curve for a metallic SWNT is shown. Graphene and SWNTs have the same configuration but they have different conformations; they have unrolled and rolled graphene sheets. The gate voltage generates the mobile surface charges ("electrons," "holes"). Upon application of a bias voltage, the "holes" will move and generate additional charge currents:

$$j_{\text{"holes"}} = env_F \tag{12.4}$$

with v_F being the Fermi speed. The mobility defined by

$$\mu \equiv \frac{\sigma}{ne} \tag{12.5}$$

reaches 15 000 $\text{cm}^{-2}\,\text{V}^{-1}\,\text{s}^{-1}$ in the experiments. This μ-behavior is observed independent of temperature T between 10 and 100 K.

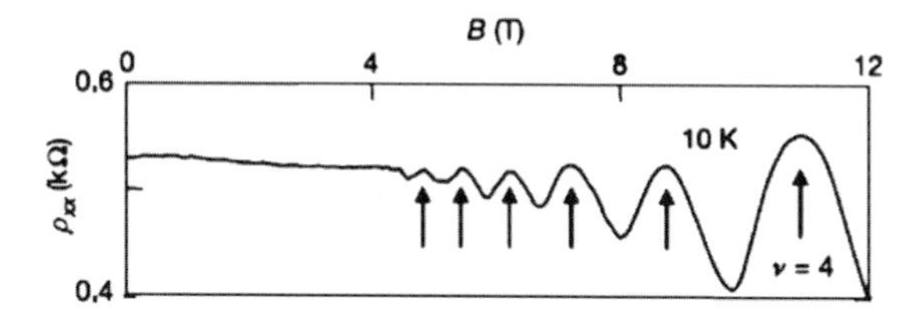

Figure 12.2 The SdH oscillations in graphene at gate voltage $V_g = -60$ V, temperature $T = 10$ K. The longitudinal resistivity ρ_{xx} (kΩ) is plotted as a function of magnetic field B (T). After [1, Figure 2].

We offer a physical explanation of the σ–V_g curve in Figure 12.1a.

The currents near the origin are due to the supercurrents given by (see (8.67))

$$j_{\text{super}} = -2en_0 \frac{2}{\pi}(v_1 - v_2) , \tag{12.6}$$

where n_0 is the condensed c-boson density, and v_1 (v_2) are Fermi speeds of the "electron" ("hole"). The system is in a superconducting state in the experimental temperature range: 10–100 K. Thus, the curve should be temperature-independent. It is very important to find the superconducting temperature T_c for the system. We predict that the critical temperature T_c is essentially the same for both graphene and metallic SWNT. The σ–V_g curve appears to have a perfect right–left symmetry. This indicates that the carriers are only "holes", and not "holes" and "electrons." Only "holes" can move along the surface boundary as explained earlier. This can be checked by Hall effect measurements.

Figure 12.2 is reproduced after [1, Figure 2]. The magnetoresistivity ρ_{xx} at $T = 10$ K and $V_g = -60$ V is plotted as a function of magnetic field $B(T)$, exhibiting SdH oscillations. The ρ_{xx}–B curve is remarkably similar to that in Figure 10.5, the lowest curve, where the SdH oscillations in the GaAs/AlGaAs heterojunction are shown. We can interpret the data in the same manner.

The 2D SdH oscillations have the following features:

(a) The absence of the Landau diamagnetism. This appears as a flat background.
(b) The oscillation period is $\varepsilon_F/(\hbar\omega_c)$, where $\omega_c \equiv eB/m^*$, $m^* =$ cyclotron mass, is the cyclotron frequency.
(c) The envelope of the oscillations decreases like $[\sinh(2\pi^2 M^* k_B T/(\hbar e B))]^{-1}$, where M^* is the magnetotransport mass distinct from the cyclotron mass m^*.

The SdH oscillations in graphene are temperature-dependent. The features evident at 10 K nearly disappear at 140 K in the experiments (not shown).

Figure 12.3 is reproduced after [1, Figure 4]. The longitudinal magnetoresistivity ρ_{xx} and the Hall conductivity σ_{xy} in graphene at $B = 14$ T and $T = 4$ K are plotted as a function of the conduction electron density n ($\sim 10^{12}$ cm^{-2}).

The plateau values of the Hall conductivity σ_{xy} are quantized in the units of

$$\frac{4e^2}{h} \tag{12.7}$$

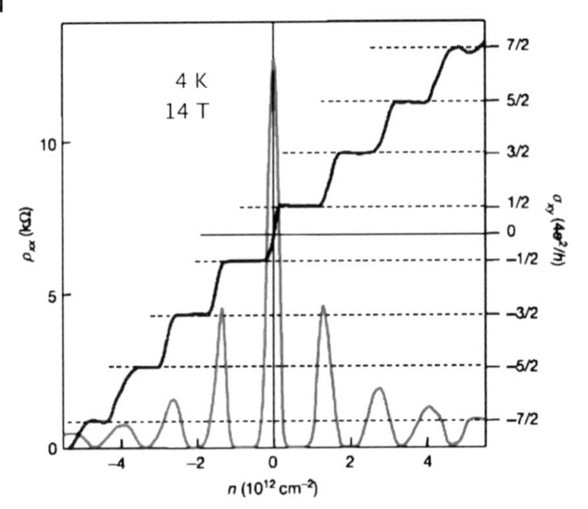

Figure 12.3 QHE in graphene. After Novoselov *et al.* [1].

within experimental errors. The longitudinal resistivity ρ_{xx} reaches zero at the middle of the plateaus. These two are the major signatures of the QHE. The strengths of the (superconducting) states decrease with increasing filling factor ν, see (11.29), as observed in this figure. We will further discuss the features in Figure 12.3 later.

In 2007 Novoselov *et al.* [4] reported the discovery of a room-temperature QHE. We reproduced their data in Figure 12.4 after [7, Figure 1]. The Hall resistivity ρ_{xy} for "electrons" and "holes" indicates precise quantization within experimental errors in units of h/e^2 at magnetic field 45 T and temperature 300 K. This is an extraordinary jump in the observation temperatures since the QHE in the GaAs/AlGaAs heterojunction was reported below 1 K. Figure 12.4 is similar to

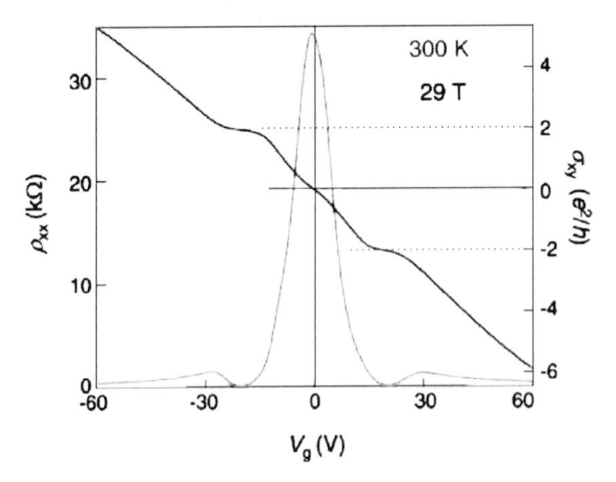

Figure 12.4 Room-temperature QHE in graphene after Novoselov *et al.* [7]. Hall conductivity $\sigma_{xy}\,(e^2/h)$ (dark) and resistance ρ_{xx} (light) as a function of gate voltage (V_g) at temperature 300 K and magnetic field 29 T. Positive values of V_g induce "electrons," and negative values of V_g induce "holes," in concentrations $n = (7.2 \times 10^{10}\ \text{cm}^{-2}\,\text{V}^{-1})V_g$.

Figure 12.3 although the abscissas are different, one in gate voltage and the other in carrier density, and hence the physical conditions are different. We give an explanation below.

The QHE behavior observed for graphene is remarkably similar to that for the GaAs/AlGaAs. The physical conditions are different, however, since the gate voltage and the applied magnetic field are varied in the experiments. The present authors regard the QHE in GaAs/AlGaAs as the superconductivity induced by the magnetic field. Briefly, the magnetoresistivity for a QHE system reaches zero (superconducting) and the accompanying Hall resistivity reveals a plateau (Meissner effect). The QHE state is not easy to destroy because of the *superconductivity energy gap* in the composite boson (c-boson) excitation spectrum. If an extra magnetic field is applied to the system at optimum QHE state (the center of the plateau), then the system remains in the same superconducting state by expelling the extra field. If the field is reduced, then the system stays in the same state by sucking in extra field fluxes, thus generating a Hall resistivity plateau. In the graphene experiments, the gate voltage applied perpendicular to the plane is varied. A little extra voltage relative to the gate voltage at the center of zero resistivity polarizes the system without changing the superconducting state. Hence, the system remains in the same superconducting state, keeping zero resistivity and constant (flat) Hall resistivity. This state has an extra electric field energy:

$$\frac{A}{2}\varepsilon_0(\Delta E)^2 , \tag{12.8}$$

where A is the sample area, ε_0 the dielectric constant, and ΔE is the extra electric field, positive or negative, generated by the sample charge. If the gate voltage is further increased (or decreased), then it will eventually destroy the superconducting state, and the resistivity will rise from zero. This explains the flat ρ_{xy} plateau and the rise in resistivity from zero.

The original authors found that the quantization in σ_{xy} is exact within experimental accuracy (0.2%) as shown in Figure 12.5 which is reproduced after [7, Figure 1c], where the Hall resistance R_{xy} at 45 T and 300 K is plotted as a function of the electron density. The quantization in R_{xy} appears at $h/(2e^2)$, which is a little strange since the most visible quantization for GaAs/AlGaAs appears at h/e^2.

From the QHE behaviors in Figures 12.3 and 12.5, we observe that the quantization occurs at a set of points:

$$\frac{h}{e^2}\frac{(2P+1)}{2} \qquad P = 0, 1, 2, \ldots \tag{12.9}$$

Let us first consider the case: $P = 0$. The QHE requires a Bose–Einstein condensation (BEC) of c-bosons. Its favorable environment is near the van Hove singularities, where the Fermi surface changes its curvature sign. For graphene, this happens when the 2D Fermi surface just touches the Brillouin zone boundary and "electrons" or "holes" are abundantly generated. Following our recent work [8], we shall explain the quantization rule given by (12.9) with the assumption that the

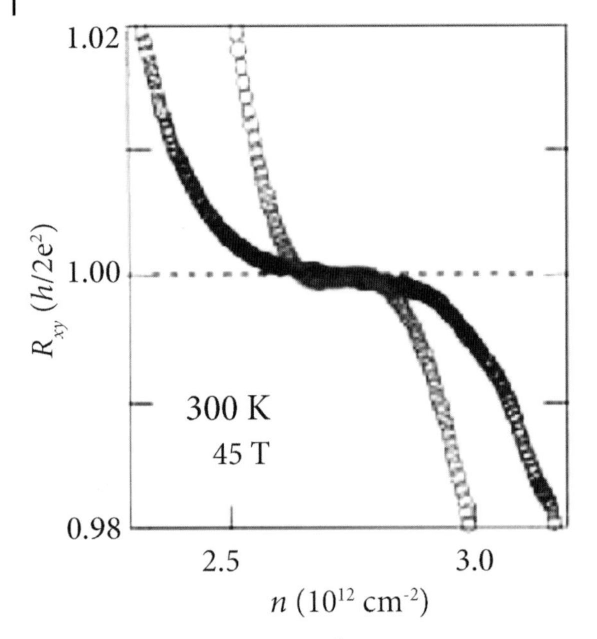

Figure 12.5 Hall resistivity $R_{xy}\,(h/e^2)$ for "electrons" (dark circles) and "holes" (light circles) shows the exact quantization within experimental errors at 45 T and 300 K, after Novoselov *et al.* [7].

c-bosons are formed from a pair of like charge c-fermions, each containing a conduction electron and two fluxons.

We postulate that each and every c-fermion has the effective charge e:

$$e^* = e \quad \text{for any c-fermion} . \tag{12.10}$$

After studying the weak-field fermionic QH state we obtain

$$n_\phi^{(Q)} = n_\text{e}/Q \tag{12.11}$$

for the density of the c-fermions with Q fluxons, where n_e is the electron density. The points of QHE lie on the straight line when ρ_H plotted as a function of the magnetic field as seen in Figure 12.4. Physically when Q is high, the LL separation is great and the c-fermion formation is more difficult. The c-boson contains two c-fermions. We calculate the Hall conductivity σ_H and obtain

$$\sigma_\text{H} \equiv \rho_\text{H}^{-1} = \frac{j}{E_\text{H}} = \frac{2en_0 v_\text{d}}{v_\text{d} n_\phi \Phi_0} = \frac{2e^2}{h} . \tag{12.12}$$

where n_0 is the c-boson density and n_ϕ the fluxon density at $\nu = 1/2$.

The QHE states with integers $P = 1, 2, \cdots$ are generated on the weaker field side. Their strength decreases with increasing P. Thus, we have obtained the rule (12.9) within the framework of the traditional fractional QHE theory [4].

In summary, we have successfully described the QHE in graphene without introducing Dirac fermions. In solid state physics we deal with "electrons" and "holes", which move in crystals and respond to the Lorentz force. These charged particles are considered as Bloch wave packets having size and charges (not a point particle). The bare point-like Dirac particle, if it exists, would be dressed with charge cloud, and it would then acquire a mass. It is very difficult to theoretically argue that Dirac fermions appear only in graphene but nowhere. The relativistic Dirac electron moves at the speed of light, c. The observed particle in graphene moves at a speed of the order of 10^8 ms^{-1}, which is much lower than the speed of light $c = 3 \times 10^8$ ms^{-1}. It is difficult to explain this difference of two orders of magnitude from first principles. The Dirac fermion model is inherently connected to the Wigner-Seitz cell model, which is rejected in favor of the rectangular unit cell model in this book. The plateau observed for the Hall conductivity σ_H is caused by the condensed c-bosons. The plateau behavior arises from the superconducting state and hence it is unlikely to be explained based on the Dirac fermion model.

References

1 Novoselov, K.S., Geim, A.K., Morozov, S.V., Jiang, D., Katsnelson, M.I., Grigorieva, I.V., Dubonos, S.V. and Firsov, A.A. (2005) *Nature*, **438**, 197.

2 Zhang, Y., Y.-Tan, W., Stormer, H.L., and Kim, P. (2005) *Nature*, **438**, 201.

3 Zhang, Y., Jiang, Z., Small, J.P., Purewal, M.S., Y. Tan, W., Fezlollahi, M., Chudow, J.D., Jaszczak, J.A., Stormer, H.L., and Kim, P. (2006) *Phys. Rev. Lett.*, **96**, 136806.

4 Bardeen, J., Cooper, L.N., and Schrieffer, J.R. (1957) *Phys. Rev.*, **108**, 1175.

5 Zhang, S.C., Hansson, T.H., and Kivelson, S. (1989) *Phys. Rev. Lett.*, **62**, 82.

6 Ezawa, Z.F. (2000) *Quantum Hall Effect*, World Scientific, Singapore.

7 Novoselov, K.S., Jiang, Z., Zhang, Y., Morozov, S.V., Stormer, H.I., Zeitler, U., Maan, J.C., Boebinger, G.S., Kim, P., and Gaim, A.K. (2007) *Science*, **315**, 1379.

8 Fujita, S. and Suzuki, A. (2013) ArXiv: 1304.7631v1 [cond-mat,mes-hall].

13
Seebeck Coefficient in Multiwalled Carbon Nanotubes

Strictly speaking, the Seebeck coefficient, S, also called the thermoelectric power, is not an electrical transport property. But as we see below, the measurement of Seebeck coefficients gives important information about the carriers in electrical transport. Because of this we discuss the Seebeck coefficient in this chapter.

On the basis of the idea that different temperatures generate different carrier densities and the resulting carrier diffusion generates a thermal electromotive force (emf), a new formula for the Seebeck coefficient (thermopower) S is obtained:

$$S = \frac{2\ln 2}{3}\frac{\varepsilon_F k_B \mathcal{D}_0}{qn} ,$$

where k_B is the Boltzmann constant, q, n, ε_F, and $\mathcal{D}_0$ are charge, carrier density, Fermi energy, and the density of states at ε_F, respectively. Ohmic and Seebeck currents are fundamentally different in nature, and hence, cause significantly different transport behaviors. For example, the Seebeck coefficient S in copper (Cu) is positive, while the Hall coefficient is negative. In general, the Einstein relation between the conductivity and the diffusion coefficient does not hold for a multicarrier metal. Multiwalled carbon nanotubes are superconductors. The Seebeck coefficient S in multiwalled carbon nanotubes is shown to be proportional to the temperature T above the superconducting temperature T_c based on the model of Cooper pairs as carriers. The S below T_c follows a temperature behavior:

$$\frac{S}{T} \propto \left(\frac{T_g'}{T}\right)^{1/2} ,$$

where T_g' is constant at the lowest temperatures.

13.1
Introduction

In 2003 Kang *et al.* [1] observed a logarithmic temperature (T)-dependence for the Seebeck coefficient S in multiwalled carbon nanotubes (MWNTs) at low temperatures. Their data are reproduced in Figure 13.1 after [1, Figure 2], where S/T are

Electrical Conduction in Graphene and Nanotubes, First Edition. S. Fujita and A. Suzuki.

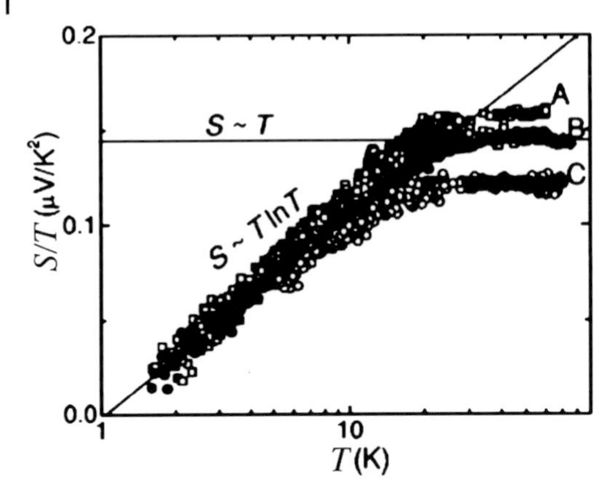

Figure 13.1 Low-temperature Seebeck coefficient S of MWNTs plotted as S/T on a logarithmic temperature scale (reproduced from [1, Figure 2]).

plotted on a logarithmic temperature scale. Above 20 K S is proportional to T:

$$S \propto T\,, \quad T > 20\,\mathrm{K}\,. \tag{13.1}$$

Below 20 K the curves follow a logarithmic behavior:

$$S \sim T \ln T\,, \quad T < 20\,\mathrm{K}\,. \tag{13.2}$$

The data are shown for three samples with different doping levels: A, B, and C. If a system of free electrons with a uniform distribution of impurities is considered, then the Seebeck coefficient, also called the thermoelectric power, S is temperature-independent which will be shown in Section 13.6. Hence, the behavior of T in both (13.1) and (13.2) is unusual. If the Cooper pairs (pairons) [2] are charge carriers and other conditions are met, then both (13.1) and (13.2) are explained microscopically, which is shown in this chapter.

The extended data up to 300 K obtained by Kang *et al.* [1] are shown in Figure 13.2, after [1, Figure 1]. In Figure 13.2a the S of MWNTs is shown, indicating a clear suppression of S from linearity below 20 K as seen in the lower-right inset. In Figure 13.2b, the Seebeck coefficient S of highly oriented single-crystal pyrolytic graphite (HOPG) is shown. This S is negative ("electron"-like) at low temperatures and become positive ("hole"-like) and constant above 150 K:

$$S_{\mathrm{graphite}} = \begin{cases} < 0, & T < 150\,\mathrm{K} \\ \text{constant} > 0, & T > 150\,\mathrm{K} \end{cases}. \tag{13.3}$$

The "electron" ("hole") is a quasielectron which has an energy higher (lower) than the Fermi energy *and* which circulates counterclockwise (clockwise) when viewed from the tip of the applied magnetic field vector. "Electrons" ("holes") are excited on the positive (negative) side of the Fermi surface with the convention that the positive normal vector at the surface points in the energy-increasing direction. Graphite

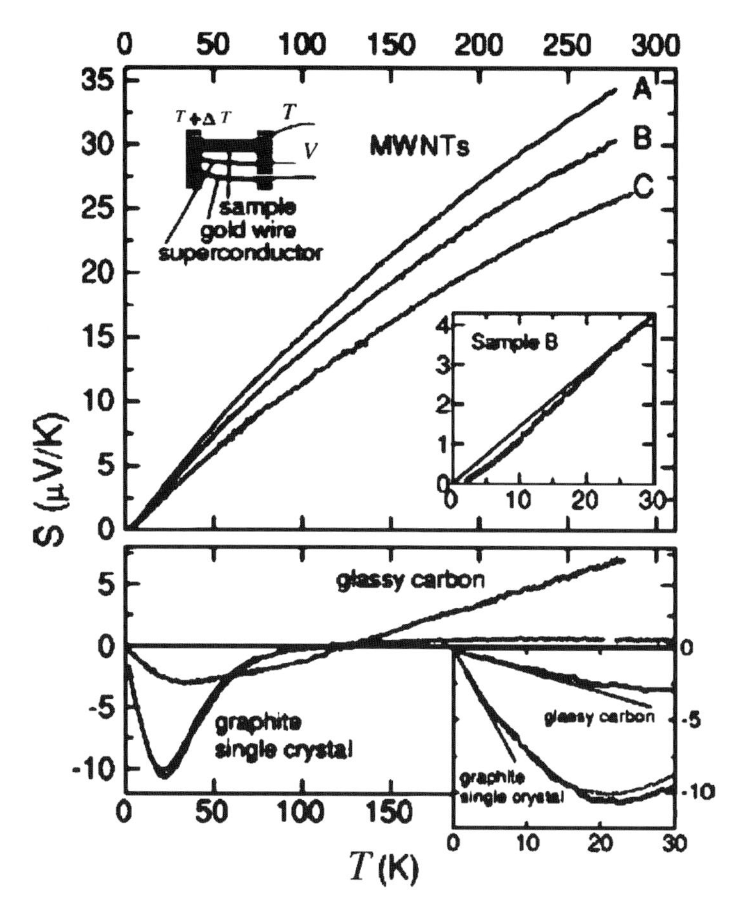

Figure 13.2 (a) The temperature dependence of thermoelectric power of MWNTs at several doping levels. The suppression of TEP from linearity at low temperatures is clearly shown in the lower-right inset (the line represents a linear T dependence). (b) The thermoelectric power of single-crystal HOPG and glassy carbons. No suppression can be recognized for both as $T \to 0$ (reproduced from [1, Figure 1]).

is composed of ABAB-type graphene layers. The different T-behaviors for graphite (3D) and MWNTs (2D) should arise from the different carriers. We will show that the majority carriers in graphene and graphite are "electrons" while the majority carriers in MWNT are "holes" based on the rectangular unit cell model, which is shown in Sections 13.5 and 13.6. Note that conduction electrons are denoted by quotation marked "electrons" ("holes") whereas generic electrons are denoted without quotation marks.

13.2 Classical Theory of the Seebeck Coefficient in a Metal

We now take a system of free electrons with mass m and charge $-e$ with a uniform distribution of impurities which act as scatterers. We assume that a free classical electron system in equilibrium is characterized by the ideal gas condition so that the average electron energy ε depends on the temperature T only:

$$\varepsilon = \varepsilon(n, T) = \varepsilon(T) , \tag{13.4}$$

where n is the electron density. The electric current density $\boldsymbol{j}$ is given by

$$\boldsymbol{j} = (-e)n\boldsymbol{v} , \tag{13.5}$$

where $\boldsymbol{v}$ is the average velocity. We assume that the density n is constant in space and time. If there is a temperature gradient, then there will be a current as shown below. We assume first a one-dimensional (1D) motion. The velocity field v depends on the temperature T, which varies in space.

Assume that the temperature T is higher at $x + \Delta x$ than at x:

$$T(x + \Delta x) > T(x) . \tag{13.6}$$

Then

$$v[n, T(x + \Delta x)] - v[n, T(x)] = \frac{\partial v(n, T)}{\partial T}\frac{\partial T}{\partial x}\Delta x . \tag{13.7}$$

The diffusion and heat conduction occur locally. We may choose Δx to be a mean free path:

$$l = v\tau , \tag{13.8}$$

which is constant in our system. Then the current j is, from (13.5),

$$j = (-e)n\frac{\partial v}{\partial T}l\frac{\partial T}{\partial x} . \tag{13.9}$$

When a metallic bar is subjected to a voltage (V) or temperature (T) difference, an electric current is generated. For small voltage and temperature gradients we may assume a linear relation between the electric current density $\boldsymbol{j}$ and the gradients:

$$\boldsymbol{j} = \sigma(-\nabla V) + A(-\nabla T) = \sigma\boldsymbol{E} - A\nabla T , \tag{13.10}$$

where $\boldsymbol{E} \equiv -\nabla V$ is the electric field and σ the conductivity. If the ends of the conducting bar are maintained at different temperatures, no electric current flows. Thus, from (13.10), we obtain

$$\sigma\boldsymbol{E}_S - A\nabla T = 0 , \tag{13.11}$$

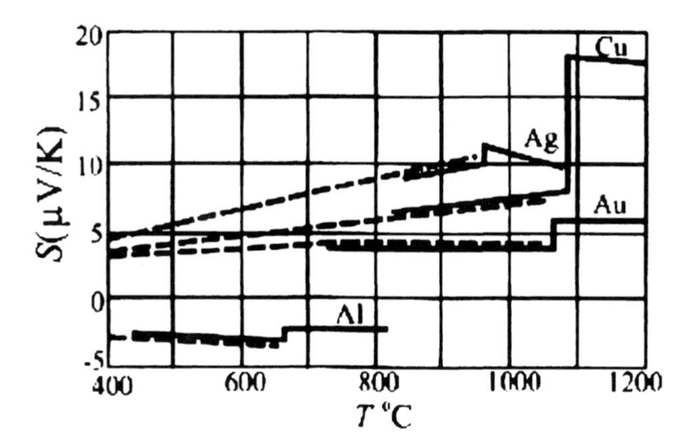

Figure 13.3 High-temperature Seebeck coefficients above 400 °C for Ag, Al, Au, and Cu. The solid and dashed lines represent two experimental data sets. Taken from [3]. The discontinuities arise from melting.

where $\boldsymbol{E}_S$ is the field generated by the Seebeck electromotive force (emf). The *Seebeck coefficient* S is defined through

$$\boldsymbol{E}_S = S\nabla T\,, \quad S \equiv \frac{A}{\sigma}\,. \tag{13.12}$$

The conductivity σ is positive, but the Seebeck coefficient S can be positive or negative. The measured Seebeck coefficient S in Al at high temperatures (400–670 °C) is negative, while the S in noble metals (Copper (Cu), silver (Ag), and gold (Au)) are positive as shown in Figure 13.3.

On the basis of the classical idea that different temperatures generate different electron drift velocities, we obtain the Seebeck coefficient (Problem 13.3.1):

$$S = -\frac{c_V}{3ne}\,, \tag{13.13}$$

where c_V is the heat capacity per unit volume and n the electron density. Using (13.8)–(13.10), we obtain

$$A = (-e)n\frac{\partial v}{\partial T}v\tau\,. \tag{13.14}$$

The conductivity σ is given by the Drude formula:

$$\sigma = \frac{ne^2\tau}{m}\,. \tag{13.15}$$

Thus, the Seebeck coefficient S is, using (13.14) and (13.15),

$$\begin{aligned} S = \frac{A}{\sigma} &= -\frac{1}{ne}m\frac{\partial v}{\partial T}\frac{l}{\tau} = -\frac{1}{ne}m\frac{\partial v^2}{\partial T} \\ &= -\frac{1}{ne}\frac{\partial}{\partial T}\frac{1}{2}mv^2 = -\frac{1}{ne}\frac{\partial \varepsilon}{\partial T} = -\frac{1}{ne}c\,, \end{aligned} \tag{13.16}$$

where

$$c \equiv \frac{\partial \varepsilon}{\partial T} \tag{13.17}$$

is the heat capacity per electron.

Our theory can be extended simply for 3D motion. The equipartition theorem holds for the classical electrons:

$$\left\langle \frac{1}{2} m v_x^2 \right\rangle = \left\langle \frac{1}{2} m v_y^2 \right\rangle = \left\langle \frac{1}{2} m v_z^2 \right\rangle = \frac{1}{2} k_B T \, , \tag{13.18}$$

where the angular brackets mean the equilibrium averages. Hence, the average energy is

$$\varepsilon \equiv \frac{1}{2} m v^2 = \frac{1}{2} \left(v_x^2 + v_y^2 + v_z^2 \right) = \frac{3}{2} k_B T \, . \tag{13.19}$$

We obtain

$$A = -en \frac{1}{2} \frac{\partial v^2}{\partial t} \tau \, . \tag{13.20}$$

Using this, we obtain the Seebeck coefficient for 3D motion as

$$S = \frac{A}{\sigma} = -\frac{c_V}{3ne} = -\frac{k_B}{2e} \, , \tag{13.21}$$

where

$$c_V \equiv \frac{\partial \varepsilon}{\partial T} = \frac{3}{2} k_B \tag{13.22}$$

is the heat capacity per electron. The heat capacity per unit volume, c_V, is related to the heat capacity per electron, c, by

$$c_V = nc \, . \tag{13.23}$$

Setting c_V equal to $3nk_B/2$ in (13.13), we obtain the *classical formula* for S:

$$S_{\text{classical}} = -\frac{k_B}{2e} = -0.43 \times 10^{-4}\ \text{V K}^{-1} = -43\ \mu\text{V K}^{-1} \, . \tag{13.24}$$

Observed Seebeck coefficients in metals at room temperature are of the order of microvolts per degree (see Figure 13.3), a factor of ten smaller than $S_{\text{classical}}$. If we introduce the heat capacity computed using Fermi statistics [4] (Problem 13.3.2)

$$c_V = \frac{1}{2} \pi^2 n k_B \frac{k_B T}{\varepsilon_F} = \frac{1}{2} \pi^2 n k_B \frac{T}{T_F} \, , \tag{13.25}$$

where $T_F(\varepsilon_F)$ is the Fermi temperature (energy), we obtain

$$S_{\text{semiquantum}} = -\frac{\pi}{6} \frac{k_B}{e} \frac{k_B T}{\varepsilon_F} \, , \tag{13.26}$$

which is often quoted in materials handbook [3]. Formula (13.26) remedies the difficulty with respect to the magnitude. But the correct theory must explain the two possible signs of S besides the magnitude. Fujita, Ho, and Okamura [5] developed a quantum theory of the Seebeck coefficient. We follow this theory below.

13.3 Quantum Theory of the Seebeck Coefficient in a Metal

Let us recall that "electrons" ("holes") are thermally excited above (below) the Fermi energy. This means that the existence of "holes" by itself is a quantum effect.

We assume that the carriers are conduction electrons ("electron", "hole") with charge q ($-e$ for "electrons," $+e$ for "holes") and effective mass m^*. Assuming a one-component system, the Drude conductivity σ is given by

$$\sigma = n\frac{q^2}{m^*}\tau\,, \tag{13.27}$$

where n is the carrier density and τ the mean free time. We observe from (13.27) that σ is always positive irrespective of whether $q = -e$ or $+e$. The Fermi distribution function f is

$$f(\varepsilon;\beta,\mu) = \frac{1}{e^{\beta(\varepsilon-\mu)}+1}\,, \quad \beta \equiv (k_B T)^{-1}\,, \tag{13.28}$$

where μ is the chemical potential whose value at 0 K equals the Fermi energy ε_F. The voltage difference $\Delta V = LE$, with L being the sample length, generates the chemical potential difference $\Delta\mu$, the change in f, and consequently, the electric current. Similarly, the temperature difference ΔT generates the change in f and the current.

At 0 K the Fermi surface is sharp and there are no conduction electrons ("electrons," "holes"). At a finite T, "electrons" ("holes") are thermally excited near the Fermi surface if the curvature of the surface is negative (positive), see Figures 13.4 and 13.5. We assume a high Fermi degeneracy:

$$T_F \gg T\,. \tag{13.29}$$

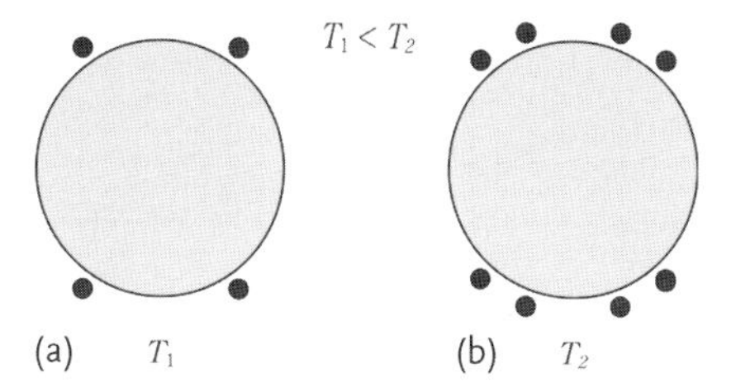

Figure 13.4 (a,b) More "electrons" (dots) are excited above the Fermi surface (solid line) at the high-temperature end: $T_2 (> T_1)$. The shaded area denotes the electron-filled states. "Electrons" diffuse from (b) to (a).

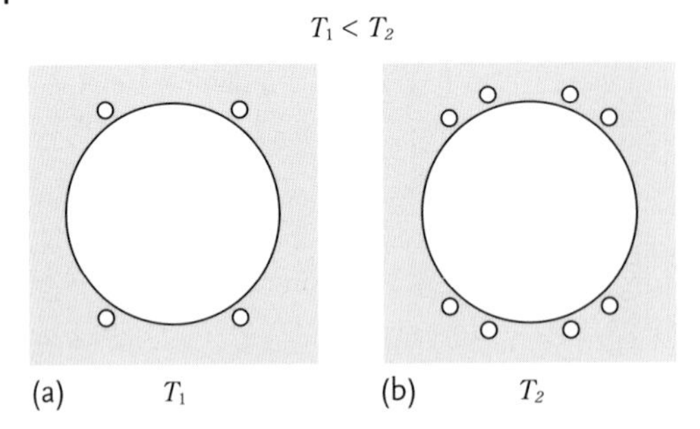

Figure 13.5 (a,b) More "holes" (open circles) are excited below the Fermi surface at the high-temperature end: $T_2 (> T_1)$. "Holes" diffuse from (b) to (a).

Consider first the case of "electrons." The number of thermally excited "electrons," N_x, having energies greater than the Fermi energy ε_F is defined and calculated as (Problem 13.3.3)

$$N_x \equiv \int_{\varepsilon_F}^{\infty} d\varepsilon \mathcal{D}(\varepsilon) \frac{1}{e^{\beta(\varepsilon-\mu)}+1} \cong \mathcal{D}_0 \int_{\varepsilon_F}^{\infty} d\varepsilon \frac{1}{e^{\beta(\varepsilon-\mu)}+1}$$
$$= -\mathcal{D}_0 \beta^{-1} \left[\ln \left(1 + e^{-\beta(\varepsilon-\mu)} \right) \right]_{\varepsilon_F}^{\infty} = \ln 2 k_B T \mathcal{D}_0 \,, \tag{13.30}$$

where $\mathcal{D}_0 \equiv \mathcal{D}(\varepsilon_F)$ is the density of states at $\varepsilon = \varepsilon_F$. The excited "electron" density $n \equiv N_x/\mathbb{V}$, where $\mathbb{V}$ is the sample volume, is higher at the high-temperature end, and the particle current runs from the high- to the low-temperature end. This means that the electric current runs towards (away from) the high-temperature end in an "electron" ("hole")-rich material. After using formula (13.12), we find

$$\begin{aligned} S &< 0 \quad \text{for "electrons"} \\ S &> 0 \quad \text{for "holes"} \,. \end{aligned} \tag{13.31}$$

The Seebeck current arises from the thermal diffusion. We assume Fick's law:

$$j = q j_{\text{particle}} = -q D \nabla n \,, \tag{13.32}$$

where D is the diffusion constant, which is computed from the kinetic-theoretical formula:

$$D = \frac{1}{d} v l = \frac{1}{d} v_F^2 \tau \,, \quad v = v_F \,, \quad l = v\tau \,, \tag{13.33}$$

where d is the dimension in this chapter. The density gradient ∇n is generated by the temperature gradient ∇T, and is given by

$$\nabla n = \frac{\ln 2}{d} k_B D_0 \nabla T \,, \quad D_0 \equiv \frac{\mathcal{D}_0}{\mathbb{V}} \,, \tag{13.34}$$

where (13.30) is used. Using the last three equations and (13.10), we obtain

$$A = \ln 2 q v_F^2 k_B \mathcal{D}_0 \tau \,. \tag{13.35}$$

Using (13.10), (13.27), and (13.35), we obtain (Problem 13.3.4)

$$S = \frac{A}{\sigma} = \frac{2\ln 2}{d}\left(\frac{1}{qn}\right)\varepsilon_F k_B \mathcal{D}_0 . \tag{13.36}$$

The mean free time τ cancels out from the numerator and denominator.

The derivation of our formula (13.36) for the Seebeck coefficient S was based on the idea that the Seebeck emf arises from the thermal diffusion. We used the high Fermi degeneracy condition (13.29): $T_F \gg T$. The relative errors due to this approximation *and* due to the neglect of the T-dependence of μ are both of the order $(k_B T/\varepsilon_F)^2$. Formula (13.36) can be negative *or* positive, while the materials handbook formula (13.26) has a negative sign. The average speed v for highly degenerate electrons is equal to the Fermi velocity v_F (independent of T). In Ashcroft and Mermin's book [4], the origin of a positive S in terms of a mass tensor $\mathsf{M} = \{m_{ij}\}$ is discussed. This tensor M is real and symmetric, and hence, it can be characterized by the principal masses $\{m_j\}$. The formula for S obtained by Ashcroft and Mermin [4, Equation (13.62)], can be positive or negative but is hard to apply in practice. In contrast our formula (13.36) can be applied straightforwardly. Besides our formula for a one-carrier system is T-independent, while Ashcroft and Mermin's formula is linear in T.

Formula (13.36) is remarkably similar to the standard formula for the Hall coefficient of a one-component system:

$$R_H = (qn)^{-1} . \tag{13.37}$$

Both Seebeck and Hall coefficients are inversely proportional to charge q, and hence, they give important information about the sign of the carrier charge. In fact the measurement of the S of a semiconductor can be used to see if the conductor is n-type or p-type (with no magnetic measurements). If only one kind of carrier exists in a conductor, then the Seebeck and Hall coefficients must have the same sign as observed in alkali metals.

Let us consider the electric current caused by a voltage difference. The current is generated by the electric force that acts on *all* electrons. The electron's response depends on its mass m^*. The density (n) dependence of σ can be understood by examining the current-carrying steady state in Figure 13.6b. The electric field $\boldsymbol{E}$ displaces the electron distribution by a small amount $\hbar^{-1} q E \tau$ from the equilibrium distribution in Figure 13.6a. Since all the conduction electrons are displaced, the conductivity σ depends on the particle density n. The Seebeck current is caused by the density difference in the thermally excited electrons near the Fermi surface, and hence, the thermal diffusion coefficient A depends on the density of states at the Fermi energy, $\mathcal{D}_0$ (see (13.35)). We further note that the diffusion coefficient D does not depend on m^* directly (see (13.33)). Thus, the Ohmic and Seebeck currents are fundamentally different in nature.

For a single-carrier metal such as sodium which forms a body-centered cubic (bcc) lattice, where only "electrons" exist, both R_H and S are negative. The *Einstein relation* between the conductivity σ and the diffusion coefficient D holds:

$$\sigma \propto D . \tag{13.38}$$

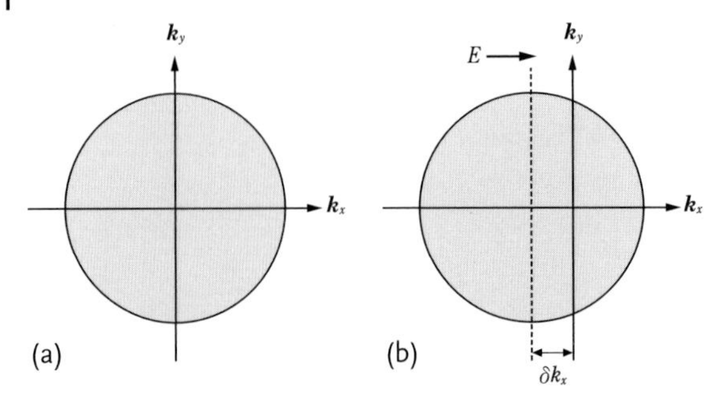

Figure 13.6 As the electric field $\boldsymbol{E}$ points in the positive x-direction, the steady-state electron distribution in (b) is generated by a translation of the equilibrium distribution in (a) by the amount $\delta \boldsymbol{k} = -eE\tau/\hbar$.

Using (13.27) and (13.33), we obtain

$$\frac{D}{\sigma} = \frac{v_F^2 \tau/3}{nq^2\tau/m^*} = \frac{2}{3}\frac{\varepsilon_F}{nq^2}\,, \tag{13.39}$$

which is a material constant. The Einstein relation is valid for a single-carrier system.

The relation does not hold in general for multicarrier systems. The ratio D/σ for a two-carrier system containing "electrons" (1) and "holes" (2) is given by (Problem 13.3.5)

$$\frac{D}{\sigma} = \frac{1/3 v_1^2 \tau_1 + 1/3 v_2^2 \tau_2}{(n_1 q_1^2/m_1^*)\tau_1 + (n_2 q_2^2/m_2^*)\tau_2}\,, \tag{13.40}$$

which is a complicated function of (m_1^*/m_2^*), (n_1/n_2), (v_1/v_2), and (τ_1/τ_2). In particular the mass ratio m_1^*/m_2^* may vary significantly for a heavy fermion condition, which occurs whenever the Fermi surface just touches the Brillouin boundary, see below. An experimental check on the violation of the Einstein relation can be carried out by simply examining the T dependence of the ratio D/σ. This ratio from (13.39) is constant for a single-carrier system, while from (13.40) it depends on T since the generally T-dependent mean free times (τ_1, τ_2) arising from the electron–phonon scattering do not cancel out from numerator and denominator. Conversely, if the Einstein relation holds for a metal, the spherical Fermi surface approximation with a single effective mass m^* is valid for this single-carrier metal.

Problem 13.3.1. Derive (13.13) using kinetic theory.

Problem 13.3.2. Derive (13.25) applying the Fermi statistics for free carriers.

Problem 13.3.3. Verify (13.30). Hint:

$$\frac{\mathrm{d}}{\mathrm{d}\varepsilon}\ln(1 + \mathrm{e}^{-\beta(\varepsilon-\mu)}) = -\frac{\beta \mathrm{e}^{-\beta(\varepsilon-\mu)}}{1 + \mathrm{e}^{-\beta(\varepsilon-\mu)}}\,.$$

Problem 13.3.4. Verify (13.36).

Problem 13.3.5. Derive (13.40).

13.4 Simple Applications

We consider two-carrier metals (noble metals). Noble metals including Cu, Ag, and Au form face-centered cubic (fcc) lattices. Each metal contains "electrons" and "holes." The Seebeck coefficient S for these metals are shown in Figure 13.3. The S is positive for all:

$$S > 0 \quad \text{for Cu, Au, Ag}\,, \tag{13.41}$$

indicating that the major carriers are "holes." The Hall coefficient R_H is known to be negative:

$$R_H < 0 \quad \text{for Cu, Au, Ag}\,. \tag{13.42}$$

Clearly the Einstein relation (13.38) does not hold since the charge sign is different for S and R_H. This complication was explained by Fujita, Ho, and Okamura [5] based on the Fermi surfaces having "necks" (see Figure 13.7). The curvatures along the axes of each neck are positive, and hence, the Fermi surface is "hole"-generating. Experiments [6–8] indicate that the minimum neck area A_{111} (neck) in the k-space is 1/51 of the maximum belly area A_{111} (belly), meaning that the Fermi surface just touches the Brillouin boundary (Figure 13.7 exaggerates the neck area). The density of "hole"-like states, n_{hole}, associated with the $\langle 111 \rangle$ necks, having a heavy fermion character due to the rapidly varying Fermi surface with energy, is much greater than that of "electron"-like states, n_{electron}, associated with the $\langle 100 \rangle$ belly. The thermally excited "hole" density is higher than the "electron" density, yielding a positive S. The principal mass m_1^* along the axis of a small neck ($m_1^{*-1} = \partial^2 \varepsilon / \partial p_1^2$) is positive ("hole"-like) and extremely large. The contribution of the"hole" to the conduction is small ($\sigma \propto m^{*-1}$). Then the "electrons"

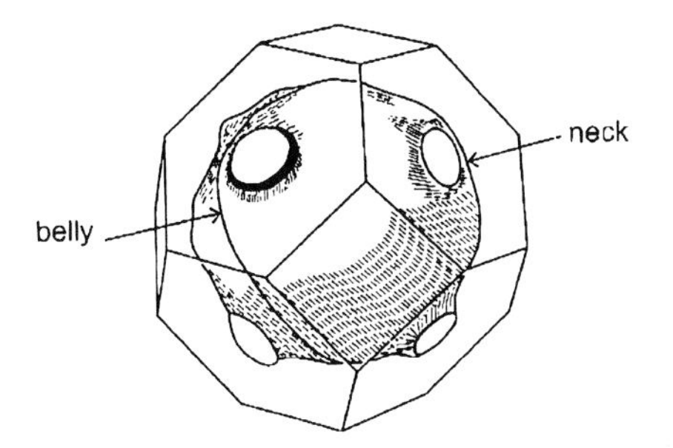

Figure 13.7 The Fermi surface of silver (fcc) has "necks," with the axes in the $\langle 111 \rangle$ direction, located near the Brillouin boundary, reproduced after [6–8].

associated with the nonneck Fermi surface dominate and yield a negative Hall coefficient R_H.

We note that the Einstein relation (13.38) does not hold in general for multicarrier systems. If the Einstein relation holds for a metal, the spherical Fermi surface approximation with a single effective mass m^* is valid.

13.5
Graphene and Carbon Nanotubes

Graphite and diamond are both made of carbons. They have different lattice structures and different properties. Diamond is brilliant and it is an insulator while graphite is black and is a good conductor. In 1991 Iijima [9] discovered carbon nanotubes in the soot created in an electric discharge between two carbon electrodes. These nanotubes ranging from 4 to 30 nm in diameter were found to have a helical multiwalled structure. The tube length is about one micron (μm). Single-wall nanotubes (SWNTs) were fabricated first by Iijima and Ichihashi [10] and by Bethune *et al.* [11] in 1993. The tube size is about 1 nm in diameter and a few microns in length. The scroll-type tube is called a multiwalled carbon nanotube (MWNT). The tube size is about 10 nm in diameter and a few microns (μm) in length. An unrolled carbon sheet is called *graphene,* which has a honeycomb lattice structure as shown in Figure 13.8.

We consider graphene which forms a 2D honeycomb lattice. The normal carriers in the transport of electrical charge are "electrons" and "holes." Following Ashcroft and Mermin [4], we assume the semiclassical (wavepacket) model of a conduction electron. It is necessary to introduce a k-vector:

$$\boldsymbol{k} = k_x \hat{\boldsymbol{e}}_x + k_y \hat{\boldsymbol{e}}_y + k_z \hat{\boldsymbol{e}}_z \tag{13.43}$$

since the k-vector is involved in the semiclassical equation of motion:

$$\hbar \dot{\boldsymbol{k}} \equiv \hbar \frac{\mathrm{d}\boldsymbol{k}}{\mathrm{d}t} = q(\boldsymbol{E} + \boldsymbol{v} \times \boldsymbol{B}) \,, \tag{13.44}$$

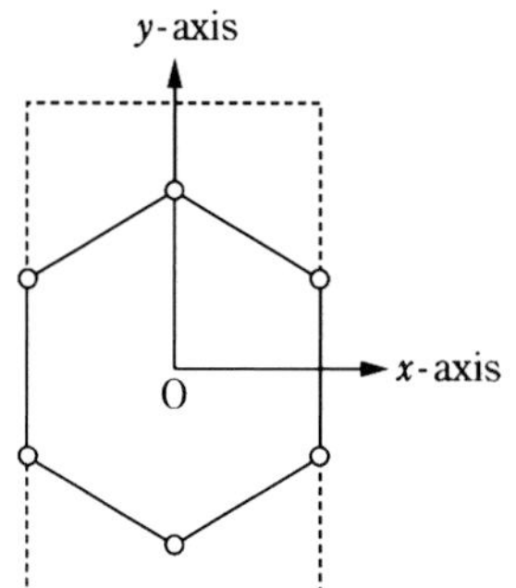

Figure 13.8 A honeycomb lattice and the Cartesian (rectangular) unit cell (dotted line) of graphene. The rectangular unit cell contains four C^+ ions represented by open circles (○).

where $\boldsymbol{E}$ and $\boldsymbol{B}$ are the electric and magnetic fields, respectively. The vector

$$v \equiv \frac{1}{\hbar}\frac{\partial \varepsilon}{\partial \boldsymbol{k}} \tag{13.45}$$

is the particle velocity, where ε is the particle energy. For some crystals such as simple cubic, face-centered cubic, body-centered cubic, tetragonal, and orthorhombic crystals, the choice of the orthogonal (x, y, z)-axes and the unit cells are obvious. Two-dimensional crystals such as graphene can also be treated similarly, with only the z-component being dropped. We will show that graphene has "electrons" and "holes" based on the rectangular unit cell model.

We assume that the "electron" ("hole") wavepacket has a charge $-e$ $(+e)$ and the size of a unit carbon hexagon, generated above (below) the Fermi energy ε_F. We will show that (a) the "electron" and "hole" have different charge distributions and different effective masses, (b) that the "electrons" and "holes" are thermally activated with different energy gaps $(\varepsilon_1, \varepsilon_2)$, and (c) that the "electrons" and "holes" move in different easy channels along which they travel.

The positively charged "hole" tends to stay away from the positive C^+ ions, and hence its charge is concentrated at the center of the hexagon. The negatively charged electron tends to stay close to the C^+ hexagon and its charge is therefore concentrated near the C^+ hexagon. In our model, the "electron" and "hole" both have sizes and charge distributions, and they are not point particles. Hence, their masses m_1 ("electron") and m_2 ("hole") must be different from the gravitational mass $m = 9.11 \times 10^{-28}$ g. Because of the different internal charge distributions, the "electrons" and "holes" have different effective masses m_1 and m_2. The "electron" may move easily with a smaller effective mass in the direction [110 c-axis] $\equiv$ [110] than perpendicular to it as we see presently. Here, we use the conventional Miller indices for the hexagonal lattice with omission of the c-axis index. For the description of the electron in terms of the mass tensor, it is necessary to introduce Cartesian coordinates, which do not match with the crystal's natural (triangular) axes. We may choose the unit cell as shown in Figure 13.8. Then the Brillouin zone boundary in the k space is a rectangle with side lengths $(2\pi/b, 2\pi/c)$. The "electron" (wavepacket) may move up or down in [110] to the neighboring hexagon sites passing over one C^+. The positively charged C^+ acts as a welcoming (favorable) potential valley center for the negatively charged "electron," while the same C^+ acts as a hindering potential hill for the positively charged "hole." The "hole" can, however, move easily horizontally without ever meeting the hindering potential hills. Then, the easy channel directions for the "electrons" and "holes" are [110] and [001], respectively.

The thermally activated electron densities are then given by [13]

$$n_j(T) = n_j \mathrm{e}^{-\varepsilon_j/(k_B T)}\,, \tag{13.46}$$

where $j = 1$ and 2 represent the "electron" and "hole," respectively. The prefactor n_j is the density at the high-temperature limit.

13.6 Conduction in Multiwalled Carbon Nanotubes

MWNTs are open-ended. Hence, each pitch is likely to contain an irrational number of carbon hexagons. Then, the electrical conduction of MWNTs is similar to that of metallic SWNTs [14].

Phonons are excited based on the same Cartesian unit cell as the conduction electrons in the carbon wall. The phonon-exchange interaction binds Cooper pairs, also called pairons [2].

The conductivity σ based on the pairon carrier model is calculated as follows. The pairons move in 2D with the linear dispersion relation [2]:

$$\varepsilon_{\mathrm{p}} = c^{(j)} p \,, \tag{13.47}$$

$$c^{(j)} = \frac{2}{\pi} v_{\mathrm{F}}^{(j)} \,, \tag{13.48}$$

where $v_{\mathrm{F}}^{(j)}$ is the Fermi velocity of the "electron" ($j = 1$) ("hole" ($j = 2$)).

Consider first "electron"-pairs. The velocity $\boldsymbol{v}$ is given by (omitting superscript)

$$\boldsymbol{v} = \frac{\partial \varepsilon_{\mathrm{p}}}{\partial \boldsymbol{p}} \quad \text{or} \quad v_x = \frac{\partial \varepsilon_{\mathrm{p}}}{\partial p} \frac{\partial p}{\partial p_x} = c \frac{p_x}{p} \,, \tag{13.49}$$

where we used (13.47) for the pairon energy ε_{p} and the 2D momentum,

$$p \equiv \left(p_x^2 + p_y^2\right)^{1/2} \,. \tag{13.50}$$

The equation of motion along the electric field E in the x-direction is

$$\frac{\partial p_x}{\partial t} = q' E \,, \tag{13.51}$$

where q' is the charge $\pm 2e$ of a pairon. The solution of (13.51) is given by

$$p_x = q' E t + p_x^{(0)} \,, \tag{13.52}$$

where $p_x^{(0)}$ is the initial momentum component. The current density j_{p} is calculated from (charge q') $\times$ (number density n_{p}) $\times$ (average velocity $\overline{v}$). The average velocity $\overline{v}$ is calculated by using (13.49) and (13.52) with the assumption that the pairon is accelerated only for the mean free time τ *and* the initial-momentum-dependent terms are averaged out to zero. We then obtain

$$j_{\mathrm{p}} = q' n_{\mathrm{p}} \overline{v} = q' n_{\mathrm{p}} c \frac{\overline{p_x}}{p} = q'^2 n_{\mathrm{p}} \frac{c}{p} E \tau \,. \tag{13.53}$$

For stationary currents, the partial pairon density n_{p} is given by the Bose distribution function $f(\varepsilon_{\mathrm{p}})$:

$$n_{\mathrm{p}} = f(\varepsilon_{\mathrm{p}}) \equiv \left[\exp\left(\beta \varepsilon_{\mathrm{p}} - \alpha\right) - 1\right]^{-1} \,, \tag{13.54}$$

where e^{α} is the fugacity. Integrating the current j_{p} over all 2D p-space, and using Ohm's law $j = \sigma E$, we obtain for the conductivity σ:

$$\sigma = (2\pi\hbar)^{-2} q'^2 c \int \mathrm{d}^2 p \, p^{-1} f(\varepsilon_{\mathrm{p}}) \tau \,. \tag{13.55}$$

In the low temperatures we may assume the Boltzmann distribution function for $f(\varepsilon_\text{p})$:

$$f(\varepsilon_\text{p}) \simeq \exp\left(\alpha - \beta\varepsilon_\text{p}\right) \, . \tag{13.56}$$

We assume that the relaxation time (inverse collision frequency) arises from the phonon scattering so that

$$\tau = (aT)^{-1} \, , \quad a = \text{constant} \, . \tag{13.57}$$

After performing the p-integration we obtain from (13.55)

$$\sigma = \frac{2}{\pi} \frac{e^2 k_\text{B}}{a\hbar^2} e^{\alpha} \, , \tag{13.58}$$

which is temperature-independent. If there are "electron" and "hole" pairons, they contribute additively to the conductivity. These pairons should undergo a Bose–Einstein condensation at lowest temperatures.

13.7 Seebeck Coefficient in Multiwalled Carbon Nanotubes

We are now ready to discuss the Seebeck coefficient S of MWNTs. First, we will show that the S is proportional to the temperature T above the superconducting temperature T_c.

We start with the standard formula for the charge current density:

$$j = q' n \overline{v} \, , \tag{13.59}$$

where $\overline{v}$ is the average velocity, which is a function of temperature T and the particle density n:

$$\overline{v} = v(n, T) \, . \tag{13.60}$$

We assume a steady state of the system in which the temperature T varies only in the x-direction while the density is kept constant. The temperature gradient $\partial T/\partial x$ generates a current (Problem 13.7.1):

$$j = q' n \frac{\partial \overline{v}(n, T)}{\partial T} \frac{\partial T}{\partial x} \Delta x \, . \tag{13.61}$$

The thermal diffusion occurs locally. We may choose Δx to be a mean free path:

$$\Delta x = l = v\tau \, . \tag{13.62}$$

The current density, j_p, at the 2D pairon momentum p, which is generated by the temperature gradient $\partial T/\partial x$, is thus given by

$$j_\text{p} = q' n_\text{p} \overline{v}_x(n_\text{p}, T) = q' n_\text{p} \frac{\partial \overline{v}}{\partial T} \frac{\partial T}{\partial x} v\tau \, . \tag{13.63}$$

Integrating (13.63) over all 2D p-space and comparing with (13.10), we obtain (Problem 13.7.2)

$$A = (2\pi\hbar)^{-2} q' \frac{\partial \overline{v}}{\partial T} \int \mathrm{d}^2 p \, v_x \, f(\varepsilon_\mathrm{p}) \tau = (2\pi\hbar)^{-2} q' \frac{\partial \overline{v}}{\partial T} c \int \mathrm{d}^2 p \frac{p_x}{p} f(\varepsilon_\mathrm{p}) \tau \,. \tag{13.64}$$

We compare this integral with the integral in (13.55). It has an extra factor in p and therefore generates an extra factor T when the Boltzmann distribution function is adopted for $f(\varepsilon_\mathrm{p})$. Thus, we obtain (Problem 13.7.3), using (13.55) and (13.64),

$$S = \frac{A}{\sigma} \propto T \,. \tag{13.65}$$

We next consider the system below the superconducting temperature T_c. The supercurrents arising from the condensed pairons generate no thermal diffusion. But noncondensed pairons can be scattered by impurities and phonons, and contribute to a thermal diffusion. Because of the zero-temperature energy gap

$$\varepsilon_\mathrm{g} \equiv k_\mathrm{B} T_\mathrm{g} \,, \tag{13.66}$$

generated by the supercondensate, the population of the noncondensed pairons is reduced by the *Boltzmann–Arrhenius factor*

$$\exp\left(-\frac{\varepsilon_\mathrm{g}}{k_\mathrm{B} T}\right) = \exp\left(-\frac{T_\mathrm{g}}{T}\right) \,. \tag{13.67}$$

This reduction applies only for the conductivity (but not for the diffusion). Hence, we obtain the Seebeck coefficient (Problem 13.7.4):

$$S = \frac{A}{\sigma} \propto \frac{T}{\exp(-T_\mathrm{g}/T)} \,. \tag{13.68}$$

In the experiment in [1] a MWNT bundle containing hundreds of individual nanotubes was used. Both circumference and pitch have distributions. Hence, the energy gap $\varepsilon_\mathrm{g}(= k_\mathrm{B} T_\mathrm{g})$ has a distribution.

Kang *et al.* [1] measured the conductance G, which is proportional to the conductivity σ, of the MWNT samples. Their data are reproduced in Figure 13.9, after [1, Figure 3], where the conductance G as a function of temperature is plotted on a logarithmic scale. The G arising from the conduction electron in each MWNT carries an Arrhenius-type exponential

$$\exp\left(-\frac{\varepsilon_\mathrm{a}}{k_\mathrm{B} T}\right) = \exp\left(-\frac{T_\mathrm{a}}{T}\right) \,, \quad \varepsilon_\mathrm{a} \equiv k_\mathrm{B} T_\mathrm{a} \tag{13.69}$$

where ε_a is the activation energy. This energy ε_a has a distribution since the MWNTs have varied circumferences and pitches. The temperature behavior of G for the bundle of MWNTs is seen to be represented by

$$-\ln G \sim \left(\frac{T_\mathrm{a}}{T}\right)^{-1/2} \tag{13.70}$$

in the range: 5–20 K. The electron-activation energy ε_a and the zero-temperature pairon energy gap ε_g are different from each other. But they have the same orders of magnitude and both are temperature-independent. We assume that the distributions are similar. We may then replace $\exp(-T_g/T)$ in (13.68) by $(T_g'/T)^{1/2}$, obtaining the Seebeck coefficient for a bundle of MWNTs

$$S_{\text{bundle}} = \frac{A}{\sigma} \propto T \left(\frac{T_g'}{T}\right)^{1/2} , \tag{13.71}$$

or

$$\ln S_{\text{bundle}} \sim T \ln T \quad \text{below } 20\,\text{K} , \tag{13.72}$$

which is observed in Figure 13.1.

The data in Figure 13.1 clearly indicates a phase change at the temperature

$$T_0 = 20\,\text{K} . \tag{13.73}$$

We now discuss the connection between this T_0 and the superconducting temperature T_c. We deal with a thermal diffusion of the MWNT bundle. The diffusion occurs most effectively for the most dissipative samples which correspond to those with the lowest superconducting temperatures. Hence, the T_0 observed can be interpreted as the superconducting temperature of the most dissipative samples.

In contrast the conduction is dominated by the least dissipative samples having the highest T_c. Figure 13.9 shows a clear deviation of G around 120 K from the experimental law: $G \sim \ln T$. We may interpret this as an indication of the limit of the superconducting states. We then obtain

$$T_c \sim 120\,\text{K} \tag{13.74}$$

for the good samples.

By considering moving pairons we obtained the T-linear behavior of the Seebeck coefficient S above the superconducting temperature T_c and the $T \ln T$-behavior

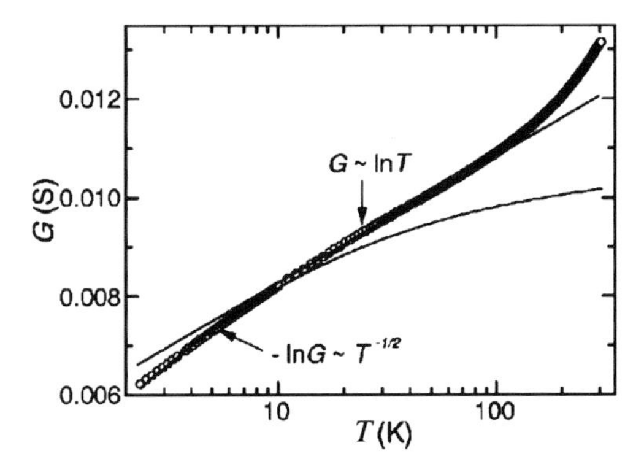

Figure 13.9 The conductance G of the multiwalled carbon nanotube samples as a function of temperature (after [1, Figure 3]).

of S at the lowest temperatures. The energy gap ε_g vanishes at T_c. Hence, the temperature behaviors should be smooth and monotonic as observed in Figure 13.1. This supports our interpretation of the data based on the superconducting phase transition. The doping changes the pairon density and the superconducting temperature. Hence, the data for A, B, and C in Figure 13.1 are reasonable.

Problem 13.7.1. Assume that the temperature T is higher at $x + \Delta x$ than at x: $T(x + \Delta x) > T(x)$. Noting that the velocity field (average velocity) v depends on the local temperature $T(x)$, show that

$$v(T(x + \Delta x)) - v(T(x)) = \frac{\partial v(n, T)}{\partial T} \frac{\partial T}{\partial x} \Delta x$$

holds for $v = v(T(x))$. Applying this formula, derive (13.61).

Problem 13.7.2. Derive (13.64).

Problem 13.7.3. Show (13.65) that the Seebeck coefficient is proportional to temperature T.

Problem 13.7.4. Derive (13.68).

References

1. Kang, N., Lu, L., Kong, W.J., Hu, J.S., Yi, W., Wang, Y.P., Zhang, D.L., Pan, Z.W., and Xie, S.S. (2003) *Phys. Rev. B*, **67**, 033404.
2. Fujita, S., Ito, K., and Godoy, S. (2009) *Quantum Theory of Conducting Matter*, Springer, New York, pp. 77–79.
3. Rossiter, P.L. and Bass, J.(1994) *Metals and Alloys*: in *Encyclopedia of Applied Physics*, vol. 10, VCH Publishing, Berlin, pp. 163–197.
4. Ashcroft, N.W. and Mermin, N.D. (1976) *Solid State Physics*, Saunders, Philadelphia, pp. 46–47, pp. 217–218, pp. 256–258, pp. 290–293.
5. Fujita, S., Ho, H-C., and Okamura, Y. (2000) *Int. J. Mod. Phys. B*, **14**, 2231.
6. Roaf, D.J. (1962) *Philos. Trans. R. Soc.*, **255**, 135.
7. Schönberg, D. (1962) *Philos. Trans. R. Soc.*, **255**, 85.
8. Schönberg, D. and Gold, A.V. (1969) Physics of Metals 1: *Electrons* (ed. J.M. Ziman), Cambridge University Press, UK, p. 112.
9. Iijima, S. (1991) *Nature*, **354**, 56.
10. Iijima,. S. and Ichihashi, T. (1993) *Nature*, **363**, 603.
11. Bethune,D.S., Kiang, C.H., Vries, M.S. de, Gorman, G., Savoy, R., Vazquez, J., and Beyers, R. (1993) *Nature*, **363**, 605.
12. Saito, R., Fujita, M., Dresselhaus, G., and Dresselhaus, M.S. (1992) *Appl. Phys. Lett.*, **60**, 2204.
13. Fujita, S. and Suzuki, A. (2010) *J. Appl. Phys.*, **107**, 013711.
14. Fujita, S., Takato, Y., and Suzuki, A. (2011) *Mod. Phys. Lett. B*, **25**, 223.

14
Miscellaneous

14.1
Metal–Insulator Transition in Vanadium Dioxide

Vanadium dioxide (VO_2) undergoes a metal–insulator transition (MIT) at 340 K with the structural change from a tetragonal (tet) to monoclinic (mcl) crystal as the temperature is lowered. The conductivity σ drops at MIT by four orders of magnitude. The low-temperature monoclinic phase is known to have a lower ground state energy. The existence of a k-vector $\boldsymbol{k}$ is a prerequisite for the conduction since the $\boldsymbol{k}$ appears in the semiclassical equation of motion for the conduction electron (wavepacket). Each wavepacket is, by assumption, composed of plane waves proceeding in a $\boldsymbol{k}$ direction perpendicular to the plane. The tetragonal $(VO_2)_3$ unit cells are periodic along the crystal's x-, y-, and z-axes, and hence there are three-dimensional k-vectors. There are one-dimensional $\boldsymbol{k}$ along the c-axis for a monoclinic crystal. We argue that this decrease in the dimensionality of the k-vectors is the cause of the conductivity drop.

14.1.1
Introduction

In 1959 Morin reported his discovery of a metal–insulator transition in vanadium dioxide (VO_2) [1]. The compound VO_2 forms a monoclinic crystal on the low-temperature side and a tetragonal crystal on the high-temperature side. When heated, VO_2 undergoes an insulator–metal transition around 340 K, with resistivity drops by four orders of magnitude as T is raised. The phase change carries a hysteresis similar to a ferro–paramagnetic phase change. The origin of the phase transition has been attributed by some authors to Peierls instability driven by strong electron–phonon interaction [2], or to Coulomb repulsion and electron localization due to the electron–electron interaction on a Mott–Hubbard picture by other authors [3–5].

A simpler view on the MIT is presented here. We assume that the electron wavepacket is composed of superposable plane waves characterized by the k-vectors. The superposability is a basic property of the Schrödinger wavefunction in free space. A mcl crystal can be generated from an orthorhombic crystal (orc)

Electrical Conduction in Graphene and Nanotubes, First Edition. S. Fujita and A. Suzuki.

by distorting the rectangular faces perpendicular to the c-axis into parallelograms. Material plane waves proceeding along the c-axis exist since the (x, y) planes containing materials (atoms) perpendicular to the z-axis are periodic. It then has one-dimensional (1D) k-vectors along the c-axis. In the x–y plane there is an oblique net whose corners are occupied by V's for mcl VO_2. The position vector $\boldsymbol{R}$ of every V can be represented by integers (m, n), if we choose

$$\boldsymbol{R}_{mn} = m\boldsymbol{a}_1 + n\boldsymbol{a}_2 , \tag{14.1}$$

where $\boldsymbol{a}_1$ and $\boldsymbol{a}_2$ are nonorthogonal base vectors. In the field theoretical formulation the field point $\boldsymbol{r}$ is given by

$$\boldsymbol{r} = \boldsymbol{r}' + \boldsymbol{R}_{mn} , \tag{14.2}$$

where $\boldsymbol{r}'$ is the point defined within the standard unit cell. Equation (14.2) describes the 2D lattice periodicity but does *not* establish k-space as shown earlier, in Section 5.2, Chapter 5.

If we omit the kinetic energy term, then we can still use (14.1) and obtain the ground state energy (except the zero-point energy). The reduction in the dimensionality of the k-vectors from 3D to 1D is the cause of the conductivity drop in the MIT.

The MIT proceeds by domains since the insulator (mcl) phase has the lower degrees of symmetry. Strictly speaking, the existence of 1D k-vectors allows the mcl material to have a small conductivity. This happens for VO_2, see below.

Wu *et al.* [6] measured the resistance R of individual nanowires $W_xV_{1-x}O_2$ with tungsten (W) concentration x ranging up to 1.14%. The nanowires are grown with the wire axis matching the c-axis of the high-temperature rutile structure. The transition temperature T_0 decreases from 340 K, passing room temperature, to 296 K as the concentration x changes from 0 to 1.14%. The temperature dependence of R for the low-temperature phase is semiconductor-like. That is, the resistance R decreases with increasing temperature. The behavior can be fitted with the Arrhenius law:

$$\sigma \propto R^{-1} \propto \exp\left(-\frac{\varepsilon_a}{k_B T}\right) , \tag{14.3}$$

where ε_a is the activation energy. The activation energy ε_a is about 300 meV.

Whittaker *et al.* [7] measured the resistance R in a VO_2 nanowire and observed that (a) the resistance R for the low-temperature (mcl) phase shows an Arrhenius-type dependence as shown in (14.3), while (b) the resistance R for the high-temperature (tet) phase appears to be T-independent. These different behaviors may arise as follows. The currents in (a) run along the nanowire axis, which is also the easy c-axis of the mcl crystal. Hence, Arrhenius behavior for the conductivity is observed for case (a). For case (b), the currents run in three dimensions. Only those electrons near the Fermi surface are excited and participate in the transport. Then, the density of excited electrons, n_x, is related to the total electron density n_0 by

$$n_x = c n_0 \frac{k_B T}{\varepsilon_a} , \tag{14.4}$$

where c is a number close to unity. The factor T cancels out with the T-linear phonon scattering rate for the conductivity. The T-dependence of the exponential factor is small since the activation energy ε_a is much greater than the observation temperature scale measured in units of 10 °C. Thus, the conductivity is nearly constant. There is a sudden drop of resistance around 300 K, when the two phases separate.

In summary the MIT in VO_2 directly arises from the lattice structure change between the tet and the mcl crystal. The tet (mcl) crystal has 3D (1D) k-vectors. The reduction in the dimensionality of the k-vectors is the cause of the conductivity drop.

14.2 Conduction Electrons in Graphite

Graphite is composed of graphene layers stacked in the manner ABAB··· along the c-axis. We may choose a Cartesian unit cell as shown in Figure 14.1.

As Figure 14.1 shows the rectangle (solid line) in the A plane (black) contains six C's wholly within and four C's at the sides. The side C's are shared by neighbors. Hence, the total number of C's is $6 \times 1 + 4 \times 1/2 = 8$. The rectangle in the B plane (gray) contains five C's within, four C's at the sides, and four C's at the corners. The total number of C's is $5 \times 1 + 4 \times 1/2 + 4 \times 1/4 = 8$. The unit cell therefore contains 16 C's. The two rectangles are stacked vertically with the interlayer separation, $c_0 = 3.35$ Å much greater than the nearest-neighbor distance between two C's,

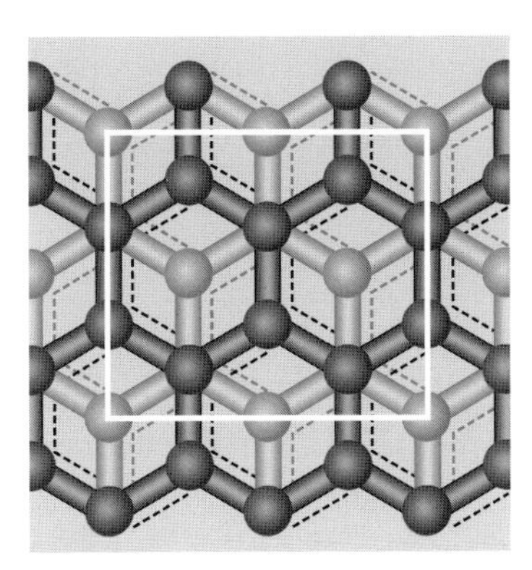

Figure 14.1 The Cartesian unit cell (white solid lines) viewed from the top for graphite. The unit cell of graphite has two layers of graphene. The carbons (circles) in the A (B) planes are shown in black (gray).

$a_0 = 1.42$ Å. The unit cell has three side-lengths:

$$b_1 = 3a_0 , \quad b_2 = 2\sqrt{3}a_0 , \quad b_3 = 2c_0 . \tag{14.5}$$

The center of the unit cell is empty. Clearly, the system is periodic along the orthogonal directions with the three periods (b_1, b_2, b_3) given in (14.5). We may assume that both the "electron" and "hole" have the same unit cell size. Thus, the system is orthorhombic with the sides (b_1, b_2, b_3), $b_1 \neq b_2$, $b_1 \neq b_3$, $b_2 \neq b_3$.

The negatively charged "electrons" (with charge $-e$) in graphite are welcomed by the positively charged C^+ when moving in a vertical direction just as in graphene. That is, the easy directions of movement for the "electrons" are vertical. The easy directions for the "holes" are horizontal. There are no hindering hills for "holes" moving horizontally. Hence, just as for graphene, the "electron" in graphite has a lower activation energy ε than the "hole":

$$\varepsilon_1 < \varepsilon_2 . \tag{14.6}$$

Because of this, the "electrons" are the majority carriers in graphite.

Graphite and graphene have very different unit cells including dimensionality. Thus, the electrical transport behaviors must be significantly different. There are "electrons" and "holes" in graphite. We predict that graphite is a superconductor with a superconducting temperature of the order of 1 K.

14.3 Coronet Fermi Surface in Beryllium

Divalent beryllium (Be) forms a hexagonal close packed (hcp) crystal. The Fermi surface in the second zone constructed in the Nearly Free Electron Model (NFEM) is represented by the "monster" (Figure 3.8a) and the actually observed "coronet" (Figure 3.8b) as shown in Figure 3.8. The figures are drawn based on the Wigner–Seitz model. Part of the coronet Fermi surface can be fitted with the quadratic-in-k dispersion relation:

$$E = \frac{1}{2m_1}p_1^2 + \frac{1}{2m_2}p_2^2 + \frac{1}{2m_3}p_3^2 \tag{14.7}$$

with two negative effective masses and one positive effective mass. The coronet encloses unoccupied states. The effective masses (m_1, m_2, m_3) may be directly obtained from the cyclotron resonance data, using Shockley's formula [8]:

$$\frac{\omega}{eB} = \sqrt{\frac{m_2 m_3 \cos^2(\mu, x_1) + m_3 m_1 \cos^2(\mu, x_2) + m_1 m_2 \cos^2(\mu, x_3)}{m_1 m_2 m_3}} \tag{14.8}$$

where $\cos(\mu, x_j)$ is the directional cosine between the field direction μ and the x_j-direction. The Cartesian coordinate system (x_1, x_2, x_3) can be chosen along the orthogonal unit cells for the hcp crystal. It is a challenge to obtain the values of (m_1, m_2, m_3) after choosing the orthogonal axes appropriately.

14.4 Magnetic Oscillations in Bismuth

Magnetic oscillations, de Haas–van Alphen and Shubnikov–de Haas oscillations, were discovered in bismuth (Bi), which forms a rhombohedral (rhl) crystal. A rhl crystal can be obtained by stretching the three body-diagonal distances from a simple cubic crystal as discussed in Sections 5.2 and 5.3. If an orthogonal unit cell with the Cartesian axes along the body-diagonal passing six corner atoms is chosen, then the system is periodic along the x-, y-, and z-axes passing the center. Thus, the system can be regarded as an orc, which has a 3D k-space. It is a challenge to obtain the effective masses (m_1, m_2, m_3) in Bi from the cyclotron resonance and other measured data.

References

1 Morin, F.J. (1959) *Phys. Rev. Lett.*, **3**, 34.
2 Goodenough, J.B. (1971) *J. Solid State Chem.*, **3**, 490.
3 Mott, N.F. (1968) *Rev. Mod. Phys.*, **40**, 677.
4 Wentzcovitch, R.M, Schulz, W.W., and Allen, P.B. (1994) *Phys. Rev. Lett.*, **72**, 3389.
5 Rice, T.M. Launois, H. and Pouget, J.P. (1994) *Phys. Rev. Lett.*, **73**, 3042.
6 Wu, T-L, Whitaker, L., Banerjee, S., and Sambandamurthy, G. (2011) *Phy. Rev. B*, **83**, 073101.
7 Whittaker, L., Patridge, C.J., and Banerjee, S. (2011) *J. Phys. Chem. Lett.*, **2**, 745.
8 Shockley, W. (1953) *Phys. Rev.*, **90**, 491.

Appendix

A.1
Second Quantization

The most remarkable fact about a system of fermions is that no more than one fermion can occupy a quantum particle state (Pauli's exclusion principle). For bosons no such restriction applies. That is, any number of bosons can occupy the same state. We shall discuss the second-quantization formalism in which creation and annihilation operators associated with each quantum state are used. This formalism is extremely useful in treating many-boson and/or many-fermion systems. Zero-mass bosons such as photons and phonons can be created or annihilated. These dynamical processes can only be described in second quantization.

A.1.1
Boson Creation and Annihilation Operators

Use of creation and annihilation operators $(a^\dagger, a)$ was demonstrated in Chapter 4, Section 4.3 for the treatment of a harmonic oscillator. Second quantization can be used for a general many-body system, as we see below.

The quantum state for a system of bosons (or fermions) can most conveniently be represented by a set of occupation numbers $\{n'_a\}$, where n'_a are the numbers of bosons (or fermions) occupying the quantum particle states a. This representation is called the *occupation-number representation* or simply the *number representation*. For bosons, the possible values for n'_a are zero, one, or any positive integers:

$$n'_a = 0, 1, 2, \ldots \quad \text{for bosons.} \tag{A1}$$

The many-boson state can best be represented by the distribution of particles (balls) in the states (boxes) as shown in Figure A.1, where we choose the 1D momentum states, $p_j = 2\pi\hbar j/L$, as examples.

Let us introduce operators n_a, whose eigenvalues are given by $0, 1, 2, \ldots$ Since (A1) is meant for each and every state a independently, we assume that

$$[n_a, n_b] \equiv n_a n_b - n_b n_a = 0 \,. \tag{A2}$$

Electrical Conduction in Graphene and Nanotubes, First Edition. S. Fujita and A. Suzuki.

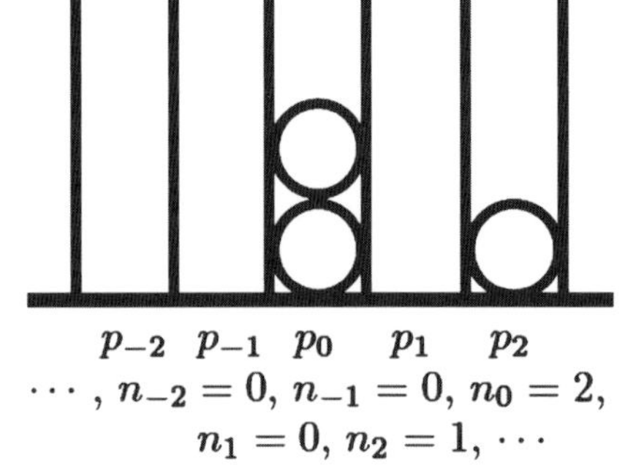

Figure A.1 A many-boson state is represented by a set of boson numbers $\{n_j,\}$ occupying the state $\{p_j\}$.

It is convenient to introduce complex dynamic variables η and $\eta^\dagger$ instead of directly dealing with the number operators n. We attach labels $a, b, \ldots$ and assume that η and $\eta^\dagger$ satisfy the following *Bose commutation rules*:

$$\left[\eta_a, \eta_b^\dagger\right] = \delta_{ab}, \quad [\eta_a, \eta_b] = \left[\eta_a^\dagger, \eta_b^\dagger\right] = 0. \tag{A3}$$

Let us set

$$\eta_a^\dagger \eta_a \equiv n_a = n_a^\dagger, \tag{A4}$$

which is Hermitian. The eigenvalue equation is

$$n|n'\rangle = n'|n'\rangle, \tag{A5}$$

where n' is the eigenvalue of n. We showed earlier in Section 4.3 that the operator n has as eigenvalues all nonnegative integers:

$$n' = 0, 1, 2, \ldots \tag{A6}$$

The corresponding normalized eigenkets are given by

$$(n'!)^{-1/2}(\eta^\dagger)^{n'}|n'\rangle. \tag{A7}$$

Let $|\phi_a\rangle$ be a normalized eigenket of n_a belonging to the eigenvalue 0 so that

$$n_a|\phi_a\rangle = \eta_a^\dagger \eta_a|\phi_a\rangle = 0. \tag{A8}$$

By multiplying all these kets $|\phi_a\rangle$ together, we construct a normalized eigenket:

$$|\Phi_0\rangle \equiv |\phi_a\rangle|\phi_b\rangle \ldots, \tag{A9}$$

which is a simultaneous eigenket of all n belonging to the eigenvalue zero. This ket is called the *vacuum ket*. It has the following property:

$$\eta_a|\Phi_0\rangle = 0 \quad \text{for any } a. \tag{A10}$$

From (A7) we see that if $n'_1, n'_2, \ldots$ are any nonnegative integers, then

$$(n'_1!n'_2!\ldots)^{-1/2}\left(\eta_1^\dagger\right)^{n'_1}\left(\eta_2^\dagger\right)^{n'_2}\ldots|\Phi_0\rangle \equiv |n'_1, n'_2, \ldots\rangle \tag{A11}$$

is a normalized simultaneous eigenket of all n belonging to the eigenvalues $n'_1, n'_2, \ldots$ Various kets obtained by taking different n' form a complete set of kets all orthogonal to each other.

Following Dirac [1, 2], we postulate that the quantum states for N bosons can be represented by a *symmetric ket*:

$$S\left[\left|\alpha_a^{(1)}\right\rangle\left|\alpha_b^{(2)}\right\rangle\cdots\left|\alpha_g^{(N)}\right\rangle\right] \equiv |\alpha_a\alpha_b\cdots\alpha_g\rangle_S\,, \tag{A12}$$

where S is the *symmetrizing operator*

$$S \equiv \frac{1}{\sqrt{N!}}\sum_P P\,, \tag{A13}$$

and P's are permutation operators for the particle indices $(1, 2, \ldots, N)$. The ket in (A12) is not normalized but

$$(n_1!n_2!\cdots)^{-1/2}|\alpha_a\alpha_b\cdots\alpha_g\rangle_S \equiv |\{n\}\rangle \tag{A14}$$

is a normalized ket representing the same state. Comparing (A14) and Eq. (A11), we obtain

$$|\alpha_a\alpha_b\cdots\alpha_g\rangle_S = \eta_a^\dagger\eta_b^\dagger\cdots\eta_g^\dagger|\Phi_0\rangle\,. \tag{A15}$$

That is, unnormalized symmetric kets $|\alpha_a\alpha_b\ldots\alpha_g\rangle_S$ for the system can be constructed by applying N creation operators $\eta_a^\dagger\eta_b^\dagger\cdots\eta_g^\dagger$ to the vacuum ket $|\Phi_0\rangle$. So far we have tacitly assumed that the total number of bosons is fixed at N'. If this number is not fixed but is variable, we can easily extend the theory to this case.

Let us introduce a Hermitian operator N defined by

$$N \equiv \sum_a \eta_a^\dagger\eta_a = \sum_a n_a = N^\dagger\,, \tag{A16}$$

the summation extending over the whole set of boson states. Clearly, the operator N has eigenvalues $0, 1, 2, \ldots$, and the ket $|\alpha_a\alpha_b\cdots\alpha_g\rangle_S$ is an eigenket of N belonging to the eigenvalue N'. We may arrange kets in the order of N', that is a zero-particle state, one-particle states, two-particle states, $\cdots$:

$$|\Phi_0\rangle\,,\quad \eta_a^\dagger|\Phi_0\rangle\,,\quad \eta_a^\dagger\eta_b^\dagger|\Phi_0\rangle\,,\quad \cdots\,. \tag{A17}$$

These kets are all orthogonal to each other, two kets referring to the same number of bosons are orthogonal as before, and two referring to different numbers of bosons are orthogonal because they have different eigenvalues N'. By normalizing the kets, we obtain a set of kets like (A14) with no restriction on $\{n'\}$. These kets form the basic kets in a representation where $\{n_a\}$ are diagonal.

A.1.2
Observables

We wish to express *observable physical quantities* (*observables*) for the system of identical bosons in terms of η and $\eta^\dagger$. These observables are, by postulate, symmetric functions of the boson variables.

An observable can be written in the form:

$$\sum_j y^{(j)} + \sum_i \sum_j z^{(ij)} + \ldots = Y + Z + \ldots , \tag{A18}$$

where $y^{(j)}$ is a function of the dynamic variables of the jth boson, $z^{(ij)}$ that of the dynamic variables of the ith and jth bosons, and so on.

We take

$$Y \equiv \sum_j y^{(j)} . \tag{A19}$$

Since $y^{(j)}$ acts only on the ket $|\alpha^{(j)}\rangle$ of the jth boson, we have

$$\begin{aligned} & y^{(j)} \left(\left|\alpha_{x_1}^{(1)}\right\rangle \left|\alpha_{x_2}^{(2)}\right\rangle \cdots \left|\alpha_{x_j}^{(j)}\right\rangle \cdots \right) \\ & = \sum_a \left(\left|\alpha_{x_1}^{(1)}\right\rangle \left|\alpha_{x_2}^{(2)}\right\rangle \cdots \left|\alpha_a^{(j)}\right\rangle \cdots \right) \left\langle \alpha_a^{(j)} \right| y^{(j)} \left| \alpha_{x_j}^{(j)} \right\rangle . \end{aligned} \tag{A20}$$

The matrix element

$$\left\langle \alpha_a^{(j)} \right| y^{(j)} \left| \alpha_{x_j} \right\rangle \equiv \langle \alpha_a | y | \alpha_{x_j} \rangle \tag{A21}$$

does not depend on the particle index j. Summing (A20) over all j and applying operator S to the result, we obtain

$$S Y \left(\left|\alpha_{x_1}^{(1)}\right\rangle \left|\alpha_{x_2}^{(2)}\right\rangle \cdots \right) = \sum_j \sum_a S \left(\left|\alpha_{x_1}^{(1)}\right\rangle \left|\alpha_{x_2}^{(2)}\right\rangle \cdot\cdot \left|\alpha_a^{(j)}\right\rangle \cdots \right) \left\langle \alpha_a^{(j)} \right| y^{(j)} \left| \alpha_{x_j}^{(j)} \right\rangle . \tag{A22}$$

Since Y is symmetric, we can replace SY by YS for the lhs. After straightforward calculations, we obtain, from (A22),

$$\begin{aligned} Y \eta_{x_1}^\dagger \eta_{x_2}^\dagger \ldots |\Phi_0\rangle &= \sum_j \sum_a \eta_{x_1}^\dagger \eta_{x_2}^\dagger \ldots \eta_{x_{j-1}}^\dagger \eta_a^\dagger \eta_{x_{j+1}}^\dagger \ldots |\Phi_0\rangle \langle \alpha_a | y | \alpha_{x_j} \rangle \\ &= \sum_a \sum_b \eta_a^\dagger \sum_j \eta_{x_1}^\dagger \eta_{x_2}^\dagger \ldots \eta_{x_{j-1}}^\dagger \eta_{x_{j+1}}^\dagger \\ &\quad \ldots |\Phi_0\rangle \delta_{b x_j} \langle \alpha_a | y | \alpha_b \rangle . \end{aligned} \tag{A23}$$

Using the commutation rules and the property (A10) we can show that (Problem A.1)

$$\eta_b \eta^\dagger_{x_1} \eta^\dagger_{x_2} \cdots |\Phi_0\rangle = \sum_j \eta^\dagger_{x_1} \eta^\dagger_{x_2} \cdots \eta^\dagger_{x_{j-1}} \eta^\dagger_{x_{j+1}} \cdots |\Phi_0\rangle \delta_{bx_j} \, . \tag{A24}$$

Using this relation, we obtain from (A23)

$$Y \eta^\dagger_{x_1} \eta^\dagger_{x_2} \cdots |\Phi_0\rangle = \sum_a \sum_b \eta^\dagger_a \eta_b \langle \alpha_a | y | \alpha_b \rangle \left(\eta^\dagger_{x_1} \eta^\dagger_{x_2} \cdots |\Phi_0\rangle \right) \, . \tag{A25}$$

Since the kets $\eta^\dagger_{x_1} \eta^\dagger_{x_2} \ldots |\Phi_0\rangle$ form a complete set, we obtain

$$Y = \sum_a \sum_b \eta^\dagger_a \eta_b \langle \alpha_a | y | \alpha_b \rangle \, . \tag{A26}$$

In a similar manner, Z in (A18) can be expressed by (Problem A.2)

$$Z = \sum_a \sum_b \sum_c \sum_d \eta^\dagger_a \eta^\dagger_b \eta_d \eta_c \langle \alpha_a \alpha_b | z | \alpha_c \alpha_d \rangle \, , \tag{A27}$$

where $\langle \alpha_a \alpha_b | z | \alpha_c \alpha_d \rangle$ is given by

$$\langle \alpha_a \alpha_b | z | \alpha_c \alpha_d \rangle \equiv \left\langle \alpha_a^{(1)} \right| \left\langle \alpha_b^{(2)} \right| z^{(12)} \left| \alpha_d^{(2)} \right\rangle \left| \alpha_c^{(1)} \right\rangle \, . \tag{A28}$$

Problem A.1. Prove (A24). Hint: Start with cases of one- and two-particle state kets.

Problem A.2. Prove (A27) by following steps similar to (A22)–(A26).

A.1.3
Fermion Creation and Annihilation Operators

In this section we treat a system of identical fermions in a parallel manner.

The quantum states for fermions, by postulate, are represented by *antisymmetric kets*:

$$|\alpha_a \alpha_b \cdots \alpha_g\rangle_A \equiv A\left(\left|\alpha_a^{(1)}\right\rangle \left|\alpha_b^{(2)}\right\rangle \cdots \left|\alpha_g^{(N)}\right\rangle \right) \, , \tag{A29}$$

where

$$A \equiv \frac{1}{\sqrt{N!}} \sum_P \delta_P P \tag{A30}$$

is the *antisymmetrizing operator*, with δ_P being $+1$ or -1, according to whether P is even or odd. Each antisymmetric ket in (A29) is characterized such that it changes its sign if an odd permutation of particle indices is applied to it, and the fermion states $a, b, \ldots, g$ are all different. Just as for a boson system, we can introduce

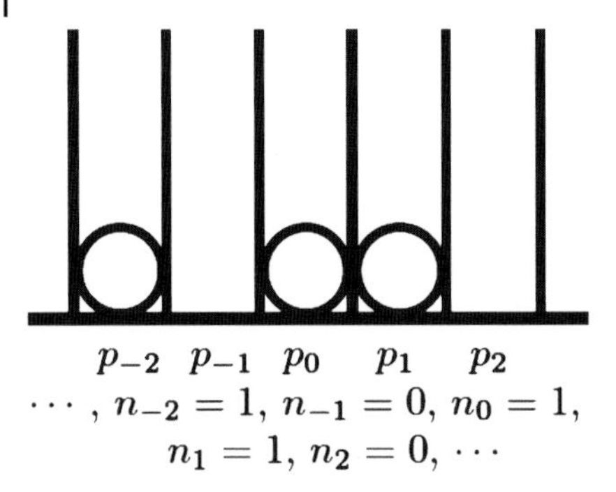

Figure A.2 A many-fermion state is represented by the set of fermion numbers $\{n_j\}$ occupying the state $p_j \equiv 2\pi\hbar j/L$. Each n_j is restricted to 0 or 1.

observables $n_1, n_2, \ldots$, each with eigenvalues 0 or 1, representing the number of fermions in the states $\alpha_1, \alpha_2, \ldots$, respectively. The many-fermion occupation-number state can be represented as shown in Figure A.2.

We can also introduce a set of linear operators $(\eta, \eta^\dagger)$, one pair $(\eta_a, \eta_a^\dagger)$ for each state α_a, satisfying the *Fermi anticommutation rules*:

$$\left\{\eta_a, \eta_b^\dagger\right\} \equiv \eta_a \eta_b^\dagger + \eta_b^\dagger \eta_a = \delta_{ab}\,, \quad \{\eta_a, \eta_b\} = \left\{\eta_a^\dagger, \eta_b^\dagger\right\} = 0\,. \tag{A31}$$

The number of fermions in the state α_a is again represented by

$$n_a \equiv \eta_a^\dagger \eta_a = n_a^\dagger\,. \tag{A32}$$

Using (A31), we obtain

$$n_a^2 \equiv \eta_a^\dagger \eta_a \eta_a^\dagger \eta_a = \eta_a^\dagger \left(1 - \eta_a^\dagger \eta_a\right) \eta_a = \eta_a^\dagger \eta_a = n_a \quad \text{or} \quad n_a^2 - n_a = 0\,. \tag{A33}$$

If an eigenket of n_a belonging to the eigenvalue n_a' is denoted by $|n_a'\rangle$, (A33) yields

$$\left(n_a^2 - n_a\right)|n_a'\rangle = \left(n_a'^2 - n_a'\right)|n_a'\rangle = 0\,. \tag{A34}$$

Since the eigenket n_a is nonzero, we obtain $n_a'(n_a' - 1) = 0$, meaning that the eigenvalues n_a' are either 0 or 1 as required:

$$n_a' = 0 \quad \text{or} \quad 1\,. \tag{A35}$$

Similarly to the case for bosons, we can show that (Problem A.3)

$$|\alpha_a \alpha_b \cdots \alpha_g\rangle_A = \eta_a^\dagger \eta_b^\dagger \cdots \eta_g^\dagger |\Phi_0\rangle \tag{A36}$$

which is normalized to unity.

Observables describing the system of fermions can be expressed in terms of operators η and $\eta^\dagger$, and the results have the same form (A26) and (A27) as for the case of bosons.

In summary, both states and observables for a system of identical particles can be expressed in terms of creation and annihilation operators. This formalism, called the *second quantization*, has some notable advantages over the usual Schrödinger formalism. First, the permutation-symmetry property of the quantum particle is represented simply in the form of Bose commutation (or Fermi anticommutation) rules. Second, observables in second quantization are defined for an arbitrary number of particles so that the formalism may apply to systems in which the number of particles is not fixed, but variable. Third, and most importantly, all relevant quantities (states and observables) can be defined referring only to the single-particle states. This property allows one to describe the motion of the many-body system in the 3D space. In fact, relativistic quantum field theory can be developed only in second quantization. Furthermore, boson creation and annihilation can only be described in second quantization.

Problem A.3. Show that (A36) is normalized.

A.1.4
Heisenberg Equation of Motion

In the Schrödinger picture (SP), the energy eigenvalue equation is

$$\mathcal{H}|E\rangle = E|E\rangle\,, \tag{A37}$$

where $\mathcal{H}$ is the Hamiltonian and E the eigenvalue. In the position representation this equation is written as

$$\mathcal{H}\left(x_1, -\mathrm{i}\hbar\frac{\partial}{\partial x_1}, x_2, -\mathrm{i}\hbar\frac{\partial}{\partial x_2}, \cdots\right)\Psi(x_1, x_2, \cdots) = E\Psi\,, \tag{A38}$$

where Ψ is the wavefunction for the system. We consider one-dimensional motion for conceptional and notational simplicity. (For three-dimensional motion, (x, p) should be replaced by $(x, y, z, p_x, p_y, p_z) = (\boldsymbol{r}, \boldsymbol{p})$.) If the number of electrons N is large, the wavefunction Ψ contains many particle variables $(x_1, x_2, \ldots)$. This complexity needed in dealing with many particle variables can be avoided if we use second quantization and the *Heisenberg picture* (HP), which will be shown in this section.

If the Hamiltonian $\mathcal{H}$ is the sum of single-particle Hamiltonians:

$$\mathcal{H} = \sum_j h^{(j)}\,, \tag{A39}$$

the Hamiltonian $\mathcal{H}$ can be represented by

$$\mathcal{H} = \sum_a \sum_b \langle\alpha_a|h|\alpha_b\rangle\eta_a^\dagger\eta_b \equiv \sum_a \sum_b h_{ab}\eta_a^\dagger\eta_b\,, \tag{A40}$$

where $\eta_a(\eta_a^\dagger)$ are annihilation (creation) operators associated with particle state a and satisfying the Fermi anticommutation rules in (A31) or the Bose commutation rules in (A3).

In the Heisenberg picture a dynamical variable $\xi(t)$ changes in time, following the *Heisenberg equation of motion*:

$$-i\hbar \frac{d\xi(t)}{dt} = [\mathcal{H}, \xi] \equiv \mathcal{H}\xi - \xi\mathcal{H} . \tag{A41}$$

Setting $\xi = \eta_a^\dagger$, we obtain

$$-i\hbar \frac{d\eta_a^\dagger}{dt} = \left[\mathcal{H}, \eta_a^\dagger\right] , \tag{A42}$$

whose Hermitian conjugate is given by

$$i\hbar \frac{d\eta_a}{dt} = \left(\left[\mathcal{H}, \eta_a^\dagger\right]\right)^\dagger = -[\mathcal{H}, \eta_a] . \tag{A43}$$

By the quantum postulate the observables ξ are Hermitian: $\xi^\dagger = \xi$. Variables η_a and $\eta_a^\dagger$ are *not* Hermitian, but both obey the *same* Heisenberg equation of motion, see (A42) and (A43).

We introduce the Hamiltonian (A40) into (A42), and calculate the commutator $[\mathcal{H}, \eta_a^\dagger]$. In such a commutator calculation, the following identities are very useful:

$$[A, BC] = [A, B]C + B[A, C] , \quad [AB, C] = A[B, C] + [A, C]B , \tag{A44}$$

$$[A, BC] = \{A, B\}C - B\{A, C\} , \quad [AB, C] = A\{B, C\} - \{A, C\}B . \tag{A45}$$

Note that the negative signs on the right-hand terms in (A45) occur when the cyclic order is destroyed for the case of the anticommutator: $\{A, B\} \equiv AB + BA$. We obtain from (A42) and (A43) (Problem A.4)

$$\begin{aligned}
-i\hbar \frac{d\eta_a^\dagger}{dt} &= \sum_c \sum_b h_{cb} \left[\eta_c^\dagger \eta_b, \eta_a^\dagger\right] \\
&= \sum_c \sum_b h_{cb} \eta_c^\dagger \left\{\eta_b, \eta_a^\dagger\right\} = \sum_c h_{ca} \eta_c^\dagger
\end{aligned}$$

or

$$-i\hbar \frac{d\eta_a^\dagger}{dt} = \sum_c h_{ac} \eta_c^\dagger . \tag{A46}$$

We take the Hermitian conjugation and obtain

$$i\hbar \frac{d\eta_a}{dt} = \sum_c \eta_c h_{ca} . \tag{A47}$$

Equation (A46) means that the change of the one-body operator $\eta_a^\dagger$ is determined by the one-body Hamiltonian h. This is one of the major advantages of working in the HP. Equations (A46) and (A47) are valid for any single-particle states $\{a\}$.

In the field operator language (A47) reads

$$i\hbar \frac{\partial \psi(\mathbf{r}, t)}{\partial t} = h\left(\mathbf{r}, -i\hbar \frac{\partial}{\partial \mathbf{r}}\right) \psi(\mathbf{r}, t) , \tag{A48}$$

which is formally identical to the Schrödinger equation of motion for a particle.

If the system Hamiltonian $\mathcal{H}$ contains an interparticle interaction

$$\mathcal{V} = \frac{1}{2}\int \mathrm{d}^3 r \int \mathrm{d}^3 r' v(\boldsymbol{r}-\boldsymbol{r}')\psi^\dagger(\boldsymbol{r},t)\psi^\dagger(\boldsymbol{r}',t)\psi(\boldsymbol{r}',t)\psi(\boldsymbol{r},t) , \tag{A49}$$

where $v(\boldsymbol{r}-\boldsymbol{r}')$ denotes the interparticle interaction potential, the evolution equation for $\psi(\boldsymbol{r},t)$ is nonlinear (Problem A.5):

$$\begin{aligned} \mathrm{i}\hbar\frac{\partial\psi(\boldsymbol{r},t)}{\partial t} &= h\left(\boldsymbol{r},-\frac{\mathrm{i}\hbar\partial}{\partial\boldsymbol{r}}\right)\psi(\boldsymbol{r},t) \\ &\quad + \int \mathrm{d}^3 r' v(\boldsymbol{r}-\boldsymbol{r}')\psi^\dagger(\boldsymbol{r}',t)\psi(\boldsymbol{r}',t)\psi(\boldsymbol{r},t) . \end{aligned} \tag{A50}$$

In quantum field theory the basic dynamical variables are particle field operators. The quantum statistics of the particles are given by the Bose commutation or the Fermi anticommutation rules satisfied by the field operators. The evolution equations of the field operators are intrinsically nonlinear when the interparticle interaction is present.

Problem A.4. Verify that the equation of motion (A46) and (A47), holds for bosons.

Problem A.5. Verify (A50). Hint: use (A46).

A.2
Eigenvalue Problem and Equation-of-Motion Method

In this Appendix we set up the energy eigenvalue problem for a quasiparticle and illustrate the equation-of-motion method to obtain the eigenvalue of a many-body Hamiltonian of quasiparticles.

A.2.1
Energy-Eigenvalue Problem in Second Quantization

Here we consider as a preliminary an electron characterized by the Hamiltonian:

$$h(x,p) = \frac{p^2}{2m} + v(x) . \tag{A51}$$

We may set up the eigenvalue equations for the position x, the momentum p, and the one-body Hamiltonian h as follows:

$$x|x'\rangle = x'|x'\rangle , \tag{A52}$$

$$p|p'\rangle = p'|p'\rangle , \tag{A53}$$

$$h|\varepsilon_\nu\rangle = \varepsilon_\nu|\varepsilon_\nu\rangle \equiv \varepsilon_\nu|\nu\rangle , \tag{A54}$$

where x', p', and ε_ν are eigenvalues.

By multiplying (A54) from the left by $\langle x|$, we obtain

$$h\left(x, -\frac{i\hbar \mathrm{d}}{\mathrm{d}x}\right)\phi_\nu(x) = \varepsilon_\nu \phi_\nu(x) , \tag{A55}$$

$$\phi_\nu(x) \equiv \langle x|\nu\rangle . \tag{A56}$$

Equation (A55) is just the Schrödinger energy-eigenvalue equation, and $\phi_\nu(x)$ is the familiar quantum wavefunction. If we know with certainty that the system is in the energy eigenstate ν, we can choose a density operator ρ_1 to be

$$\rho_1 \equiv |\nu\rangle\langle\nu| , \quad \langle\nu|\nu\rangle = 1 . \tag{A57}$$

This ρ_1 is a one-body density operator for the system h in a pure state $|\nu\rangle$.

Let us now consider $\mathrm{tr}\{|\nu\rangle\langle x|\rho_1\}$, which can be transformed as follows:

$$\mathrm{tr}\{|\nu\rangle\langle x|\rho_1\} = \sum_a \langle a|\nu\rangle\langle x|\nu\rangle\langle\nu|a\rangle = \sum_a \langle x|\nu\rangle\langle\nu|a\rangle\langle a|\nu\rangle = \langle x|\nu\rangle .$$

From this we can write the wavefunction $\phi_\nu(x)$ as follows:

$$\phi_\nu(x) = \mathrm{tr}\{|\nu\rangle\langle x|\rho_1\} = \langle x|\rho_1|\nu\rangle , \tag{A58}$$

where the symbol "tr" denotes a one-body trace. It can be seen from (A57) that the wavefunction $\phi_\nu(x)$ can be regarded as a *mixed* representation of the density operator ρ_1 in terms of the states (ν, x). In a parallel manner, we can show that the wavefunction in the momentum space $\phi_\nu(p) \equiv \langle p|\nu\rangle$ can be regarded as a mixed representation of ρ_1 in terms of energy state ν and momentum state p (Problem A.1):

$$\phi_\nu(p) = \mathrm{tr}\{|\nu\rangle\langle p|\rho_1\} = \langle p|\rho_1|\nu\rangle . \tag{A59}$$

In analogy with (A58) we introduce a *quasiwavefunction* $\Psi_\nu(p)$ through

$$\Psi_\nu(p) \equiv \mathrm{Tr}\left\{\psi_\nu^\dagger a_p \rho\right\} , \tag{A60}$$

where $\psi_\nu^\dagger$ is the energy-state creation operator, a_p the momentum-state annihilation operator, and ρ a many-body-system density operator that commutes with the Hamiltonian $\mathcal{H}$:

$$[\mathcal{H}, \rho] = 0 . \tag{A61}$$

In Equation (A60) the symbol "Tr" denotes a many-body trace. Equation (A61) is the necessary condition that ρ be a stationary density operator, which is seen at once from the quantum Liouville equation:

$$i\hbar\frac{\partial\rho}{\partial t} = [\mathcal{H}, \rho] . \tag{A62}$$

Let us consider a system for which the total Hamiltonian $\mathcal{H}$ is the sum of single-electron energies h:

$$\mathcal{H} = \sum_j h^{(j)} . \tag{A63}$$

For example, the single-electron Hamiltonian h may contain the kinetic energy and the lattice potential energy. We assume that the Hamiltonian $\mathcal{H}$ does not depend on time explicitly.

In second quantization the Hamiltonian $\mathcal{H}$ can be represented by

$$\mathcal{H} = \sum_a \sum_b \langle \alpha_a | h | \alpha_b \rangle \eta_a^\dagger \eta_b \equiv \sum_a \sum_b h_{ab} \eta_a^\dagger \eta_b , \tag{A64}$$

where η_a ($\eta_a^\dagger$) are annihilation (creation) operators, satisfying the Fermi anticommutation rules (A31).

We calculate the commutator $[\mathcal{H}, \psi_a^\dagger]$. After straightforward calculation we obtain (Problem A.2)

$$\left[\mathcal{H}, \psi_\nu^\dagger\right] = \varepsilon_\nu \psi_\nu^\dagger = \sum_\mu \psi_\mu^\dagger h_{\mu\nu} . \tag{A65}$$

Multiplying Equation (A65) by $a_p \rho$ from the right and taking a many-body trace, we obtain

$$\sum_\mu \Psi_\mu(p) h_{\mu\nu} = \varepsilon_\nu \Psi_\mu(p) , \tag{A66}$$

which is formally identical to the *Schrödinger energy-eigenvalue equation* for the one-body problem (Problem A.3):

$$\sum_\mu \phi_\nu(p) h_{\mu\nu} = \varepsilon_\nu \phi_\nu(p) , \quad \phi_\nu(p) \equiv \langle p | \nu \rangle . \tag{A67}$$

The quasiwavefunction $\Psi_\nu(p)$ can be regarded as a mixed representation of the *one-body density operator* n in terms of the states (ν, p) (Problem A.4):

$$\Psi_\nu(p) \equiv \langle p | n | \nu \rangle . \tag{A68}$$

The operator n is defined through

$$\mathrm{Tr}\left\{\eta_b \rho \eta_a^\dagger\right\} \equiv \langle \alpha_b | n | \alpha_b \rangle \equiv n_{ba} . \tag{A69}$$

These n_{ba} are called b–a elements of the *one-body density matrix.*

We reformulate (A66) for later use. Using (A61) and (A65), we can write (A66) as in the form (Problem A.5):

$$\mathrm{Tr}\left\{\left[\mathcal{H}, \psi_\nu^\dagger\right] a_p \rho\right\} = \mathrm{Tr}\left\{\psi_\nu^\dagger [a_p, \mathcal{H}] \rho\right\} = \varepsilon_\nu \Psi_\nu(p) , \tag{A70}$$

where we used the invariance of cyclic permutation under the trace.[1] The complex conjugate of (A70) is similarly given by (Problem A.6)

$$\varepsilon_\nu \psi_\nu^*(p) = \mathrm{Tr}\left\{\left[\mathcal{H}, a_p^\dagger\right] \psi_\nu \rho\right\} . \tag{A71}$$

Either (A70) or (A71) can be used to formulate the energy-eigenvalue problem. If we choose the latter, we may proceed as follows:

1. Given $\mathcal{H}$ in the momentum space, compute $[\mathcal{H}, a_p^\dagger]$; the result can be expressed as a linear function of $a_p^\dagger$;
2. Multiply the result obtained in (1) by $\psi_\nu \rho$ from the right, and take a trace; the result is a linear function of ψ^*;
3. Use (A71) in the resulting equation obtained in (2), and we obtain a linear homogeneous equation for ψ^*, which is a standard form of the energy-eigenvalue equation.

The energy-eigenvalue problem developed here is often called the *equation-of-motion method.*

Problem A.1. Verify (A59).

Problem A.2. Derive (A65). Hint: Use identities (A49).

Problem A.3. Derive (A67) from (A54).

Problem A.4. Prove (A68).

Problem A.5. Derive (A70).

Problem A.6. Verify (A71).

A.2.2
Energies of Quasielectrons (or "Electrons") at 0 K

We re-derive the energy gap equations, utilizing the equation-of-motion method [3]. Below the critical temperature T_c, where the supercondensate is present, quasielectrons move differently from those above T_c. Here we study the energies of quasielectrons at 0 K. This ground state is described in terms of the original reduced Hamiltonian in (7.45), see below.

$$\mathcal{H}_{\text{red}} = \sum_k 2\varepsilon_k^{(1)} b_k^{(1)\dagger} b_k^{(1)} + \sum_k 2\varepsilon_k^{(2)} b_k^{(2)\dagger} b_k^{(2)} - \sum_k{}' \sum_{k'}{}' \left[v_{11} b_k^{(1)\dagger} b_{k'}^{(1)} \right.$$
$$\left. + v_{12} b_k^{(1)\dagger} b_{k'}^{(2)\dagger} + v_{21} b_k^{(2)} b_{k'}^{(1)} + v_{22} b_k^{(2)} b_{k'}^{(2)\dagger} \right] . \tag{A72}$$

1) $\mathrm{Tr}\{AB\rho\} = \mathrm{Tr}\{\rho AB\} = \mathrm{Tr}\{B\rho A\}$.

By using this $\mathcal{H}_{\text{red}}$, we obtain (Problem A.7)

$$\left[\mathcal{H}_{\text{red}}, c_{p\uparrow}^{(1)\dagger}\right] = \varepsilon_p^{(1)} c_{p\uparrow}^{(1)\dagger} - \left[v_{11}\sum_k{}' b_k^{(1)\dagger} + v_{12}\sum_k{}' b_k^{(2)}\right] c_{-p\downarrow}^{(1)}, \tag{A73}$$

$$\left[\mathcal{H}_{\text{red}}, c_{-p\downarrow}^{(1)\dagger}\right] = -\varepsilon_p^{(1)} c_{-p\downarrow}^{(1)\dagger} - \left[v_{11}\sum_k{}' b_k^{(1)} + v_{12}\sum_k{}' b_k^{(2)\dagger}\right] c_{p\uparrow}^{(1)\dagger}. \tag{A74}$$

These two equations indicate that the dynamics of quasielectrons described in terms of c's are affected by stationary pairons described in terms of b's.

Now let us find the energy of a quasielectron. We follow the equation-of-motion method. We multiply (A73) from the right by $\psi_\nu^{(1)}\rho_0$, where $\psi_\nu^{(1)}$ is the "electron" energy-state annihilation operator and

$$\rho_0 \equiv |\Psi\rangle\langle\Psi| \equiv \prod_k{}' \left(u_k^{(1)} + v_k^{(1)} b_k^{(1)\dagger}\right) \prod_{k'}{}' \left(u_{k'}^{(2)} + v_{k'}^{(2)} b_{k'}^{(2)\dagger}\right) |0\rangle\langle\Psi| \tag{A75}$$

is the density operator describing the supercondensate, and take a grand ensemble trace denoted by TR. After using (A70), the lhs can be written as

$$\text{TR}\left\{\left[\mathcal{H}_{\text{red}}, c_{p\uparrow}^{(1)}\right] \psi_\nu^{(1)}\rho_0\right\} = \text{TR}\left\{E_{\nu,p}^{(1)} c_{p\uparrow}^{(1)\dagger} \psi_\nu^{(1)}\rho_0\right\} \equiv E_p^{(1)} \psi_\uparrow^{(1)*}(\boldsymbol{p}), \tag{A76}$$

where we dropped the subscript ν; the quasielectron is characterized by momentum $\boldsymbol{p}$ and energy $E_p^{(1)}$. The first term on the rhs simply yields $\varepsilon_p^{(1)}\psi_\uparrow^{(1)*}(\boldsymbol{p})$. Consider now

$$\text{TR}\left\{b_k^{(1)\dagger} c_{-p\downarrow}^{(1)} \psi_\nu^{(1)}\rho_0\right\} \equiv \text{TR}\left\{b_k^{(1)\dagger} c_{-p\downarrow}^{(1)} \psi_\nu^{(1)} |\Psi\rangle\langle\Psi|\right\}. \tag{A77}$$

The state $|\Psi\rangle$ is normalized to unity, and it is the only system state at 0 K. Hence we obtain

$$\text{TR}\left\{b_k^{(1)\dagger} c_{-p\downarrow}^{(1)} \psi_\nu^{(1)}\rho_0\right\} = \langle\Psi| b_k^{(1)\dagger} c_{-p\downarrow}^{(1)} \psi_\nu^{(1)} |\Psi\rangle. \tag{A78}$$

We assume here that $\boldsymbol{k} \neq \boldsymbol{p}$, since the state must change after a phonon exchange. We examine the relevant matrix element and obtain (Problem A.8)

$$\langle 0| \left(u_k^{(1)} + v_k^{(1)} b_k^{(1)}\right) b_k^{(1)\dagger} \left(u_k^{(1)} + v_k^{(1)} b_k^{(1)\dagger}\right) |0\rangle = u_k^{(1)} v_k^{(1)}. \tag{A79}$$

We can therefore write

$$\text{TR}\left\{b_k^{(1)\dagger} c_{-p\downarrow}^{(1)} \psi_\nu^{(1)}\rho_0\right\} = u_k^{(1)} v_k^{(1)} \psi_\downarrow^{(1)}(-\boldsymbol{p}), \tag{A80}$$

$$\psi_\downarrow^{(1)}(-\boldsymbol{p}) \equiv \text{TR}\left\{c_{-p\downarrow}^{(1)} \psi_\nu^{(1)}\rho_0\right\}. \tag{A81}$$

Collecting all contributions, we obtain from (A73)

$$E_p^{(1)} \psi_\uparrow^{(1)*}(\boldsymbol{p}) = \varepsilon_p^{(1)} \psi_\uparrow^{(1)*}(\boldsymbol{p}) - \left[v_{11} {\sum_k}' u_k^{(1)} v_k^{(1)} + v_{12} {\sum_k}' u_k^{(2)} v_k^{(2)} \right] \psi_\downarrow^{(1)}(-\boldsymbol{p}) . \tag{A82}$$

Using (7.62) and (7.63), we obtain

$$\Delta_1 \equiv v_{11} {\sum_k}' u_k^{(1)} v_k^{(1)} + v_{12} {\sum_k}' u_k^{(2)} v_k^{(2)} . \tag{A83}$$

We can therefore simplify (A82) to

$$E_p^{(1)} \psi_\uparrow^{(1)*}(\boldsymbol{p}) = \varepsilon_p^{(1)} \psi_\uparrow^{(1)*}(\boldsymbol{p}) - \Delta_1 \psi_\downarrow^{(1)}(-\boldsymbol{p}) . \tag{A84}$$

Similarly we obtain from (A74)

$$E_{-p}^{(1)} \psi_\downarrow^{(1)}(-\boldsymbol{p}) = -\varepsilon_p^{(1)} \psi_\downarrow^{(1)}(-\boldsymbol{p}) - \Delta_1 \psi_\uparrow^{(1)*}(\boldsymbol{p}) . \tag{A85}$$

Energy $E_p^{(1)}$ can be interpreted as the *positive* energy required to create an up-spin unpairing electron at $\boldsymbol{p}$ in the presence of the supercondensate. The energy $E_{-p}^{(1)}$ can be regarded as the *positive* energy required to remove a down-spin electron from the paired state $(\boldsymbol{p}\uparrow, -\boldsymbol{p}\downarrow)$. These two energies are equal to each other

$$E_p^{(1)} = E_{-p}^{(1)} = E_p^{(1)} > 0 . \tag{A86}$$

In the stationary state (A84) and (A85) must hold simultaneously, thus yielding

$$\begin{vmatrix} E_p^{(1)} - \varepsilon_p^{(1)} & \Delta_1 \\ \Delta_1 & E_p^{(1)} + \varepsilon_p^{(1)} \end{vmatrix} = 0 , \tag{A87}$$

whose solutions are $E_p^{(1)} = \pm(\varepsilon_p^{(1)2} + \Delta_1^2)^{1/2}$. Since $E_p^{(1)} > 0$, we obtain

$$E_p^{(1)} = \left(\varepsilon_p^{(1)2} + \Delta_1^2 \right)^{1/2} . \tag{A88}$$

The theory developed here can be applied to the "hole" in a parallel manner. We included this case in (A88). Our calculation confirms our earlier interpretation that $E_p^{(1)}$ is the energy of the quasielectron. In summary, unpaired electrons are affected by the presence of the supercondensate, and their energies are given by (A88).

Problem A.7. Derive (A73) and (A74).

Problem A.8. Verify (A78).

A.3
Derivation of the Cooper Equation (7.34)

We derive the Cooper equation (7.34) in this Appendix.

Let us consider a Cooper pair. Second-quantized operators for a pair of "electrons" (i.e., "electron" pairons) are defined by

$$B_{12}^{\dagger} \equiv B_{k_1\uparrow k_2\downarrow}^{\dagger} \equiv c_1^{\dagger} c_2^{\dagger} , \quad B_{34} = c_4 c_3 . \tag{A89}$$

Odd-numbered "electrons" carry up-spins $\uparrow$ and even-numbered carry down-spins $\downarrow$. The commutators among B and $B^{\dagger}$ can be computed using the commutators among B and $B^{\dagger}$ along with the Fermi anticommutation rules, and they are given by (Problem A.1)

$$[B_{12}, B_{34}] \equiv B_{12} B_{34} - B_{34} B_{12} = 0 , \tag{A90}$$

$$B_{12}^2 \equiv B_{12} B_{12} = 0 , \tag{A91}$$

$$\left[B_{12}, B_{34}^{\dagger}\right] = \begin{cases} 1 - n_1 - n_2 & \text{if} \quad \boldsymbol{k}_1 = \boldsymbol{k}_3 \quad \text{and} \quad \boldsymbol{k}_2 = \boldsymbol{k}_4 \\ c_2 c_4^{\dagger} & \text{if} \quad \boldsymbol{k}_1 = \boldsymbol{k}_3 \quad \text{and} \quad \boldsymbol{k}_2 \neq \boldsymbol{k}_4 \\ c_1 c_3^{\dagger} & \text{if} \quad \boldsymbol{k}_1 \neq \boldsymbol{k}_3 \quad \text{and} \quad \boldsymbol{k}_2 = \boldsymbol{k}_4 \\ 0 & \text{otherwise} , \end{cases} \tag{A92}$$

where

$$n_1 \equiv c_{k_1\uparrow}^{\dagger} c_{k_1\uparrow} , \quad n_2 \equiv c_{k_2\downarrow}^{\dagger} c_{k_2\downarrow} \tag{A93}$$

are the number operators for electrons.

Let us now introduce the *relative* and *net* (CM) *momenta* $(\boldsymbol{k}, \boldsymbol{q})$ such that

$$\boldsymbol{k} \equiv \frac{1}{2}(\boldsymbol{k}_1 - \boldsymbol{k}_2) , \quad \boldsymbol{q} \equiv \boldsymbol{k}_1 + \boldsymbol{k}_2 ; \quad \boldsymbol{k}_1 = \boldsymbol{k} + \frac{1}{2}\boldsymbol{q} , \quad \boldsymbol{k}_2 = -\boldsymbol{k} + \frac{1}{2}\boldsymbol{q} . \tag{A94}$$

Alternatively we can represent *pairon annihilation* and *creation operators* by

$$B'_{kq} \equiv B_{k_1\uparrow k_2\downarrow} \equiv c_{-k+q/2\downarrow} c_{k+q/2\uparrow} , \quad B'^{\dagger}_{kq} \equiv c_{k+q/2\uparrow}^{\dagger} c_{-k+q/2\downarrow}^{\dagger} . \tag{A95}$$

The prime on B will be dropped hereafter. In the $\boldsymbol{k}$–$\boldsymbol{q}$ representation the commutation relations are re-expressed as

$$[B_{kq}, B_{k'q'}] = 0 , \quad [B_{kq}]^2 = 0 , \tag{A96}$$

$$\left[B_{kq}, B_{k'q'}^{\dagger}\right] = \begin{cases} 1 - n_{k+q/2\uparrow} - n_{-k+q/2\downarrow} & \text{if} \quad \boldsymbol{k} = \boldsymbol{k}' \quad \text{and} \quad \boldsymbol{q} = \boldsymbol{q}' \\ c_{-k+q/2\downarrow} c_{-k'+q'/2\downarrow}^{\dagger} & \text{if} \quad \boldsymbol{k} + \frac{\boldsymbol{q}}{2} = \boldsymbol{k}' + \frac{\boldsymbol{q}'}{2} \\ & \text{and} \quad -\boldsymbol{k} + \frac{\boldsymbol{q}}{2} \neq -\boldsymbol{k}' + \frac{\boldsymbol{q}'}{2} \\ c_{k+q/2\uparrow} c_{k'+q'/2\uparrow}^{\dagger} & \text{if} \quad \boldsymbol{k} + \frac{\boldsymbol{q}}{2} \neq \boldsymbol{k}' + \frac{\boldsymbol{q}'}{2} \\ & \text{and} \quad -\boldsymbol{k} + \frac{\boldsymbol{q}}{2} = -\boldsymbol{k}' + \frac{\boldsymbol{q}'}{2} \\ 0 & \text{otherwise} . \end{cases} \tag{A97}$$

If we drop the "hole" contribution from the generalized BCS Hamiltonian in Equation (7.23), we obtain the *Cooper Hamiltonian* $\mathcal{H}_C$:

$$\mathcal{H}_C = \sum_{\substack{k \\ \varepsilon_k > 0}} \sum_{s} c^\dagger_{ks} c_{ks} - v_0 \sum_k{}' \sum_{k'}{}' \sum_q{}' B^\dagger_{kq} B_{k'q} \,, \quad (_0 \equiv v_{11}) \,, \tag{A98}$$

where the prime on the summation means the restriction:

$$0 < \varepsilon\left(\left|k + \frac{q}{2}\right|\right) \,, \quad \varepsilon\left(\left|-k + \frac{q}{2}\right|\right) < \hbar\omega_D \,. \tag{A99}$$

The Hamiltonian $\mathcal{H}_C$ can be expressed in terms of pair operators $(B, B^\dagger)$:

$$\begin{aligned} \mathcal{H}_C = &\sum_k{}' \sum_q{}' \left[\varepsilon\left(\left|k + \frac{q}{2}\right|\right) + \varepsilon\left(\left|-k + \frac{q}{2}\right|\right)\right] B^\dagger_{kq} B_{kq} \\ &- v_0 \sum_k{}' \sum_{k'}{}' \sum_q{}' B^\dagger_{kq} B_{k'q} \,. \end{aligned} \tag{A100}$$

Using (A96) and (A97), we obtain (Problem A.2)

$$\begin{aligned} \left[\mathcal{H}_C, B^\dagger_{kq}\right] = &\left[\varepsilon\left(\left|k + \frac{q}{2}\right|\right) + \varepsilon\left(\left|-k + \frac{q}{2}\right|\right)\right] B^\dagger_{kq} \\ &- v_0 \sum_{k'}{}' B^\dagger_{k'q}(1 - n_{k+q/2\uparrow} - n_{-k+q/2\downarrow}) \,. \end{aligned} \tag{A101}$$

If we represent the energies of pairons by w_ν and the associated pair annihilation operator by ϕ_ν, $\mathcal{H}_C$ can be expressed by

$$\mathcal{H}_C = \sum_\nu w_\nu \phi^\dagger_\nu \phi_\nu \,. \tag{A102}$$

This equation is similar to (A45) in Appendix A.1 with the only difference that here we deal with *pair energies* and *pair-state operators*. We multiply (A101) by $\phi_\nu \rho_{gc}$ from the right and take a grand ensemble trace:

$$\begin{aligned} w_\nu a_{kq} \equiv w_q a_{kq} &= \mathrm{Tr}\left\{\left[\mathcal{H}_C, B^\dagger_{kq}\right] \phi_\nu \rho_{gc}\right\} \\ &= \left[\varepsilon\left(\left|k + \frac{q}{2}\right|\right) + \varepsilon\left(\left|-k + \frac{q}{2}\right|\right)\right] a_{kq} \\ &\quad - v_0 \sum_{k'}{}' \left\langle B^\dagger_{k'q} \left(1 - n_{k+q/2\uparrow} - n_{-k+q/2\downarrow}\right) \phi_\nu \right\rangle \,, \end{aligned} \tag{A103}$$

where a_{kq} is defined by

$$a_{kq,\nu} \equiv \mathrm{Tr}\left\{B^\dagger_{kq} \phi_\nu \rho_{gc}\right\} \equiv a_{kq} \,. \tag{A104}$$

The energy w_ν can be characterized by $\boldsymbol{q}$, and we have $w_\nu \equiv w_q$. In other words, excited pairons have net momentum $\boldsymbol{q}$ and energy w_q. We shall omit the subscripts ν in the pairon wavefunction: $a_{kq,\nu} \equiv a_{kq}$. The angular brackets mean the grand canonical ensemble average of an observable $\mathcal{A}$:

$$\langle \mathcal{A} \rangle \equiv \mathrm{Tr}\{\mathcal{A}\rho_{\mathrm{gc}}\} \equiv \frac{\mathrm{Tr}\{\mathcal{A}\exp[\beta(\mu\mathcal{N}-\mathcal{H})]\}}{\mathrm{Tr}\exp[\beta(\mu\mathcal{N}-\mathcal{H})]} , \tag{A105}$$

where $\mathcal{H}$ and $\mathcal{N}$ represent the Hamiltonian and the number operator, respectively.

In the bulk limit: $N \to \infty$, $V \to \infty$ while $n = N/V =$ finite, where N represents the number of electrons, and k-vectors become continuous. Denoting the wavefunction in this limit by $a(\boldsymbol{k}, \boldsymbol{q})$ and using a factorization approximation, we obtain from (A103)

$$w_q a(\boldsymbol{k},\boldsymbol{q}) = \left\{\varepsilon\left(\left|\boldsymbol{k}+\frac{\boldsymbol{q}}{2}\right|\right) + \varepsilon\left(\left|-\boldsymbol{k}+\frac{\boldsymbol{q}}{2}\right|\right)\right\} a(\boldsymbol{k},\boldsymbol{q}) - \frac{v_0}{(2\pi\hbar)^3} \int' \mathrm{d}^3k' a(\boldsymbol{k}',\boldsymbol{q}) \left\{1 - f_{\mathrm{F}}\left[\varepsilon\left(\left|\boldsymbol{k}+\frac{\boldsymbol{q}}{2}\right|\right)\right] - f_{\mathrm{F}}\left[\varepsilon\left(\left|-\boldsymbol{k}+\frac{\boldsymbol{q}}{2}\right|\right)\right]\right\} , \tag{A106}$$

$$\langle n_{\mathrm{p}} \rangle = \frac{1}{\exp(\beta\varepsilon_{\mathrm{p}})+1} \equiv f_{\mathrm{F}}(\varepsilon_{\mathrm{p}}) , \tag{A107}$$

where f_{F} is the Fermi distribution function. The factorization is justified since the coupling between electrons and pairons is weak.

In the low-temperature limit ($T \to 0$ or $\beta \to \infty$),

$$f_{\mathrm{F}}(\varepsilon_{\mathrm{p}}) \to 0 , \quad (\varepsilon_{\mathrm{p}} > 0) . \tag{A108}$$

We then obtain

$$w_q a(\boldsymbol{k},\boldsymbol{q}) = \left\{\varepsilon\left(\left|\boldsymbol{k}+\frac{\boldsymbol{q}}{2}\right|\right) + \varepsilon\left(\left|-\boldsymbol{k}+\frac{\boldsymbol{q}}{2}\right|\right)\right\} a(\boldsymbol{k},\boldsymbol{q}) - \frac{v_0}{(2\pi\hbar)^3} \int' \mathrm{d}^3k' a(\boldsymbol{k}',\boldsymbol{q}) . \tag{A109}$$

This equation is identical to Cooper's equation, Eq. (1) of his 1956 Physical Review [4]. For a 2D system, replacing $\mathrm{d}^3k/(2\pi\hbar)^3$ in (A109) for the 3D case by $\mathrm{d}^2k/(2\pi\hbar)^2$ we obtain the Cooper equation (7.34) for the 2D case.

In the above derivation we obtained the Cooper equation in the zero-temperature limit. Hence, the energy of the pairon, w_q, is temperature-independent.

Problem A.1. Derive (A90), (A91) and (A92).

Problem A.2. Derive (A101).

A.4
Proof of (7.94)

The number operator for the pairons in the state $(\boldsymbol{k}, \boldsymbol{q})$ is

$$n_{kq} \equiv B^{\dagger}_{kq} B_{kq} = c^{\dagger}_{k+q/2} c^{\dagger}_{-k+q/2} c_{-k+q/2} c_{k+q/2} \,, \tag{A110}$$

where we omitted the spin indices. We write n^2_{kq} explicitly, transform the middle factors and obtain

$$\begin{aligned} n^2_{kq} &= c^{\dagger}_{k+q/2} c^{\dagger}_{-k+q/2} c_{-k+q/2} c_{k+q/2} c^{\dagger}_{k+q/2} c^{\dagger}_{-k+q/2} c_{-k+q/2} c_{k+q/2} \\ &= c^{\dagger}_{k+q/2} c^{\dagger}_{-k+q/2} \left(1 - c^{\dagger}_{-k+q/2} c_{-k+q/2}\right) \\ &\quad \times \left(1 - c^{\dagger}_{k+q/2} c_{k+q/2}\right) c_{-k+q/2} c_{k+q/2} = n_{kq} \,, \end{aligned} \tag{A111}$$

where we used (A36) (the Fermi commutation rules), and (A38) in Appendix A.1. Hence, we obtain

$$\left(n^2_{kq} - n_{kq}\right) |n'_{kq}\rangle = \left(n'^2_{kq} - n'_{kq}\right) |n'_{kq}\rangle = 0 \,. \tag{A112}$$

Since $|n'_{kq}\rangle \neq 0$, we obtain

$$n'_{kq} = 0 \quad \text{or} \quad 1 \,. \tag{A113}$$

We now introduce

$$B_q \equiv \sum_k B_{kq} \,, \tag{A114}$$

and obtain (Problem A.1)

$$[B_q, n_q] = \sum_k \left(1 - n_{k+q/2} - n_{-k+q/2}\right) B_{kq} = B_q \,, \quad \left[n_q, B^{\dagger}_q\right] = B^{\dagger}_{q'} \,. \tag{A115}$$

Although the occupation number n_q is not connected with B_q as $n_q \neq B^{\dagger}_q B_q$, the eigenvalues n'_q of n_q satisfying (A114) can be shown straightforwardly to yield

$$n'_q = 0, 1, 2, \cdots \,, \tag{A116}$$

with the eigenstates

$$|0\rangle \,, \quad |1\rangle = B^{\dagger}_q |0\rangle \,, \quad |2\rangle = B^{\dagger}_q B^{\dagger}_q |0\rangle, \cdots \,. \tag{A117}$$

The derivation of the boson occupation numbers $n' = 0, 1, 2, \cdots$, from $[\eta, n] = \eta$, $n \equiv \eta^{\dagger}\eta$, follows the steps after (A10), and ending with (A17) in Appendix A.1. We may follow the same steps. By setting $\boldsymbol{q} = 0$ in this equation, we obtain (7.85).

Problem A.1. Prove (A115).

A.5 Statistical Weight for the Landau States

The statistical weight $\mathcal{W}$ for the Landau states in 3D and 2D are calculated in this appendix.

A.5.1 The Three-Dimensional Case

Poisson's sum formula [5, 6] is

$$\sum_{n=-\infty}^{\infty} f(2\pi n) = \frac{1}{2\pi} \sum_{m=-\infty}^{\infty} F(m) \equiv \frac{1}{2\pi} \sum_{m=-\infty}^{\infty} \int_{-\infty}^{\infty} \mathrm{d}\tau F(\tau) \mathrm{e}^{-im\tau} , \tag{A118}$$

where F is the Fourier transform of f, and the sum $\sum_{n=-\infty}^{\infty} f(2\pi n + t)$, $0 \le t < 2\pi$, is periodic with the period 1. The sum is by assumption uniformly convergent.

We write the sum in (9.47) as

$$2\sum_{n=0}^{\infty} \sqrt{\varepsilon - (2n+1)\pi} = (\varepsilon - \pi)^{1/2} + \phi(\varepsilon; 0) , \tag{A119}$$

$$\phi(\varepsilon; x) \equiv \sum_{n=-\infty}^{\infty} (\varepsilon - \pi - 2\pi|n + x|)^{1/2} . \tag{A120}$$

Note that $\phi(\varepsilon; x)$ is periodic in x with the period 1, and it can, therefore, be expanded in a Fourier series. After the Fourier series expansion, we set $x = 0$ and obtain (A119). By taking the real part ($\Re$) of (A119) and using (A118) and (9.47), we obtain

$$\left[A\frac{(\hbar\omega_c)^{3/2}}{\sqrt{2\pi}}\right]^{-1} \mathcal{W}(E) = \frac{1}{\pi}\int_0^{\varepsilon} \mathrm{d}\tau(\varepsilon - \tau)^{1/2} + \frac{2}{\pi}\sum_{m=1}^{\infty}(-1)^m \times \int_0^{\varepsilon} \mathrm{d}\tau(\varepsilon - \tau)^{1/2} \cos m\tau , \tag{A121}$$

where we assumed

$$\varepsilon \equiv \frac{2\pi E}{\hbar\omega_c} \gg 1 \tag{A122}$$

and neglected π against ε. The integral in the first term in (A121) yields $(2/3)\varepsilon^{3/2}$, leading to $\mathcal{W}_0$ in (9.51). The integral in the second term can be written after integrating by parts, changing the variable ($m\varepsilon - m\tau = t$), and using $\sin(A - B) = \sin A \cos B - \cos A \sin B$ as

$$\frac{1}{2m^{3/2}}\left(\sin m\varepsilon \int_0^{m\varepsilon} \mathrm{d}t \frac{\cos t}{\sqrt{t}} - \cos m\varepsilon \int_0^{m\varepsilon} \mathrm{d}t \frac{\sin t}{\sqrt{t}}\right) . \tag{A123}$$

We use asymptotic expansion for $m\varepsilon = x \gg 1$:

$$\int_0^x \mathrm{d}t \frac{\sin t}{\sqrt{t}} \sim \sqrt{\frac{\pi}{2}} - \frac{\cos x}{\sqrt{x}} - \cdots , \tag{A124}$$

$$\int_0^x \mathrm{d}t \frac{\cos t}{\sqrt{t}} \sim \sqrt{\frac{\pi}{2}} + \frac{\sin x}{\sqrt{x}} - \cdots . \tag{A125}$$

The second terms in the expansions lead to $\mathcal{W}_\mathrm{L}$ in (9.52), where we used $\sin^2 A + \cos^2 A = 1$ and

$$\sum_{m=1}^{\infty} \frac{(-1)^{m-1}}{m^2} = \frac{\pi^2}{12} . \tag{A126}$$

The first terms lead to the oscillatory term $\mathcal{W}_\mathrm{osc}$ in (9.53).

A.5.2
The Two-Dimensional Case

We write the sum in (10.35) as

$$2\sum_{n=0}^{\infty} \Theta\left[\varepsilon - (2n+1)\pi\right] = \Theta(\varepsilon - \pi) + \psi(\varepsilon; 0) , \tag{A127}$$

$$\psi(\varepsilon; x) \equiv \sum_{n=-\infty}^{\infty} \Theta(\varepsilon - \pi - 2\pi|n + x|) . \tag{A128}$$

Note that $\psi(\varepsilon; x)$ is periodic in x and can therefore be expanded in a Fourier series. After the Fourier expansion, we set $x = 0$ and obtain (A127). By taking the real part ($\Re$) of (A127) and using (A118), we obtain

$$\Re\{(\mathrm{A127})\} = \frac{1}{\pi}\int_0^{\infty} \mathrm{d}\tau \Theta(\varepsilon - \tau) + \frac{2}{\pi}\sum_{\nu=1}^{\infty}(-1)^{\nu} \int_0^{\infty} \mathrm{d}\tau \Theta(\varepsilon - \tau)\cos \nu\tau , \tag{A129}$$

where we assumed $\varepsilon \equiv 2\pi E/\hbar\omega_\mathrm{c} \gg 1$ and neglected π against ε. The integral in the first term in (A129) yields ε. The integral in the second term is

$$\int_0^{\infty} \mathrm{d}\tau \Theta(\varepsilon - \tau)\cos \nu\tau = \frac{1}{\nu}\sin \nu\varepsilon . \tag{A130}$$

Thus, we obtain

$$\Re\{(\text{A127})\} = \frac{1}{\pi}\varepsilon + \frac{2}{\pi}\sum_{\nu=1}^{\infty}\frac{(-1)^{\nu}}{\nu}\sin\nu\varepsilon\,. \tag{A131}$$

Using (10.35) and (A131), we obtain

$$\begin{aligned}\mathcal{W}(E) &= \mathcal{W}_0 + \mathcal{W}_{\text{osc}}\\ &= C(\hbar\omega_{\text{c}})\left(\frac{\varepsilon}{\pi}\right) + C\hbar\omega_{\text{c}}\frac{2}{\pi}\sum_{\nu=1}^{\infty}\frac{(-1)^{\nu}}{\nu}\sin\left(\frac{2\pi\nu E}{\hbar\omega_{\text{c}}}\right)\,,\end{aligned} \tag{A132}$$

which establishes (10.38)–(10.40).

A.6 Derivation of Formulas (11.16)–(11.18)

Let us start with a BCS-like Hamiltonian (11.8). Dropping the "holes" from the Hamiltonian $\mathcal{H}$, we obtain a Hamiltonian associated with "electrons" as

$$\begin{aligned}\mathcal{H}_{\text{e}} &= \sum_{k}{}' \varepsilon_{k}^{(1)} n_{k}^{(1)} + \sum_{k}{}' \varepsilon_{k}^{(3)} n_{k}^{(3)} - \sum_{q}{}'\sum_{k}{}'\sum_{k'}{}' v_0 B_{k'q}^{(1)\dagger} B_{kq}^{(1)}\\ &= \sum_{k}\sum_{q}\left(\varepsilon_{|k+\frac{q}{2}|} + \varepsilon^{(3)}_{|-k+\frac{q}{2}|}\right) B_{kq}^{\dagger}B_{kq} - v_0\sum_{q}{}'\sum_{k}{}'\sum_{k'}{}' B_{k'q}^{\dagger}B_{kq}\,,\end{aligned} \tag{A133}$$

where $n_{k}^{(j)} = c_{k}^{(j)\dagger}c_{k}^{(j)}$ is the number operator for the "electron" (1) (fluxon (3)), $B_{kq}^{(1)\dagger} \equiv c_{k+q/2}^{(1)\dagger}c_{-k+q/2}^{(1)\dagger}$ is the pair operator for the "electron" and we suppressed the "electron." Note that the prime on the summation in (A133) means the restriction: $0 < \varepsilon_{k} < \hbar\omega_{\text{D}}$, ω_{D} = Debye frequency. Using the anticommutation rules (11.9) we obtain

$$\begin{aligned}\left[\mathcal{H}_{\text{e}}, B_{kq}^{\dagger}\right] &= \left(\varepsilon_{|k+\frac{q}{2}|} + \varepsilon^{(3)}_{|-k+\frac{q}{2}|}\right) B_{kq}^{\dagger}\\ &\quad - v_0\sum_{k'}{}' B_{k'q}^{\dagger}\left(1 - n_{k+\frac{q}{2}} - n^{(3)}_{-k+\frac{q}{2}}\right)\,.\end{aligned} \tag{A134}$$

The Hamiltonian $\mathcal{H}_{\text{e}}$ is bilinear in $(B, B^{\dagger})$, and can therefore be diagonalized: $\mathcal{H}_{\text{e}} = \sum_{\mu} w_{\mu}\phi_{\mu}^{\dagger}\phi_{\mu}$, where w_{μ} is the energy and ϕ_{μ} the annihilation operator. We multiply (A134) by ϕ_{μ} from the right, take a grand canonical ensemble average, denoted by angular brackets, and get

$$w_\mu \Psi_\mu(\mathbf{k}, \mathbf{q}) = \left(\varepsilon_{|\mathbf{k}+\frac{\mathbf{q}}{2}|} + \varepsilon^{(3)}_{|-\mathbf{k}+\frac{\mathbf{q}}{2}|}\right) \Psi_\mu(\mathbf{k}, \mathbf{q}) - \frac{v_0}{(2\pi\hbar)^2} \times \int' \mathrm{d}^2 k' \Psi_\mu(\mathbf{k}', \mathbf{q}) \left\{1 - f_\mathrm{F}\left(\varepsilon_{|-\mathbf{k}'+\frac{\mathbf{q}}{2}|}\right) - f_\mathrm{F}\left(\varepsilon^{(3)}_{|-\mathbf{k}'+\frac{\mathbf{q}}{2}|}\right)\right\}, \tag{A135}$$

where $\langle n_\mathrm{p} \rangle = f_\mathrm{F}(\varepsilon_\mathrm{p})$ is the Fermi distribution function. The reduced wavefunction $\Psi_\mu(\mathbf{k}, \mathbf{q}) \equiv \langle B^\dagger_{\mathbf{k}\mathbf{q}} \phi_\mu \rangle = \langle \boldsymbol{\mu} | \hat{n} | \mathbf{k}, \mathbf{q} \rangle$ can be regarded as the mixed representation of the *reduced* density operator $\hat{n}$ defined through $\langle \mathbf{k}', \mathbf{q}' | \hat{n} | \mathbf{k}, \mathbf{q} \rangle \equiv \langle B^\dagger_{\mathbf{k},\mathbf{q}} B_{\mathbf{k}',\mathbf{q}'} \rangle$. The fc-boson energy ω_μ can be specified by $(N_\mathrm{L}, \mathbf{q})$, and it will be denoted by ω_0 since it is N_L-independent. As $T \to 0$, $f_\mathrm{F}(\varepsilon_\mathrm{p}) \to 0$. Dropping the fluxon energy and replacing $\mathbf{q}/2$ by $\mathbf{q}$, we obtain (11.16). We solve this equation, assuming $\varepsilon_\mathrm{F} \gg \hbar\omega_\mathrm{D}$. Using a Taylor series expansion, we obtain (11.17) to the linear in q.

Now we derive (11.18). The BEC occurs when the chemical potential μ vanishes at a finite T. The critical temperature T_c can be determined from

$$n = (2\pi\hbar)^{-2} \int \mathrm{d}^2 p [\mathrm{e}^{\beta_\mathrm{c}\varepsilon} - 1]^{-1}, \quad \beta_\mathrm{c} \equiv (k_\mathrm{B} T_\mathrm{c})^{-1}. \tag{A136}$$

After expanding the integrand in powers of $\mathrm{e}^{-\beta_\mathrm{c}\varepsilon}$ and using $\varepsilon = cp$, we obtain

$$n = 1.654 (2\pi)^{-1} \left(\frac{k_\mathrm{B} T_\mathrm{c}}{\hbar c}\right)^2, \tag{A137}$$

yielding a general formula (11.18) for 2D BEC.

References

1 Dirac, P.A.M. (1958) *Principles of Quantum Mechanics*, 4th edn, Oxford University Press, London, UK, pp. 207–210.
2 Dirac, P.A.M. (1928) *Proc. Roy. Soc.*, **A114**, 243.
3 Schrieffer, J.R. (1964) *Theory of Superconductivity*, Benjamin, New York.
4 Cooper, L.N. (1956) *Phys. Rev.*, **104**, 1189.
5 Morse, P.M. and Feshbach, H. (1953) *Methods of Theoretical Physics*, McGraw-Hill, New York, pp. 466–467.
6 Courant, R. and Hilbert, D. (1953) *Methods of Mathematical Physics*, vol. 1, Interscience-Wiley, New York, pp. 76–77.

Index

Symbols
2D charge-density wave, 210
2D massless boson, 136
2D quasifree electron system, 188
2D superconductor, 105, 158
3D superconductor, 105
g-factor, 166
$\boldsymbol{k}$–$\boldsymbol{q}$ representation, 150, 267
k-vector, 34, 35, 37, 46, 57, 92
T-linear dependence of the heat capacity, 41
T-linear heat capacity, 40, 131
T-linear law, 131
T-linear phonon scattering rate, 249

A
activated-state temperature behavior, 92
activation (or excitation) energy ε_3, 94
activation energies (ε_1, ε_2), 163
activation energy, 85, 91, 94, 96, 244, 250
activation energy ε_3, 86, 88, 96, 98, 141
addition law, 16
Angle-Resolved Photoemission Spectroscopy (ARPES), 138, 139, 214
angular frequency, 35, 178
angular momentum, 166, 180
anisotropic magnetoresistance, 189
annihilation (creation) operator, 69, 259
annihilation electron field operator, 70
anticommutation rule, 211, 273
antielectron, 158
antisymmetric ket, 257
antisymmetrizing operator, 257
Arrhenius law, 88, 248
Arrhenius plot, 96
Arrhenius slope, 96
Arrhenius-law Boltzmann factor, 216
Arrhenius-type exponential, 244
asymptotic expansion, 272
azimuthal angle, 152

B
ballistic electron, 144
ballistic electron model, 163
ballistic transport, 141
band edge, 31
band index, 37, 46
band theory of electrons, 37
bare lattice potential, 36
basic properties of superconductors, 99, 104
bcc lattice, 6, 37
BCS energy gap equation, 125
BCS formula for the critical temperature, 138
BCS ground state energy, 109
BCS theory, 108, 109
BCS theory of superconductivity, 208
BCS-like Hamiltonian, 211
BCS-like theory, 221
BCS Hamiltonian, 108
BEC in 2D, 214
BEC of the pairons, 158
BEC temperature, 137
binding energy, 114
Bloch electron, 37, 47, 48, 180
Bloch electron dynamics, 57, 77, 87, 91, 93
Bloch electron state, 46
Bloch electron's wavelength, 46
Bloch energy bands, 37
Bloch state, 37
Bloch system, 119
Bloch wave, 93
Bloch wavefunction, 29, 30, 35
Bloch's theorem, 27, 33–37, 46, 80, 148
Bohr magneton, 166
Bohr–Sommerfeld quantization rule, 180
Boltzmann distribution function, 97, 146, 243
Boltzmann equation, 19, 21
Boltzmann equation for a homogeneous stationary system, 199

Boltzmann equation for an electron–impurity system, 21
Boltzmann equation method, 11
Boltzmann–Arrhenius factor, 244
Bose commutation rules, 67, 69, 133, 254
Bose distribution function, 6, 97, 133, 155, 242
Bose-condensed state, 144
Bose–Einstein Condensation (BEC), 97, 98, 133, 136, 141, 146, 155, 203, 243
boson, 5
boson occupation number, 270
boson speed, 214
bosonic pairon model, 163
bosonic second-quantized operator, 69
bound electron pair, 114
Bravais lattice vector, 34
Bravais vector, 46, 78
Bravais vector for the sq lattice, 79
Brillouin boundary, 28, 35, 38, 56, 238, 239
Brillouin zone, 38, 39, 57, 80, 148, 158
Brillouin zone boundary, 241
Brillouin zone of copper, 38
bulk limit, 42, 45, 55, 156, 269

C

canonical variables (Q_κ, P_κ), 54
carbon, 1
carbon hexagon, 92, 98
carbon nanotube, 1, 77
carrier charge, 86, 95
carrier density, 86, 95
carriers in the SdH oscillations, 221
Cartesian axes, 48
Cartesian coordinate system, 49
Cartesian coordinates, 147
Cartesian frame of coordinates, 34
Cartesian unit cell, 146, 249
cause of both QHE and superconductivity, 208
cause of superconductivity, 56, 107
cause of the QHE, 207, 208
center of mass (CM) momentum, 112
center of mass (CM) of any composite, 210
center of oscillation, 170
changing potential field, 65
channeling electrons, 93
charge current density, 15, 243
charge distribution, 158
charging energy, 144
chemical potential, 134, 156, 172, 274
C-hexagon, 84, 93
classical electron, 15
closed loop superconductor, 102
closed orbit in k-space, 189
closed k-orbit, 180
closed r-orbit, 180
CM momentum, 210
CM of the holes (wavepacket), 147
CM of the pairons, 163
CM of the "electrons", 147
collision rate, 13
collision term, 19
collision time, 14, 97
commutation relation in the $\boldsymbol{k}$–$\boldsymbol{q}$ representation, 150
commutation relations, 113, 267
commutation relations for pair operators, 119
commutators among B and $B^\dagger$, 267
complex dynamical variable, 66
composite (c-) boson, 208, 210, 211
composite (c-) boson (fermion), 207
composite (c-) fermion, 208, 210, 211
composite boson excitation spectrum, 225
composite particle, 6
compound superconductor, 106, 107
condensation energy, 142, 144
condensation of massless bosons in 2D, 136
condensed pairon, 137, 144
conduction electron, 4, 93, 177, 216
conduction electron density, 192
conduction in graphene, 93
conduction in the wall, 98
conductivity, 97, 232
conductivity of a SWNT, 85, 94
conductivity of carbon NTs, 87
constant-energy surface, 49
Cooper Hamiltonian, 268
Cooper pair, 75, 91, 96, 98, 107, 130, 180, 208, 211, 267
Cooper pair (pairon), 85, 95, 96, 108, 110
Cooper pair (pairon) carrier model, 141
Cooper pair flux quantum, 102
Cooper-like equation, 212
Cooper's equation, 116, 151, 267, 269
coronet Fermi Surface, 250
Coulomb (charging) energy, 142
Coulomb field energy, 143
Coulomb force between a pair of electrons, 72
Coulomb interaction, 36, 72, 75
creation (annihilation) operator, 69
creation electron field operator, 70
creation operator for "electron" (1) and "hole" (2) pairon, 149
creation operator for zero-momentum pairon $B^\dagger_{\boldsymbol{k}0} \equiv b^\dagger_{\boldsymbol{k}}$, 153
creation operators for "electron" (1) and "hole" (2) pairon, 112

critical field, 107
critical magnetic field, 105
critical temperature, 86, 99, 105, 106, 109, 137, 157, 182, 206, 209, 214, 274
cross section of the r-orbit, 180
crystal's natural (triangular) axes, 147
cuprate superconductor, 107
current density, 17, 97, 204
current density of patrons, 242
current relaxation rate, 21
curvature inversion, 56
cyclotron frequency, 170, 181, 200
cyclotron mass, 191, 196
cyclotron motion, 170, 200
cyclotron resonance, 171

D

DC Josephson effect, 102
de Haas–van Alphen (dHvA) oscillations, 175, 177, 184, 200
de Haas–van Alphen (dHvA) oscillations in silver, 179
de Haas–van Alphen (dHvA) oscillations in susceptibility, 177
Debye continuum model, 56
Debye distribution, 56
Debye energy, 139
Debye frequency, 108, 114, 138, 211
decay rate, 183, 200
deformation potential approximation, 66
delta-function replacement formula, 197
density gradient, 236
density of conduction electrons, 17
density of excited bosons, 134, 156
density of normal modes, 55, 56
density of scatterers, 16
density of state in the momentum space, 42
density of states (DOS), 40, 41, 55, 148, 153, 172, 236
density of states at the Fermi energy, 154, 237
density of states for electrons with down-spins, 168
density of states for electrons with up- and down-spins, 168
density of states for electrons with up-spins, 168
density of states in energy, 41
density of states per spin at the Fermi energy, 109
density of states per unit volume, 44
density of zero-momentum bosons, 157
density-wave mode, 65
diamagnetic moment, 172, 174
diamagnetic susceptibility for a metal, 174
diamond, 1
differential conductance, 161
differential cross section, 19
diffusion coefficient, 11
diffusion constant, 236
Dingle temperature, 200
Dirac delta function, 174
Dirac fermion moving with a linear dispersion relation, 221
Dirac picture (DP), 72
Dirac's theory, 209
directional cosine, 250
dispersion relation, 30
dissipationless flows, 137
divalent metal, 39
dominant carrier in graphene, 84
down-spin, 267
down-spin electron, 167
dressed electron, 191, 197, 198
drift velocity, 17, 215
Drude formula, 85, 89, 95, 233, 235
Dulong–Petit's law, 54
dynamic response factor, 66

E

effective charge, 215
effective electron density, 216
effective lattice potential, 36
effective mass, 31, 48, 86, 95, 183, 187
effective mass m_3^*, 85, 94
effective mass of an electron, 152
effective masses (m_1, m_2), 149, 158, 163
effective masses (m_1, m_2, m_3), 42
effective phonon-exchange interaction, 75
effective potential field, 36
Ehrenfest–Oppenheimer–Bethe (EOB) rule, 6, 210
eigenvalue equation, 261
eigenvalue problem, 261
Einstein relation, 237, 238
electric conduction in SWNT, 141
electric current, 17
electric current density, 232
electrical conduction in SWNTs, 3
electrical conductivity, 15, 16
electrical conductivity of NTs, 85, 86
electron, 3, 7, 17, 84, 86, 91–94, 98, 158, 163, 177, 221
electron (fluxon)–phonon interaction, 210
electron (hole) wavepacket, 77, 147
electron carrier model, 141
electron density (field), 69

electron density deviation, 66
electron density of states per spin at the Fermi energy, 114
electron effective mass, 148
electron energy, 210
electron energy gap, 157
electron flux quantum, 178, 180
electron in a Landau state, 171
electron mass, 75
electron pair (—pairon), 158
electron speed, 16
electron spin resonance, 167
electron variables, 259
electron-gas system, 74
electronic heat capacity, 40, 104
electron-pair operators $(b_k, b_k^\dagger)$, 133
electrons moving in graphene walls, 85
electron's response, 66
electron–electron interaction, 207
electron–fluxon composite, 203, 211
electron–impurity scattering, 16
electron–phonon interaction, 57, 65, 163
electron–phonon scattering, 16
electron–transverse phonon interaction, 70
electrostatic potential shift, 145
elementary excitation, 126
ellipsoidal constant-energy surface, 49
ellipsoidal Fermi surface, 178, 179
ellipsoidal surface, 42
enclosed magnetic flux, 180
energy bands, 29
energy gap, 104, 123, 127
energy gap at 0 K, 109
energy gap between the moving and the stationary fc-bosons, 215
energy gap equation, 128
energy of the moving pairon, 146
energy of the pairon, 269
energy of zero-point motion, 5
energy-dependent current relaxation rate, 24
energy-dependent relaxation rate, 196
energy-eigenvalue equation for a harmonic oscillator, 170
energy-momentum (or dispersion) relation, 46
energy-state creation operator, 262
envelope of the oscillations, 183
equation of motion for a pairon, 97
equation-of-motion method, 128, 261
equations of motion for a Bloch electron, 46
equations of motion for a harmonic oscillator, 66
equipartition theorem, 54
exact pairon wavefunction, 151
excited electrons near the Fermi surface, 187
excited pairons, 269
exponential decay rate, 187
extremum condition for | ΨĂB, 122

F

face-centered cubic (fcc) lattice, 38, 40
factorization approximation, 269
fc-boson, 214, 216
fc-boson condensed at a finite momentum, 216
fc-boson density, 216
fc-boson energy, 274
fc-boson number, 213
fc-boson, having the linear dispersion relation, 215
fcc lattice structure, 38
Fermi anticommutation rules, 70, 108, 112, 133, 211, 258, 259
Fermi degeneracy, 173, 185
Fermi distribution function, 6, 40, 41, 172, 196, 235, 269
Fermi distribution function for free electrons, 21
Fermi energy, 6, 40, 108, 147, 168, 181, 235
Fermi liquid model, 27, 36, 37, 46, 108, 131
Fermi sea, 114
Fermi speed, 146, 158, 213
Fermi sphere, 38, 41
Fermi statistics of electrons, 208
Fermi surface, 37, 39, 41, 117, 148, 152, 178
Fermi surface for a superconductor, 131
Fermi surface of Cu, 38
Fermi surface of Na, 38
Fermi velocity, 65, 96, 137, 242
Fermi-liquid state, 208
fermion, 5
fermion–antifermion symmetry, 158
Fermi–Dirac statistic, 37
Feynman diagram, 72
Fick's law, 11, 236
field effect (gate voltage) study, 84, 93
filling factor (Landau-level occupation ratio), 205
finite size effect, 85, 93
first Brillouin zone, 28, 38, 40
flux quanta, 102, 180
flux quanta (fluxons), 191, 207
flux quantization, 101, 209
flux quantization for the Cooper pair, 180
fluxon, 209
fluxon number operator $n_{ks}^{(3)}$, 211
fluxon–phonon interaction strength, 210

force term, 19
Fourier's law, 12
four-valence electron conduction, 93
fractional charge, 208
fractional LL occupation ratio (filling factor), 208
fractional quantum Hall effect (QHE), 203, 205–207
fractional ratio, 203
free boson model, 214
free electron model, 174, 181
free electron model in 3D, 152
free electrons in equilibrium, 168
free energy, 181, 186
free energy for a system of free electrons, 172
free massless boson, 214
free pairon model, 137
free-electron Fermi sphere, 37
frequency integral, 55
Fröhlich interaction Hamiltonian, 65, 70, 71
f-sum rule, 136, 214
fugacity, 97, 134, 135, 156, 157, 242
full-spin boson, 209
fundamental (f) c-boson (fc-boson), 212
fundamental composite (c-) boson, 212
fundamental quantum nature, 208

G

galvanomagnetic phenomena, 171
gapless semiconductor, 77, 87, 160
gas constant, 135
gate voltage effect, 92, 96
gate voltage shift, 145
generalized BCS Hamiltonian, 56, 113, 268
generalized energy gap equation, 124
Ginzburg–Landau theory, 105
grand canonical ensemble average, 269
grand ensemble trace, 268
graphene, 2, 4, 77, 78, 87, 91, 93, 147, 148, 221, 240, 241
graphene layer, 249
graphene sheet, 91, 93, 141, 146, 158, 163
graphene wall, 91, 94
graphite, 1, 4, 231, 240
gravitational mass, 6
ground pairon, 109, 128
ground state energy, 122, 127, 154, 212
ground state energy of the system of fc-bosons, 213
ground state of the Bloch system, 121
ground state wave function, 122
group velocity, 30, 47
group velocity of the Bloch wavepacket, 30

H

half-spin fermion, 209
Hall coefficient, 18, 86, 148, 205, 237
Hall effect, 17
Hall effect measurements, 204
Hall field, 203
Hall resistivity, 203, 207, 208, 225
Hall resistivity in GaAs/AlGaAs, 206, 208
Hall resistivity plateau, 205, 206, 209, 225
Hall voltage, 18
Hall's experiment, 17
Hamiltonian for an electron gas system, 71
Hamiltonian for an electron–phonon system, 71
Hamiltonian for moving pairons, 150
Hamiltonian of a free electron in $\boldsymbol{B}$, 170
Hamiltonian of a simple harmonic oscillator, 4
Hamilton's equations of motion, 47
harmonic approximation, 54, 55
harmonic equation of motion, 66
Harrison's model, 40
heat capacity, 40, 53–55, 136, 233
heat capacity per electron, 234
heat capacity per unit volume, 234
Heaviside step function, 185
Heisenberg equation of motion, 260
Heisenberg picture (HP), 259, 260
Heisenberg uncertainty relation, 5
Heisenberg's uncertainty principle, 38
helical angle, 84, 93
helical line, 84, 85, 93
helicity, 84, 85, 93
hexagonal close packed (hcp) crystal, 39, 250
high-temperature superconductivity (HTSC), 107, 208
Hohenberg's theorem, 136, 214
hole, 3, 7, 17, 84–86, 91–94, 96, 98, 158, 163, 177, 221
hole axial transport, 96
hole channel current in a SWNT, 92
hole current, 85
hole mass, 85
hole mass in the carbon wall, 85
hole pair (+pairon), 158
holes, 85
holes moving in graphene walls, 85
homogeneous superconductor, 104
honeycomb lattice for graphene, 163
honeycomb lattice structure, 2
hyperboloidal Fermi surface, 179

I

impurity scattering rate, 86, 95

incompressible quantum fluid state, 205
independent electron model, 37
instantaneous Coulomb interaction, 72
integer QHE, 206, 207, 212
integer QHE plateau, 208
integer quantum Hall effect (QHE), 203, 205
integral number of carbon hexagons, 91
interaction Hamiltonian, 70
interboson distance, 214
interelectronic Coulomb interaction, 36
internal energy density, 135
interpairon distance, 137
interparticle interaction potential, 261
inverse collision frequency, 243
inversion (mirror) symmetry, 148
inversion symmetry, 80
ion contribution, 16

J

Jain's theory of fractional hierarchy, 203
Josephson effects, 102
Josephson interference, 103
Josephson junction, 102
Josephson tunneling, 102

K

kinetic momentum, 171, 184, 197, 198
kinetic theory of gas dynamics, 11
Kronig–Penney model, 30

L

Lagrangian in the harmonic approximation, 54
Landau diamagnetism, 165, 171, 174, 186
Landau energy, 172
Landau Level (LL), 170, 171
Landau Level (LL) degeneracy, 198
Landau oscillator quantum number, 172
Landau states, 170, 171, 184, 271
Landau susceptibility, 174
Landau-level occupation ratio (filling factor), 205
lattice, 4
lattice axes, 34
lattice dynamics, 53, 55, 57
lattice force, 47
lattice momentum, 46
lattice periodic potential, 46
lattice vibration, 55
lattice-ion mass, 75
Laughlin wavefunction, 206, 208
linear dispersion relation, 96, 133, 138, 139, 155, 163, 242
linear dispersion relation for the 3D Cooper pair, 213
linear dispersion relation in two dimensions, 146
linear energy-momentum (dispersion) relation, 153
linear energy-momentum (dispersion) relation for the center of mass motion, 146
linear heat capacity, 177
linear operators ($\eta, \eta^\dagger$), 258
Liouville operator, 72
LL degeneracy, 216
localized Bloch wavepacket, 46
longitudinal elastic wave, 56
longitudinal phonon, 70
longitudinal wave, 65
longitudinal wave mode, 65
long-range order, 105
Lorentz force, 17, 48, 177
lowest bound energy, 114

M

magnetic energy, 168
magnetic flux, 101, 103, 215
magnetic flux line, 100, 101, 107
magnetic moment, 101, 165, 166, 180
magnetic moment per unit area, 186
magnetic oscillations, 186, 188, 196
magnetic oscillations in bismuth (Bi), 251
magnetic pressure, 100
magnetic susceptibility, 177, 187
magnetization, 169, 177, 182, 183, 186, 187
magnetoconductivity, 187
magnetogyric ratio, 165, 166
magnetomechanical ratio, 166
magnetoresistance (MR), 189
magnetoresistivity, 225
magnetotransport mass, 191, 196, 197, 200
main characteristic of metallic conduction, 142
major axes of the ellipsoid, 48
majority carriers in graphite, 250
majority carriers in nanotubes, 84
many-body perturbation method, 71
many-body trace, 262
many-body-system density operator, 262
Markovian approximation, 74, 75
mass conservation law, 136
massless boson, 214
Matthiessen's rule, 15, 16
mcl crystal, 81, 249
mean free path, 14
mean free time, 16
Meissner effect, 100, 101, 107, 127, 158, 205, 225
metallic (semiconducting) SWNT, 141, 163

metallic compound, 106
metallic single-wall carbon nanotubes, 141
metallic SWNT, 3, 142, 157
metal–insulator transition (MIT), 247
Miller indices, 147
MIT in VO_2, 249
mixed representation, 274
mixed representation of one-body density operator, 263
mixed state, 107
mobility, 222
mode index, 66
molar heat capacity, 135, 136
molar heat capacity at constant density (volume), 135
molar heat capacity for a 2D massless boson, 136
momentum distribution function, 19
momentum-state annihilation operator, 262
momentum-state electron operator, 70
monoclinic phase, 247
monovalent fcc metal, 131
monovalent metal, 33, 36–38, 131
motional diamagnetism, 171
Mott's vrh theory, 96
Mott–Hubbard picture, 247
moving (noncondensed) fc-boson, 213
moving fc-boson, 216
moving pairon, 149, 151, 154
moving patron, 245
multivalent metal, 38
multiwalled carbon nanotube (MWNT), 1, 77, 85, 229, 230, 240, 242
MWNT bundle, 244

N
nanotube, 87
nearly free electron model (NFEM), 38
neck and belly orbits, 179
neck Fermi surface, 86, 221
net (CM) momentum, 267
net momentum of a pair of electrons, 72
neutral supercondensate, 131
new band model, 87
Newton's equation of motion, 15, 48
NFEM (Nearly Free Electron Model), 38, 250
noble metal, 38, 179
noncondensed fc-boson, 216
non-Ohmic behavior, 141, 142, 163
normal coordinates, 54
normal current, 104, 110, 145
normal curvature, 48
normal mode, 54
normal modes of oscillations, 54
normal modes of oscillations for a solid, 65
normal momenta, 54
normal Ohmic conduction, 95
normal-mode frequencies, 55
number density, 15, 16, 147
number density of zero-momentum bosons, 134, 156
number of zero-momentum bosons, 134
number operator, 211, 254, 270
number operator for electron (1) (hole (2)), 150
number operator for pairons having net momentum q, 154
number operator for the pairons in the state $(\boldsymbol{k}, \boldsymbol{q})$, 154
number operators for electrons, 267
number operators for electrons and holes, 112
number representation, 253
numbers of the electrons with up- and down-spins, 168

O
observable, 256
occupation number, 253, 270
occupation numbers of pairons having a CM momentum $\boldsymbol{q}$, 155
occupation-number representation, 253
Ohmic behavior, 143
Ohm's law, 15, 97, 204, 242
one-body density operator, 262, 263
one-body Hamiltonian, 261
one-body operator, 260
one-body trace, 262
one-electron-picture approximation, 37
one-pairon states, 122
one-phonon exchange process, 73
Onsager's flux quantization hypothesis, 191
Onsager's formula, 177, 178, 181, 209
Onsager's magnetic flux quantization, 180
open orbits in the k-space, 189
orthogonal unit cell, 79
orthorhombic (orc) crystal, 82, 251
oscillation period, 183, 187
oscillatory density of states, 221
oscillatory magnetization, 187
oscillatory statistical weight, 181, 187
oxide layer, 102

P
pair annihilation operator, 268
pair energies, 268
pair operators, 153, 211, 268
pair wavefunction, 116, 151
pair-annihilate hole-type c-boson pairs, 212

pair-create electron-type c-boson pairs, 212
pairon, 97, 107, 130, 142, 155, 242
pairon (Cooper pair), 85, 95, 96, 108
pairon (Cooper pair) carrier model, 142
pairon annihilation operator, 112, 150
pairon carrier model, 242
pairon density, 97, 158
pairon density of states, 118
pairon energy, 242
pairon ground state energy, 109, 115, 157
pairon momentum, 243
pairon net momentum, 133, 155
pairon occupation-number states, 155
pairon operator, 112, 150, 267
pairon speed, 145
pairon wavefunction, 116, 269
pairon–phonon scattering cross section, 145
pair-state operators, 268
particle density, 243
partition function per electron, 171
patron, 242
Pauli exclusion principle, 5, 155
Pauli magnetization, 186, 187
Pauli paramagnetic susceptibility, 175
Pauli paramagnetism, 165, 167, 169, 174
Pauli's exclusion principle, 38, 207, 253
Peierls instability, 247
penetration depth, 101
periodic boundary condition, 57
periodic lattice potential, 34
periodic oscillation of the statistical weight, 187
permutation operator, 255
permutation-symmetry property of the quantum particle, 259
phase, 178, 180
phase change, 104, 105
phonon, 16, 53, 55
phonon energy, 75, 210
phonon exchange, 151, 158
phonon momentum, 75
phonon scattering, 97, 145
phonon scattering rate, 86, 95, 145
phonon-exchange attraction, 71, 85, 91, 95, 108, 110, 116, 208, 221
phonon-exchange attraction between the electron and the flux quantum (fluxon), 221
phonon-exchange effect, 73
phonon-exchange interaction, 74, 75
physical vacuum state, 119
pitch, 84, 91–93, 98, 141
pitch angle, 98, 142
pitch in a metallic SWNT, 157
Planck constant, 181
Planck distribution function, 55, 145, 146
plane wave, 35
plateau height, 205
plateau stability, 205
Poisson's sum formula, 173, 186, 271
position representation, 259
potential energy, 53
potential field energy of a magnetic dipole, 166
primitive vectors, 78
principal axes of curvatures, 49
principal axis, 49
principal mass, 49
principal-axis transformation, 54

Q

QH state, 205
QHE at filling factor $\nu = 2$, 221
QHE in GaAs/AlGaAs, 225
QHE in graphene, 221, 225
QHE state, 217, 225
quadratic dispersion relation, 48, 183
quadrivalent metal, 148
quantization of cyclotron motion, 170
quantum Hall (QH) state, 205
Quantum Hall Effect (QHE), 188, 203
Quantum Hall Effect (QHE) oscillations, 221
quantum Liouville equation, 72, 262
quantum Liouville operator, 73
quantum number, 35
quantum postulate, 260
quantum statistical postulate, 5
quantum statistical theory, 110
quantum statistics of the particles, 261
quantum wavepacket, 47
quantum zero-point motion, 34
quasielectron, 109, 147
quasielectron energy gap, 109
quasifree electron, 184
quasifree electron Hamiltonian, 197
quasifree electron model, 6, 183, 187
quasiparticle dispersion relations, 138
quasiparticle energy, 127
quasiwavefunction, 262

R

radius of a MWNT tube, 85
rectangular cell model, 91, 92
reduced density operator, 274
reduced generalized BCS Hamiltonian, 119
reduced Hamiltonian, 153, 154
reduced wavefunction for the stationary fc-bosons, 213

reflection (mirror) symmetry, 80
relative and net momenta, 112, 150
relative momentum, 267
relaxation (collision) time, 86, 95
relaxation rate, 16, 24, 86, 95
relaxation time, 24, 97, 243
resistance, 86
resistivity, 204
right-hand screw rule, 165
rigidity (shear) modulus, 82
ring supercurrent, 101
room-temperature quantum Hall effect (QHE) in graphene, 221
running wave, 30, 65

S

Sbunikov–de Haas (SdH) oscillation, 187
Sbunikov–de Haas (SdH) oscillations in GaAs/AlGaAs, 187
scanned probe microscopy (SPM), 142
scattering angle, 196
scattering cross section, 16, 196
scattering rate, 16
Schrödinger energy-eigenvalue equation, 262, 263
Schrödinger equation, 46, 170
Schrödinger equation for an electron, 34
Schrödinger picture (SP), 259
SdH oscillation period, 200
second quantization, 253, 259
second-quantization formalism, 108, 253
second-quantized operators for a pair of electrons, 267
Seebeck coefficient, 86, 229, 233, 239, 244
Seebeck coefficient (thermopower), 229
Seebeck coefficient for 3D motion, 234
Seebeck coefficient for a bundle of MWNTs, 245
Seebeck coefficient in multi-walled carbon nanotubes, 229
Seebeck coefficient of highly oriented single-crystal pyrolytic graphite, 230
Seebeck coefficient of MWNTs, 243
Seebeck coefficient S in copper (Cu), 229
Seebeck current, 236, 237
Seebeck electromotive force, 233
semiclassical (wavepacket) model of a conduction electron, 240
semiclassical electron dynamics, 78
semiclassical equation of motion, 78
semiconducting SWNT, 3, 91, 93, 142
semiconductor-like T-behavior, 87
shear modulus, 83
Shockley's formula, 250
Shubnikov–de Haas (SdH) oscillations, 196, 221
simple cubic (sc) lattice, 46, 65
single-wall carbon nanotube (SWCN), 1, 91
single-wall nanotube (SWNT), 1, 77, 84, 91, 93, 98, 141, 240
sinusoidal oscillations to the free energy, 183
six basic properties of superconductors, 111
solid angle, 19
speed of sound, 65
spherical Fermi surface, 109
spin angular momentum, 166
spin anomaly, 166
spin degeneracy, 38, 41, 44, 167, 172
spin degeneracy factor, 116
spin-statistics theorem, 5
stationary density operator, 262
stationary pairon, 151
statistical weight, 172, 184, 186, 271
supercondensate, 110, 130, 131, 215
superconducting energy gap, 139
superconducting ground state, 142
superconducting properties, 208
Superconducting Quantum Interference Device (SQUID), 103
superconducting state, 86, 97, 110, 141, 142
superconducting state of HTSC, 107
superconducting temperature, 144, 243, 245
superconducting transition, 104, 110
superconductivity, 158
superconductivity energy gap in the composite boson (c-boson) excitation spectrum, 225
superconductors, 100
supercurrent, 101–104, 107, 110, 142, 144, 158, 215, 244
supercurrent density, 158, 215
supercurrent ring experiment, 99
superfluid phase, 111
superposable plane waves, 247
susceptibility, 169, 177, 179, 187
symmetric ket, 255
symmetrizing operator, 255

T

tcl crystal, 81, 83
temperature gradient, 236
tet crystal, 249
thermal activation, 77
thermal conduction, 11, 12
thermal diffusion of the MWNT bundle, 245
thermal electromotive force (emf), 229
thermal speed, 14

thermally activated electron density, 80, 149, 241
thermally activated process, 77
thermally excited electron, 40
thermoelectric power, 229, 230
third-order phase transition, 136
time-dependent perturbation theory, 73
total magnetic susceptibility, 175
total number of pairons, 109
transition between the hole-type c-fermion states, 212
transverse elastic wave, 56
transverse lattice normal mode, 70
transverse wave mode, 65
traveling normal mode, 66
traveling wave, 66
tunneling experiment, 104
two-body density operator, 72
two-electron density matrix, 74
two-pairon states, 122
type I elemental superconductor, 107
type II magnetic behavior, 107, 110

U

unit hexagon, 147
up-spin, 267
up-spin electron, 167

V

vacuum ket, 69, 254
vacuum-state ket for phonons, 73
van Hove singularities, 56, 57
van Leeuwen's theorem, 171
vanadium dioxide (VO_2), 247
variable range hopping (vrh) theoretical formula, 96
variation, 123
vector potential, 170
vertices, 107
virtual electron pair, 132
virtual exchange of phonon, 108
virtual phonon, 73

W

wave train, 30, 35
wavefunction, 170
wavelength, 65
wavepacket, 93, 247
wave-particle duality, 47
weak-coupling approximation, 73
Wigner–Seitz (WS) cell model, 77, 87, 93
Wigner–Seitz (WS) cell model for graphene, 3
WS model, 149
WS unit cell for graphene, 78

Y

Young modulus, 82, 83

Z

zero resistance, 99
zero resistivity, 206
zero-bias anomaly, 161
zero-momentum boson, 156
zero-momentum pairon, 118, 152, 154
zero-momentum pairon operator, 128
zero-pairon state, 122
zero-point energy, 55
zero-temperature BCS pairon size, 137
zero-temperature electron energy gap, 157
zero-temperature energy gap, 244
zone number, 29, 35